I0064463

# Land Snails
## of Belize
### Central America

## A Chronicle of Remarkable
## Diversity and Function

**Daniel C. Dourson**
**Ronald S. Caldwell**
**Judy A. Dourson**

All photos by Daniel Dourson unless otherwise stated

All Illustrations by Rachael Grabowski unless otherwise stated

Cartoons by AB Osborne

Copyright © 2018
All rights reserved.

Published by
Goatslug Publications
1128 Gritter Ridge Road, Stanton Kentucky, 40380, USA
Email: theroguebiologist@gmail.com

Reproduction of any section is prohibited without verbal or written permission from the authors. If used for educational purposes, no prior permission is required. Artwork or images from sources other than the authors are reprinted with written permission from the copyright holders. Permission must be obtained from each individual holder.

**Front Cover:** A Speckled snail-sucker, *Sibon nebulata* eating a banded cone, *Drymaeus emeus*, Bladen Nature Reserve, Belize, Central America .

---

This book is dedicated to the memory of Dr. Fred G. Thompson, Curator Emeritus of Malacology and Invertebrate Zoology at the Florida Museum of Natural History, Gainesville, Florida, USA. Dr. Thompson's work on New World and Paleotropical terrestrial mollusks was pioneering and exemplary.

Although I never shared field time with Fred, I spent many hours in his Florida office listening to his snail hunting stories and adventures in Mexico and Central America. Fred mentored our work in Belize and was an invaluable resource for this book. I also had the privilege of describing a species from Belize, *Eucalodium belizensis,* with Fred. We are forever grateful for his patience and wealth of information he shared. ~Dan Dourson

# Contents

# Acknowledgements

We would like to thank the following organizations and individuals for their valued contributions in the creation of this book:

**Belize Foundation for Research and Environmental Education (BFREE), Bladen Village, Toledo District, Belize, Central America** for serving as a premier research and education facility from which much of the research was conducted. BFREE co-founded the Rainforest Science Cooperative Lab at BFREE that serves as a repository for the Belize specimens. Special recognition goes to Jacob Marlin for his support of our research and his undeniable enthusiasm for land snails; Thomas Pop and Sipriano Canti for extensive sifting and sorting of leaf litter samples; others at BFREE who provided assistance: Heather Barrett, Steven Brewer, Elmer Tzalam, Wilfred Mutrie, Liberato Pop, William Garcia, Katelyn Loukes, Sharna Tolfree, Sofia Marlin, Amarta Coc, Carolina Chiac, Solana Cus, and Marcelino Pop.

**Ya'axche Conservation Trust (YCT), Punta Gorda, Belize, Central America** for support of this research. Special thanks to Marcus Tut and Marcus Cholum, YCT Rangers who were trained as research technicians and conducted the baseline snail plot research project in the Bladen Nature Reserve and Lee McLoughlin, former YCT protected areas manager for supporting this research.

**Cumberland Mountain Research Center at Lincoln Memorial University, Harrogate, Tennessee, USA** for co-funding the Rainforest Science Cooperative Lab at BFREE providing lab space and materials to conduct research and digital imaging equipment to photograph microsnails. Thanks to John Hoellman for field assistance. A very special thanks to Sara Collins, an undergraduate researcher and research assistant, who conducted her undergraduate research on land snails and as a result, spent countless hours collecting, drying, sifting, and sorting leaf litter samples. Dr. Adam Rollins also contributed to our research with field assistance, lab assistance and co-authored a paper.

**Florida Museum of Natural History, Gainesville, Florida, USA** for collecting supplies, serving as repository for the Belize collection and use of digital imaging used to photograph microsnails. Special thanks to John Slapcinsky, Mollusk Collections Manager, for logistical support at the museum and Verity Mathis, Mammals Collections Manager, for assistance in use of the digital imaging equipment.

**Natural History Museum, London, UK** for access to type specimens through a digital data base. Thanks to Jon Ablett, curator of Mollusc Division for photographing type specimens for comparison.

**Belize Forest Department** for assistance with scientific research permits necessary to conduct research in Belize. A special thanks to Rasheeda Garcia for her assistance.

**BERDS (Biodiversity and Environmental Resource Data System), Cayo District, Belize, Central America** for including data about the land snail fauna of Belize with special thanks to Jan Meerman for his interest in land snails.

Copperhead Environmental Consulting for collecting assistance during expedi-

tions into the Bladen and Piper Roby, for development of the map of ecoregions.
**Shipstern Conservation and Management Area**: Heron Moreno, Manager; Mayeli Guifarro, Joel Diaz, and Joe Alvarado for collection assistance as well as lodging and accommodation.

**Runaway Creek Nature Reserve:** Gil Boese, Director, Wilbur Martinez, Reynold Cal and Stevan Reneau for collection assistance as well as lodging and accommodation.

**Sustainability America:** Robert Browne, Director, for field assistance and logistics in collecting land snails in and around Sartenja.

**National Aggregates at Gracie Rock:** Thanks to Joe Fuzy for access to site.

**Gallon Jug Ranch:** Allan Jeal, for assistance and accommodation at Gallon Jug.

**Yalbac Ranch:** Jeff Roberson, for permission and access to the Yalbac Hills to study the land snails of the area.

**International Zoological Expeditions (IZE), Blue Creek, Toledo District, Belize Central America:** Thanks to Fred Dodd for access to Blue Creek Cave to survey for land snails.

**Tiger Sandy Bay, La Democracia, Belize District, Belize, Central America:** Special thanks to Steve Downard for access to survey for land snails near the Peccary Hills.

**Duplooy's Jungle Lodge, Cayo District, Belize Central America**: Thanks to Judy Duplooy for access to grounds to survey for land snails.

**Belize Audubon Society, St. Hermann's Cave at the Blue Hole National Park**: Special thanks to Joshua Richards, park ranger, for photos of land snails in the park.

**National Geographic Expedition March 2016 Sinkhole/Snail Team:** Dalton Jackson, James Abbott, Lauren diBaccari, Amy Batchman Dourson, Aaron Dourson, William Garcia, Thomas Pop, Judy Dourson and Dan Dourson. Special thanks to cave explorers, Kasia Biernacka and Pawel Skoworodko, Poland, for their amazing expertise and assistance in dropping into the unexplored sinkhole as well as collection assistance.

**Better in Belize Eco-Resort:** Thanks to Susanne Jefferson for opening her home and grounds to collection of land snails in this region of the country.

**Richard & Carol Farneti Foster:** Thanks for your interest in capturing feeding behavior of the carnivorous snail, *Euglandina ghiesbreghti,* and snail-sucker snakes as well as accommodation, logistical support and most importantly, your friendship throughout the years.

**Chrissy & Anita Tupper, Cheers with Tropical Twist Restaurant:** Thanks for your support, encouragement and friendship throughout our years of research.

**Reviewers**
Thanks to the following reviewers for candid, useful critique of this book: John Slapcinsky, Dr. Fred Thompson, Dr. Abraham Breure, Dr. G. Thomas Watters, and Ira Richling. We greatly appreciate your expertise, time and input.

# 1 Introduction

In general, our understanding of land snail diversity, ecology, and distribution is vastly incomplete, particularly in tropical regions of the world. Belize is no exception. The geographic area of Belize is small but belongs to a much larger region including Mexico and Central America that contains many biomes, different geological structures, complex physiographic features, and a myriad of ecological settings (Thompson 2011). According to Thompson, the number of recorded land snail species (around 1239 taxa) from this enormous area is about 35% of the actual fauna. As much as 65% remains undiscovered. The high number of estimated undescribed species is based on the fact that most of Panama, Costa Rica, Nicaragua, nearly all of Honduras, most of Guatemala, much of Belize, and 85% of Mexico have not been surveyed or at best, sparsely explored for mollusks, especially micro species less than 5 mm. This book seeks to bring forward the latest information on the land snail fauna of Belize.

**Historic Land Snail Research in Central America**
Most of what we know about the land snail fauna from Mexico and Central America is based on early eighteenth century works, sound taxonomic studies and reviews dating back to Shuttleworth, Menke, and Pfeiffer (Thompson 2011). Specimen locations are rarely straightforward. Many collections have been separated and sent to different museums while others have proven to be impossible to locate. Fischer and Crosse material is in Paris; Von Martens' collection is in Berlin but most of his early Central American material can be found in London. Much of the Pfeiffer collections were lost during WWII. The material that remains including many types described by him on the basis of Cuming's 1851 material are located in London. Morelet's collections partly reside in London and partly in Geneva. The largest collection of Central American land snails currently is found at the Florida Museum of Natural History. Since the majority of early shell collections of Central America reside in museums outside of the area, accessibility to researchers of Central America is limited.

While literature records are important contributions to the overall knowledge, specimens upon which the records are based are even more valuable as they can be studied to verify reports or make new taxonomic, ecological or biogeographical discoveries. Historic collections are indeed an important source of fundamental reference, but caution must be used. A major source of frustration comes from the misidentification of museum specimens. Just because a land snail is labeled a certain species doesn't mean it's that species! Many institutionalized specimens have been incorrectly identified by well-meaning researchers without further verification. In this book, every attempt was made to compare material collected during this study with the original holotype, paratype or syntype specimens and, in some cases, early illustrations.

Land snail collecting in Belize prior to this study has been sparse and sporadic with most surveys allied to archeology digs, around historic landmarks such as the Caracol Maya Ruins and Rio Frio Cave (Haas and Solem 1960) or along main thoroughfares. In the 1970s, land snail studies throughout Mexico and Central America including Belize were primarily conducted by Fred Thompson who has collected and described numerous species. His enthusiasm and tireless devotion to the study of gastropods in this great region has vastly increased our knowledge of Mexico and Central America land snails. This information, while valuable to a small group of malacologists and scientists, has remained out of reach to Latin American biologists including Belizeans. Consequently, there has been little domestic interest in the study of land snails in the country, for that matter, most of Central America.

**Present Land Snail Research in Belize**
Our countrywide research conducted from 2006-2016 has revealed a surprising number of species. At present, 158 native species and subspecies have been documented and confirmed within the arbitrary boundaries of Belize. Thompson (2011) reported only 24 known species. This ten year study represents a staggering 558% increase in the land snail biodiversity for the country. Of the 158 confirmed species, 25 are new to science of which 17 are described in the book. The remaining 8 undetermined species are also illustrated in the book and will require further investigation. *Eucalodium belizensis, Carychium belizeense,* and *Paradoxipoma enigmaticum* were recently described from Belize by other authors and are included.

Every expedition into this vastly under-collected area has yielded spectacular finds including the recently discovered and described Mayan Drum, *Eucalodium belizensis* Thompson & Dourson 2013, known only from the type locality in southern Belize. As a result, the Maya Mountains is considered to be one of the most important molluscan regions in Central America (pers. comm. Thompson 2010) and may exceed other regions of comparable size in terms of numbers of species and endemism. This, however, remains to be more thoroughly investigated.

Early isolation from North and South America, geophysical (soils) and ecological influences such as diverse vegetation are thought to be the driving forces for such a rich fauna (Brewer & Webb 2002). As the isthmus or land bridge of Central America closed, the once-isolated Maya Mountains became allied to the greater continent of South America resulting in a rich tapestry of both temperate and tropical land snails. These gastropods continue to diverge, creating new and unique locally endemic species. The inaccessibility of the densely forested and precipitous topography has also been a major deterrent to past collectors leaving the interior mountain ranges entirely unsampled. While other Central America countries have well-established access roads and trails across their highest peaks, the Maya Mountains have remained without these travel conduits.

Other habitats within the Maya Mountains await further investigation including the volcanic highlands, epiphytic plants found high in the jungle canopies, and rare elfin forests at the uppermost elevations (>1000 meters). Although these areas generally yield far less snail diversity and numbers of shells than the limestone sites, greater land snail endemism is expected. Deep cave systems will also add rare, endemic, and troglobitic forms of land snails not previously documented. Large isolated sinkholes (page 15) are common to Belize and represent another unsampled feature likely to harbor interesting land snail assemblages. The numerous islands occurring along the Mesoamerican Reef also remain largely unsampled.

Interspecies relationships between gastropods and other organisms (fish, birds, bats, salamanders, lizards, and snakes) may provide further justification for the study of land snails as interest in land snails is cultivated and scientific studies push deeper into unsampled regions of the country.

In the past, animal endemism in Belize was reported to be low for most taxonomic groups, the country containing few endemic species. Our research suggests otherwise, demonstrating that there are a number of endemic land snail species concentrated around the Maya Mountain Massif and in hydrologically isolated cave systems.

**Environment and Land Snail Distribution**
Without question, calcium carbonate is an essential mineral to land snails for regulation of bodily processes, reproduction, and most importantly, shell-building (Burch 1962; Fournie *et al.* 1984). Land snails obtain calcium in several ways including consuming soil particles from calcareous substrates, eating decaying leaf matter (Burch & Pearce 1990; Nation 2005), almost certainly by ingesting Physarales slime molds which precipitate amorphous calcium carbonate, and gleaning calcium from the shells and bones of deceased animals. *Triodopsis platysayoides* (an endemic West Virginia land snail) have been documented feeding on the vacant shells of land snails including its own kind (Dourson 2008).

Land snail abundance (number of shells) and land snail diversity (number of species) have long been associated with a variety of geological and ecological factors. Studies have shown for example, that terrestrial gastropods living around carbonate cliffs can exhibit large and diverse populations (Nekola 1999) but show significant declines in abundance in as little as 50 m from a calcareous source (Kalisz & Powell 2003) or limestone cliffline (Dourson 2007). Other research has demonstrated while limestone may impact abundance, it has little affect on diversity.

Land snail scarcity is often associated with low soil pH (Burch 1955), declining soil cations, specifically calcium (Petranka 1982), increasing coniferous presence (Jacot 1935; Karlin 1961), and increasing elevation (Petranka 1982). The influence of pH on land snails is thought to be indirect, the main effect of a low pH being a lowering of the amounts of soil cations, principally calcium (Karlin 1961; Cameron 1970). But low abundance in non-calcareous (acidic) areas may only give the illusion of low diversity. Douglas (2011) found that land snail diversity on acid soils covered by old growth forests at Lilley Cornett Woods in Kentucky, USA were analogous to limestone soils at Floracliff Nature Sanctuary along the Kentucky River Palisades.

In Belize, geological sources of calcium occur in limestone, dolomite, calcareous shale, sandstone, and the igneous-derived amphibolite, all which have calcium and/or magnesium in varying amounts supplying the necessary calcium for shell-building. But what about land snails thriving on non-calcareous substrate like elfin forests or in the canopy of trees? It is not clear how or where these land snails obtain sources of calcium. In regions lacking calcareous substrates, land snails may rely on abscissed leaves of trees, herbaceous plants or Physarales slime molds as a primary source for calcium. Or it could be something even more astonishing! The land snail, *Pittieria aurantiaca,* from Costa Rica (below) is an extraordinary example of secondary calcium acquisition. Using its tentacles, the gastropod taps the rear end of the lantern bug, *Phrictus quinquepartitus*. This strange behavior seems uncharacteristic for a carnivorous snail that normally hunts other land snails for their flesh. But *Pittieria aurantiaca* is not interested in eating the bug but instead is attracted to the hon-

© Piotr Naskrecki

Above image of a carpenter ant cleaning honeydew from the head of the land snail, *Pittieria aurantiaca,* ejected from the lantern bug abdomen. Observations made by entomologists Piotr Naskrecki & Kenji Nishida (2007).

© Piotr Naskrecki

eydew which is ejected from the bug's abdomen. After tapping the rear of the insect, the snail catches flying droplets of honeydew by forming a hood over the tip of the abdomen with its head and foot. The honeydew accumulates on the snail's ventral surface and is then consumed (Naskrecki & Nishida 2007).

Why would a predaceous snail want honeydew in the first place? In addition to the benefit of providing high quality carbohydrates, the phloem (sap) of the tree species that the lantern bug feeds on, *Simarouba amara*, contains high levels of calcium, the mineral needed to build shells. Carpenter ants also want in on the action but are too short and too slow to catch the honeydew ejected from the lantern bug at between 0.8 and 1.7 m/sec (2.6-5.6 ft/sec.) according to Naskrecki and Nishida

(2007). So ants climb the snail to harvest the sweet honeydew from the surface of the gastropod's head. Not surprisingly, the snail quickly becomes agitated by ants crawling over its eyes and chemoreceptors. Unable to rid itself of the pesky ants, the snail simply crawls away. This extraordinary land snail feeding strategy between mollusk and arthropod clearly demonstrates the wide range of calcium carbonate sources in nature and the length to which land snails will go to obtain calcium. It also confirms the complexity of food webs and interspecies relationships as well as how little is known about these relationships. These amazing observations and photographs were made by entomologists Piotr Naskrecki and Kenji Nishida.

McHargue & Roy (1932) in a study of several species of deciduous forest trees found that the amount of calcium in leaves expressed as a percentage of dry weight ranged from 1.64 to 7.8%, with the higher values occurring towards the end of the growing season. Other studies have shown that unlike other macronutrients that are reabsorbed by trees prior to leaf abscission, foliar calcium concentrations in deciduous trees increase throughout the growing season and peak at senescence (Guha & Mitchell 1966; Potter 1987). Calcium is a relatively immobile nutrient and the reabsorption of calcium may not be as high a priority to deciduous species as other nutrients. Further, Gosz et al. (1973) reported that dead birch leaves could form a significant pool of calcium on the forest floor, since the concentrations of calcium remained high in dead leaves 12 months after abscission. Vesterdal & Raulund-Rasmussen (1998) found that the nutrient content of the forest floor under a single species of tree varied with the soil type, but they also found variation between different species on the same soil type. The idea that certain deciduous species contain higher levels of foliar calcium than others was supported by Arthur et al. (1993), who reported that yellow birch had relatively high levels of foliar calcium. Gosz et al. (1972) found foliar calcium concentrations were higher in yellow birch than in maple or beech. Ricklefs & Matthews (1982) looked at the leaf chemistry of 34 species of broad leaved deciduous trees and found that yellow birch had higher than average calcium concentrations. Jenkins (2007) found that calcium levels in dogwood leaves growing in the Great Smoky Mountains National Park were the most significant source of calcium in acidic forests, although many have since died due to dogwood anthracnose.

Additional factors like gradient (slope), litter moisture, elevation and microhabitat (leaf litter, moss and logs) can significantly affect the presence or absence of land snails. Coney et al. (1982) found more species of land snails on steep slopes than on more moderate ones. Petranka (1982) found that 15 of the 56 land snail species found on Black Mountain, Kentucky, showed some preference for slope, with 9 species showing an affinity for increasing slope. The importance of leaf litter moisture (thought to be a factor of slope) to land snails was emphasized by Boycott (1934), Getz (1974), Pollard (1975), and others. Although aspect was reported to markedly affect microclimate (Braun 1940; Geiger 1965;) Petranka's study in 1982 found no environmental variable to be significantly correlated with aspect.

With respect to elevation, Petranka (1982) reported that pH, potassium, calcium and magnesium levels would decrease with increasing elevation and that the num-

9

ber of snail species and the number of individuals found per site would also decrease with elevation. In a study by Coney *et al.* (1982) conducted in the Hiwassee River Basin of Tennessee, the most important environmental factors influencing the presence or absence of land snail species was microhabitat (leaf litter, moss and logs, P<0.05 for 27 species), followed in decreasing order of importance by slope (P<0.05 for 15 species), rock type (P<0.05 for 13 species), stages of forest succession (P<0.05 for 12 species), soil pH (P<0.05 for 8 species), elevation (P<0.05 for 7 species) and soil moisture (P<0.05 for 6 species).

Less well-known is how the convergence of large physiographic and geophysical landscape edges (i.e. Maya Mountains and Vaca Plateau) serve to bridge distinctive regions and allied terrestrial gastropod communities (Dourson 2007; Dourson and Beverly 2009; Douglas 2011). Neighboring land masses and geology may be the conduit for the distribution and mixing of some snail faunas, acting as travel corridors for dispersal that results in remarkably high land snail diversity in comparatively small places (Dourson 2007).

A recent expedition into the Bladen Nature Reserve in Belize documented 2 species of land snails unknown to science, at the bottom of a 100 meter deep sink hole. Image © Kasia Biernacka.

# PHYSIOGRAPHIC REGIONS OF BELIZE

Wright *et al.* (1959) describes the land form of Belize as a low-lying shelf on an embayed area on the eastern side of the Central American isthmus with the southern half being up-lifted over 1000 meters forming the Maya Mountain Massif. To the north and south are sub-sidiary limestone masses. Taken together, topography is mountainous to hilly over 60% of the country with 40% low lying coastal plain occupying the remainder.

The authors divide Belize into 7 physiographic regions (see map on page 13). The following physiographic unit descriptions are drawn from Miller (1996), Marshall (2007), Hammond (1982), and Wright *et al.* (1959) in order of importance and terminology.

**1. Eastern Block-faulted Coastal Plain** (on Yucatan Platform): Lowland coastal areas of the north are characterized by broad lagoons and fed by stream networks draining carbonate terrains. Faulting has caused northeast-southwest trending rivers. Rio Hondo and New River are examples of rivers defined by faulting. The New River (Rio Nuevo) forms New River Lagoon, the largest body of freshwater in Belize, and faulting is evidenced by the New River Escarpment. To the extreme north occurs the Dry Tropical Forest Biome preserved in Ship-stern Nature Reserve. Evidence of Chixalub impact crater ejecta has been found on Albion Island.

**2. Peten Karst Plateau**: This unit includes the Yalbac Hills and continues into the uplands of eastern Guatemala. Dolines (shallow depressions), dry valleys, and karst windows occur throughout the Yalbac Hills. A series of prominent limestone escarpments occur in the west-ern part of Orange Walk District. Elevations in Yalbac Hills are 200-250 meters. Drainage is mostly surface. Upland lagoons, a favorite habitat for Morlet's crocodiles, occur. Unlike many areas of Belize, cockpit karst has not developed. Within this area, uplifting of the un-derlying Yucatan Platform has resulted in gradual dipping and faulting. This has produced a series of escarpments trending north-northeast to south-southwest. These are demarcated by Booth's River and Rio Bravo (Chan Chich). A steep escarpment, Yalbac Escarpment, occurs along the northern edge of Labouring Creek, which runs west-southwest.

**3. Peccary Hills**: Also called Manatee Hills, this unit is dominated by Tower Karst sur-rounded by Savanna. Tower Karst is the end of an evolutionary sequence of fluviokarst to cockpits to towers. These towers are well isolated from the main Maya Mountains unlike similar structures in southern Belize. The towers occur in roughly an arc from Dangriga to Belize City. Many towers can be seen along Manatee road. Gracie Rock, an example of these towers, is mentioned often in Belize scientific literature. Talus has developed at the base of some towers and bajos (wet depressions) are common, providing an important aquatic re-source during the rainy season, drying with onset of the dry season. Caves and shelters are common. Tower Karst summits can rise 200 meters above flat valley floors.

**4. Vaca Plateau**: This is the largest karst area in Belize (1000 m$^2$) and is part of a karst plat-form of Cretaceous carbonates extending into Belize from central Guatemala. There are rugged and steep local reliefs of 100-150 meters and increases in general elevation from 400 meters in the north to 700 meters in the south. Portions of this fault-bounded highland block may have escaped submergence since Permian times and thus could have provided a bio-logical refugium during times of isthmus inundation. There are two distinct sections. The northern two-thirds is a fluviokarst landscape where the dominant land forms are valleys cut by surface rivers. A cave of note in this section is Rio Frio Cave. Interest-ingly, the floor of the main passage of this cave is granite. The southern section is dominated by a cockpit karst, with less evidence of a fluvial past. Caracol archaeologi-cal site is in this section. Vaca Plateau also contains the Chiquibul Cave System, the largest hydrologically-linked cave network in Central America.

**5. Maya Mountains**: This includes Cockscomb, Bladen, and Little Quartz Ridge. The Maya Mountain Massif dominates the geologic structure of Belize and has contributed to the development of other physiographic units. The two highest points are Doyle's Delight (1124 meters) and Victoria Peak (1120 meters), both in the Cockscomb Range. This raised fault block (horst) is chiefly hard non-soluble metasediments and igneous intrusions and very ancient rock (Paleozoic). The main mass is made of quartz-rich rocks and thus low in plant nutrients. Topography of the Maya Mountain Massif is extremely rugged and steep (mean slope = 35°) and cut by numerous knolls, sinkholes, caves, streams, and rock outcrops. Geology is varied, approximately 62% is underlain by limestone, 22% extrusive volcanics, and 16% by alluvium and sedimentary formation. The southern Maya Mountains receive over 200 inches of rainfall per year. The geological, elevational, and meteorological variation all contribute to one of the most varied ecological units in Belize.

**6. Pine Ridge-Savanna Coastal Plain**: The essential difference in coastal plain sections of north and south is that east of the Maya Mountains the coastal areas are characterized by a series of small deltas. These deltas are made up of coarse clastic sediments eroded from the metamorphic and igneous interior of the Maya Mountain Massif. Pine ridge in Belize does not refer to a topographic ridge but rather a strand of pines due to the well-drained dry conditions.

**7. Hokeb Ha Fault Ridges** (or K-T Fault Ridges): We use the Maya name for Blue Creek Cave as the preferred unit name. This unit is the most southerly of the karst regions of Belize. The age is late Cretaceous to Tertiary Paleocene (K-T), and consists of long, isolated, block-fault ridges trending southwest to northeast, largest being 20 kilometers long. The longest ridge contains Blue Creek Cave (Hokeb Ha). The extent of the ridges is about 300 m$^2$ and ridges rise above the Pine Ridge-Savanna Coastal Plain some 220-380 meters. Ridges exhibit cockpit topography, with some cross-cutting valleys suggesting a former fluviokarst.

See **Physiographic Units** on next page (Map by Piper Roby of Copperhead Consulting, Paint Lick, Kentucky, USA).

# PHYSIOGRAPHIC REGIONS OF BELIZE

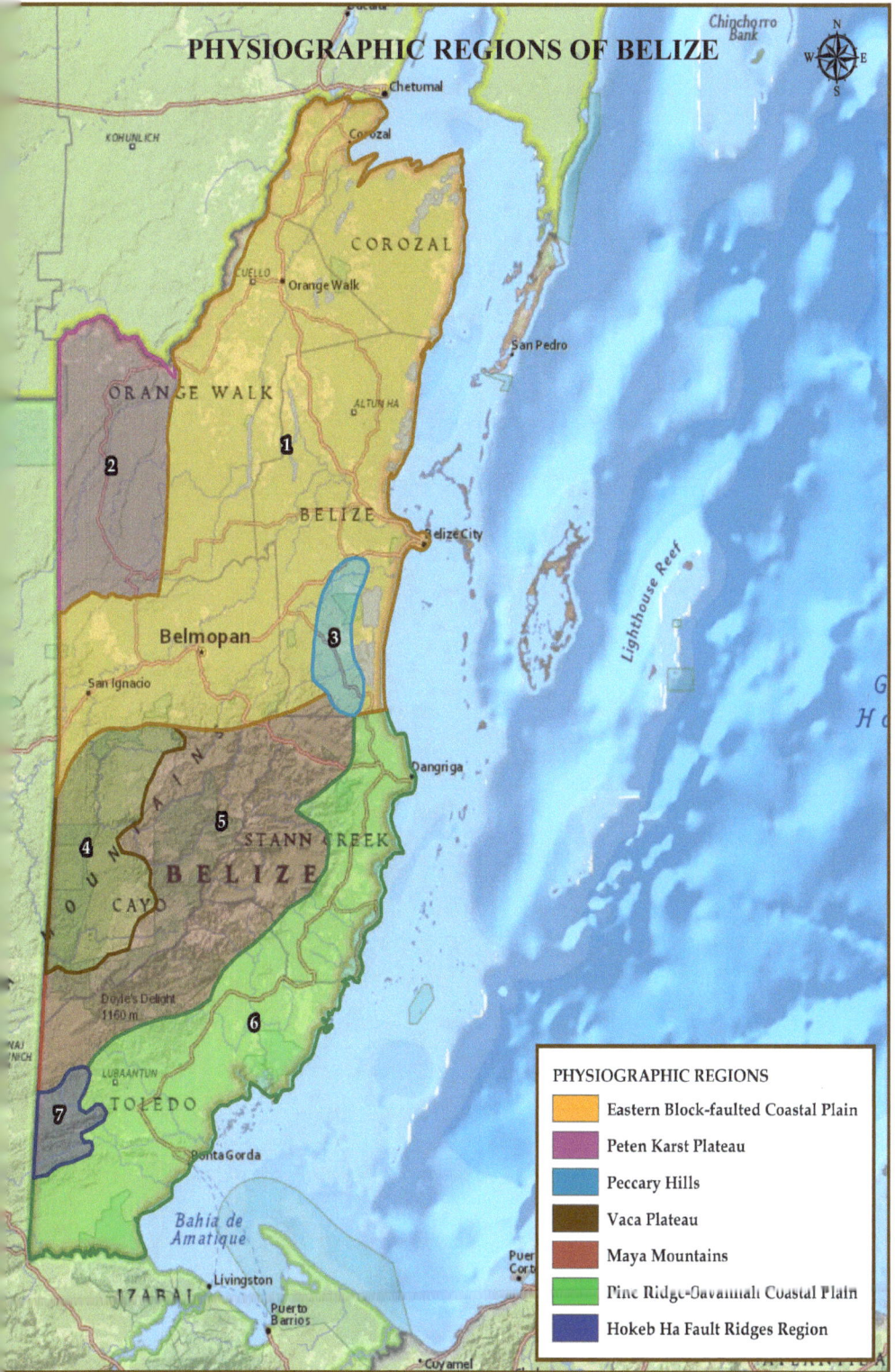

COROZAL

ORANGE WALK

**1**

**2**

BELIZE

**Belmopan**

**3**

STANN CREEK

**5**

**4**

BELIZE

CAYO

**6**

*Doyle's Delight 1160 m.*

TOLEDO

**7**

*Bahía de Amatique*

Lighthouse Reef

Chinchorro Bank

Chetumal
Corozal
Orange Walk
San Pedro
Belize City
San Ignacio
Dangriga
Punta Gorda
Livingston
Puerto Barrios

## PHYSIOGRAPHIC REGIONS

| | |
|---|---|
| | Eastern Block-faulted Coastal Plain |
| | Peten Karst Plateau |
| | Peccary Hills |
| | Vaca Plateau |
| | Maya Mountains |
| | Pine Ridge-Savannah Coastal Plain |
| | Hokeb Ha Fault Ridges Region |

**Study Area**

# Land Snail Habitats of Belize

Stephen Alvarez

Hundreds of sinkholes are found throughout Belize and are entirely unsampled

Steamy limestone knobs of extreme southern Belize (entirely unsampled)

15

# Land Snail Habitats of Belize

Volcanic outcrops covered in elfin forests of the Cockscomb Basin (unsampled)

Macal River Gorge near Black Rock Lodge (limestone outcroppings in the distance)

Limestone cliffs above the Macal River (fairly well sampled)

# Land Snail Habitats of Belize

Low karst foothills and plains of northern and southern Belize (mostly unsampled).

Areas along the Maya Mountain Divide, above Lost Valley, BNR (mostly unsampled)

Lower elevation karst hills of central Belize (mostly unsampled)

17

# Land Snail Habitats of Belize

Epiphytes of trees (unsampled)

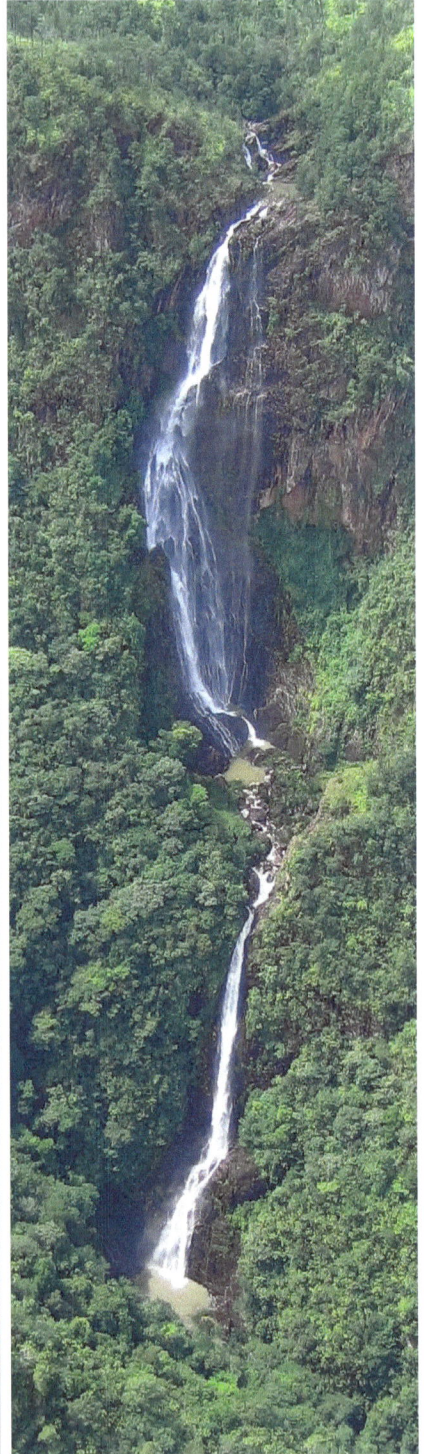

Spray zones of waterfalls (unsampled)

© Astrum Helicopters-www.astrumhelicopters.com

# Land Snail Habitats of Belize

Entrance to one of numerous caves in the Runaway Creek area (mostly unsampled)

Flowstone formations of caverns like ATM Cave, Cayo District, Belize (mostly unsampled)

19

# Land Snail Habitats of Belize

Isolated limestone knobs, Gracie Rock, Belize District, harbors several endemic land snails

Few land snail species survive in the burned pine savannas of middle and southern Belize

# 2 Land Snail Chronicles

Land snails come in every color, form and size imaginable; many beyond belief. Take, for example, terrestrial gastropods in the genus *Opisthostoma* (three species illustrated above). These eye-catching gems have taken calcium carbonate sculpturing to extremes. No less complicated than a Roman cathedral, their architecture represents one of the most spectacular achievements in convoluted evolution ever seen in a land snail. The genus is known only from limestone outcrops and caves of Borneo. (Illustrations by Jaap Verneulen).

As part of a large and diverse group of organisms known as Gastropods, land snails fall within the prodigious and immense Phylum Mollusca. Containing around 85,000 described and living species world-wide, mollusks include aquatic snails (both marine and freshwater), land snails, terrestrial slugs, sea-slugs, sea-hairs, limpets, bivalves (clams, oysters, and mussels), squids, octopuses, and the famous nautiluses. Octopuses are considered the most neurologically advanced invertebrates on earth, having feet divided into a number of prehensile and skillful tentacles capable of twisting the lid off a glass jar to access the food within. Although snails are slow moving, squids are among the fastest animals known, exceeding underwater speeds of more than 70 mph in short bursts, a result of their water-propelled jet propulsion. At 18 meters long and weighing in at around two tons, the colossal squid is the largest living invertebrate having 15-inch eyes, the biggest of any animal. Mollusks are also the longest lived, multi-cell animals, one species of bivalve, the Ocean Quahog, *Arctica islandica,* living more than 500 years.

There is scarcely a place on the dry surface of the world, outside the polar regions, where one cannot find at least a few examples of land snails (Abbott 1989). From hot, nearly waterless deserts to snow-capped mountain tops, land snails are thriving. Despite such biological and physiological limitations, land snails have developed efficient mechanisms for coping with freezing, starvation and desiccation. For example, when conditions become increasingly dry, land snails may cover the apertures of their shells with epiphragms, mucous sheets that harden, sealing in critical moisture and preventing desiccation. Other snails have calcified opercula capable of closing, also sealing in moisture. With these mechanisms, some snails can remain dormant for years, resuming activity during wet weather. Land snails in Belize are typically nocturnal animals and are largely active in warm (above 50 degrees Fahrenheit), rainy weather.

Even though mollusks rank as one of the most numerous and speciose groups of organisms on Earth, they remain largely unstudied. As a result, little is known of their importance in ecosystems. Land snails, like most invertebrates, suffer from being in a conservation "blind-spot". As snail research moves forward, our understanding of the value of these organisms is increasing. Research has shown, for example, that land snails play an important role in micronutrient cycling in terrestrial ecosystems (Dallinger et al. 2000), disperse plant seeds and fungal spores (Richter 1980) and have been shown to predict vertebrate conservation priorities (Moritz et al. 2001). Further, live snails and their vacant shells provide a food and calcium carbonate source to many systematic groups. These include but are not limited to ants, firefly larva, snail-killing flies (Foote 1959); *Cychrine* beetles, which feed chiefly on land snails (Symondson 2004), Arachnids including harvestmen, carnivorous snails, numerous species of salamanders (Petranka 1998), turtles and frogs (Burch 1962), a variety of small mammals including shrews, mice, and moles (Reid 2006), snakes (Lee 1994, Dourson et al. 2012), a variety of passerine birds (Graveland et al. 1994; Graveland 1996), thrushes, ruffed grouse and wild turkey (Martin et al. 1951), bats (Bonato et al. 2004; Thabah et al. 2007) and primates including humans.

While a building body of evidence suggests the importance of mollusks in present-day ecosystems, their historical value is less well known; namely their contributions made to existing plant communities, animals and, in particular, caves. In the past, the colossal accumulation of deceased mollusks, corals, and tiny creatures known as Foraminifera (that have calcareous skeletons), provided the necessary building material to create limestone, where caves are essentially formed. Many species uniquely adapted to caves, roost or otherwise live in these vast underworlds, a number of species occurring nowhere else. These ancient shells have also provided the necessary limestone in cement to form the foundations of our cities and homes.

Declining land snail populations can have ripple effects to surrounding ecosystems. The great tit (*Parus major*) in the Netherlands, for example, has declined precipitously with declining land snails as a result of acid rain (Graveland et al. 1994; Graveland 1996). A lack of snail shells in the bird's diet causes their egg shells to thin and break, therefore reducing reproductive success rates of the species. In North America, Hames et al. (2002) have documented a correlation between a reduced number of wood thrushes (a winter resident of Belize) and acid rain, hypothesizing a connection to reduced land snail populations.

Sensitive to changes in their environment, native land snails could provide an early warning to impending habitat deterioration, similar to the way that freshwater snails found in streams and rivers are used to determine the quality of waterways. Research has shown that when snails feed on contaminated foods such as mushrooms, green vegetation and forest litter, environmental toxins are ingested and sequestered in their tissues (Dallinger and Wieser 1984a), the midgut gland being the main accumulation site of these trace elements (Dallinger 1993). Further laboratory experiments by Dallinger and Wieser (1984) found that land snails who fed on lettuce laced with zinc, cadmium, lead and copper easily sequestered these elements. More concerning, snails quickly become poisoned when simply raised on soils contaminated with cadmium raising fear that toxins in polluted soils may be more bio-active than previously believed (Scheifler *et al.* 2003a). The environmental implications to ecosystems and the consequence to snails and higher organisms that feed on contaminated gastropods are valid as well as alarming. Rimmer *et al.* (2005) found elevated levels of mercury in the blood of Bicknell's thrush on mountains in New England and thrushes are reported to eat snails (Martin *et al.* 1951), presumably to obtain calcium for egg laying. Native land snails could therefore be used to forecast impending problems created by anthropogenic pollutants reported to be accumulating in jungles of Belize. This hypothesis however, remains untested in the country.

## Land Snails as Pests
Land snails, including slugs, are often considered agricultural pests. Yet in most cases, the problem snails and slugs are non-native introductions or exotic. Accidently introduced into Belize by way of plants, potting soils or shipping crates, these exotics can naturalize quickly and multiply. Degradation of Belize's native habitats only makes things worse by providing a conduit for the exotic species to disperse, moving into new uninfested areas of the country. Not only can these exotics cause problems in your garden and to crops, they often carry molluscan diseases and problematic parasites that can effect domestic animals, livestock, native wildlife, and even humans. Exotic slugs can be

especially damaging pests in greenhouses and agricultural lands costing millions of dollars worth of damage. In contrast, native snails and slugs rarely cause trouble as most native species become scarce or disappear entirely where the natural vegetation has been eliminated.

The giant African land snail (the size of an orange) is a major pest on agriculture crops including bananas causing millions of dollars of damage every year. This species is spreading across the globe and is now reported from several Caribbean islands including Barbados where it is a major crop nuisance. If seen in Belize, these invasive pest snails should be quickly eliminated. Caution: it is important to distinguish the differences between the exotic Giant African land snail and the much smaller, yet similar-looking native snail, Princess Cone, *Orthalicus princeps princeps* (see below).

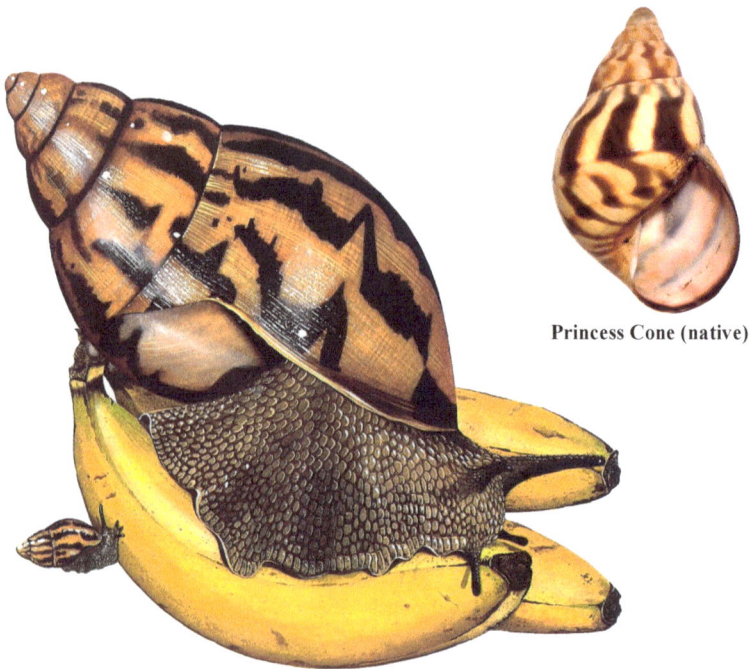

Princess Cone (native)

Giant African Land Snail (exotic), image from web

## Parasites and Snails

Although nearly every kind of mollusk is inhabited by some form of parasite, only a few gastropods are of medical or veterinary importance (Burch 1962). Of these, almost all live in freshwater environments. Snails are required hosts in the life cycle of parasitic trematode worms. A few land snails like *Cochlicopa lubrica* found in the USA are vectors of lancet liver flukes in sheep, cattle, deer, and groundhogs (Burch 1962). *Zonitoides arboreus,* a native snail to Belize, has been implicated in the spread of lungworm in domestic sheep. Many genera of land snails have been known to be intermediate hosts for *Pare-*

*laphustrongylus tenuis*, the North American deer parasite called brainworm. *Veronicellid* slugs found in Belize and the Giant African Land Snail (not yet recorded from Belize) are intermediate hosts for *Angiostrongylus costaricensis*, a rat parasite nematode also called rat lungworm that can cause eosinophilic meningitis or Morera's disease in humans. The parasite can be transferred to people when infected juvenile slugs (sometimes found on lettuce and other vegetables) are accidentally eaten raw or from contact with the slime of the Giant African Land Snail. The rat lungworm has evolved to use land snails as an intermediary host, spending part of its life cycle in a rat's blood system and brain. After a rat eats an infected snail, thousands of worms (up to 60mm long) grow in the rat's brain. In humans, the worms cannot find their way out so the worms soon die which causes an immune response and brain swelling. Symptoms range from headaches to tingling, numbness and involuntary flexing of muscles. In severe cases, sufferers may go into coma and die. The parasite has been observed in Costa Rican children since the early 1950's.

## Medical Mollusks

Mollusks support a number of contributions to human health. Marine mollusks may help fight liver cancer through the development of a drug that uses Kahalalide F, a protein extracted from a species of mollusks that eat sea slugs in the Pacific Ocean. Lethal toxins produced by cone snails, also marine snails, are used to develop a non-addictive drug called Ziconitide for patients with cancer and AIDS who suffer from chronic pain that cannot be relieved by opiates. Slime from the land snail, *Helix aspersa* (one of the commonly eaten snails referred to as escargot) is now used to treat many different types of skin disorders. Snail slime is reported to reduce scarring, repair skin damage from overexposure to the sun and reduce scarring caused by severe acne. The mucus of snails is known to contain antibacterial properties but remains largely unstudied. In Belize, the Maya use gastropods to treat a number of ailments including skin disorders, warts, glaucoma and whooping cough.

## The Diet of Land Snails

Most snails are dietary generalists (Burch and Pearce 1990), consuming a wide variety of herbaceous plant leaves or stems, decaying vegetation and leaf litter (detritus), wood or bark and fungal fruiting bodies such as mushrooms and wood-inhabiting shelf or bracket fungi (Burch and Pearce 1990; Dourson 2008). In Belize, most species in the family Spiraxidae are known carnivores, feeding on a variety of gastropods (pers. comm. Fred Thompson). Land snails sample and judge potential food by using the chemoreceptors located on the lower two tentacles (Shearer and Atkinson 2001). Once a food source is determined, the snail begins the feeding process by first touching the food with its foot and mouth. This is followed by the rasping of bits of food with the radula structure located in the mouth (Machensted and Markel 2001). The meal is then swallowed and muscular contractions move the food along the esophageal tract mixing it with saliva. Feeding episodes can last anywhere from a few minutes to nearly an hour, depending on the durability of the food being consumed (Dourson 2008). All gastropods, shelled species in particular, require a source of calcium carbonate such as limestone necessary for body functions, reproduction, but most importantly, shell building (Burch 1962).

## Reproduction

The majority of land snails are hermaphrodites, each individual having ovatestes in which both sperm and eggs are produced. When two individuals of the same species in search of propagation through scent trails find one another, they typically exchange sperm. Sperm can be stored for months to years by each individual snail until conditions are favorable for fertilization and egg laying. Eggs are deposited under logs or in moist leaf litter. Interestingly, many land snails including several native slugs of the US in the genus *Philomycus,* are characterized by the presence of "love darts".

Chase and Blanchard (2006) studied the complicated sex life of the European land snail *Cornu aspersum* (also known as escargot) that use calcareous love darts as part of their courtship. Their research revealed that the love darts are not shot but are forcefully expelled through the body wall of the partner. The dart itself appears in a variety of sizes and shapes with most species containing a single dart that is used only once. In his research on this fascinating sexual behavior, Dr. Joris Koene learned that some land snails employ multiple darts while others use the same dart to stab their partners repeatedly, as many as 3000 times. But what exactly is the function of the love dart?

Love dart →

Chase and Blanchard (2006)

Further study indicates that the dart functions after copulation to increase the reproductive fitness of the shooter. Snails are promiscuous and store sperm from multiple donors for several years before they use it to fertilize eggs. Thus, sperm donors must compete to fertilize eggs. Dart receipt promotes the safe storage of the shooter's sperm so there will be more sperm from successful shooters available for fertilization than from unsuccessful shooters. Since the female function chooses the sperm by a lottery-like mechanism, successful dart shooters sire more babies than unsuccessful dart shooters. Chase and Blanchard (2006) tested whether the dart works by simply rupturing the skin or by injecting a bioactive agent. Just before the dart is thrust into a partner, it is covered with mucus

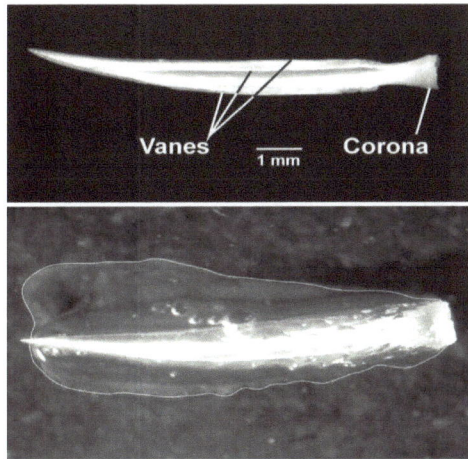

Chase and Blanchard (2006)

(figure a) from a special gland located near the dart's launching site. Koene *et al.* (2013) conducted an interesting test in which needle stabbings were substituted for dart shooting. In one mating, saline was injected through the needle, in the other mating mucus was injected. They found that the mating associated with mucus injections were responsible for more than twice the number of offspring as were mating associated with saline injections. Thus, mucus is the agent of the dart's effect on reproduction.

## Defense Strategies used by Land Snails

Land snails use a variety of strategies to protect themselves from harm. The shell is the first line of defense and works well to ward off a number of predators. Conversely, there are animals such as small mammals, birds, insects and carnivorous land snails that routinely hunt and consume snails. Snail predators such as birds and small mammals will consume the entire shell, chew through the shells to extract the snail flesh or, in the case of the beetle larva, drill a hole through the side of the shell to extract the live animal. Other defense strategies include the hairs found on *Trichodiscina* species (page…) which help camouflage already cryptic shells by picking up forest debris. Another defense strategy that has stood the test of time are lamellae barriers (or sometimes called teeth) located in and around the snail's aperture. These structures have evolved to prevent *Cychrine* beetles from entering the shell, keeping the snail safe from harm. Teeth and aperture barriers may also provide a calcium storehouse to repair damaged shells or act as pivotal points for balancing the shell during the snails movement forward. Although slugs are without protective shells, they are not defenseless. Slugs are well-endowed with copious and stickier mucus than shelled snails, the slime containing repugnant substances that repel predators. For this reason, few animals are eager to grab and dine on a gummy-slug.

## Empty Snail Shells

When snails die, the shells do not immediately decompose. Some research suggests that shells can remain intact for years (Pearce 2008) and empty shells in some limestone locations can reach exceptional numbers. But these discarded shells are anything but vacant and actually provide a secure, protected refuge for a whole host of living micro-invertebrates including pseudo scorpions, ants, millipedes, tardigrades and on occasion, other smaller land snails. Some invertebrate species even deposit eggs to be incubated in the security of shells.

## Fluorescence in Land Snails

Not to be confused with bioluminescence (the natural light observed in fireflies), fluorescence is the term used to describe the absorption of light at one wavelength and its emission in another. Only one species of land snail is known to produce bioluminescence, *Quantula striata,* from Malaysia. Research has shown, however, that the mucus of several North American family groups of land snails including Discidae and Helicodiscidae have fluorescent slime under ultraviolet light. The slime of the Flamed Tigersnail, *Anguispira alternata* (see pictures below) has a particularly brilliant, bluish fluorescence (Rawls and Yates 1971). Under laboratory conditions, the slime on the foot, body and the mucus trails of snails in the genera *Anguispira* and *Discus* glowed brightly when exposed to UV light. Other species of Polygyrids were tested and all others failed to exhibit any sign of fluorescence. The fluorescence which was observed in the specimens of the three genera noted is extremely long-lived, being as bright and as distinctive in specimens preserved for twenty years or more as it is in living snails (Rawls and Yates 1971). Interestingly, the crawling slime of *Anguispira jessica* that is typically clear did not fluoresce under UV exposure but the defense slime produced by the species under attack which is orange-yellow did fluoresce.

The function of fluorescence in land snails remains a mystery. There is speculation that fluorescent slime is simply a random act of evolution. A functional hypothesis for the fluorescence in land snails was proposed by Dourson (2012). It turns out that moonlight has a component of UV light and *Anguispira* species are most active at night. What if certain nocturnal land snail predators like snail-hunting beetles are avoidance-conditioned to the fluorescent defense slime the same way that predators are trained to steer clear of the

The orange defense slime of the Flamed tigersnail, *Anguispira alternata,* (left) fluoresces under UV light (right).

28

bright colors of coral snakes? Snail slime, especially that of *Anguispira* species, is pungent and distasteful to animals, including humans.

*Anguispira* slime has a disagreeable, numbing affect in one's mouth (Dourson, personal experience). Like the coral snake that uses color to conserve its venom, snails may be using florescence to conserve precious mucus reserves. The more snails are harassed, the more mucus is produced so losing copious slime from an attack would put the snail at risk of dehydration. It is not known if there are any land snails in Belize that possess fluorescent slime. This awaits further investigation!

# 3 The Value of Snails in Belizean Ecosystems

The greatest animal diversity on Earth belongs not to the mammals nor birds nor even reptiles. That distinction goes to the invertebrates, animals we seldom give a second thought to unless they are dining on our blood or eating our garden. They represent a staggering 95% of the total animal species recorded. For their size, they are faster, stronger, can jump farther and live longer than mammals and birds. Their life cycles are generally more complex and interesting than higher order animals. These amazing creatures are without a backbone and include the well-known group of arthropods referred to as insects (having six legs), arachnids (spiders and scorpions, having eight legs), the not-so-familiar myriapods (millipedes and centipedes having multiple legs) and the non-arthropod invertebrates (mollusks: snails and mussels).

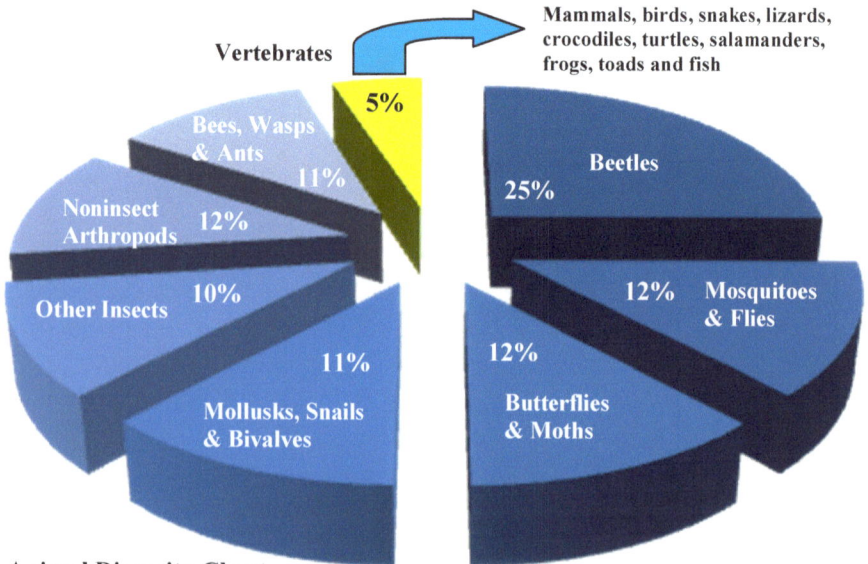

Vertebrates — Mammals, birds, snakes, lizards, crocodiles, turtles, salamanders, frogs, toads and fish

5%

Bees, Wasps & Ants 11%

Noninsect Arthropods 12%

Other Insects 10%

Mollusks, Snails & Bivalves 11%

Beetles 25%

Mosquitoes & Flies 12%

Butterflies & Moths 12%

**Animal Diversity Chart**

Clearly, along with plants, invertebrates are the driving engines of life. Is it any wonder? Invertebrates have been on Earth much longer and have survived a precarious, shifting planet. Natural lifecycles would slow down and cease to function properly without the significant and largely unquantified services provided by these organisms. Many vertebrates like birds and bats depend upon invertebrates like mollusks as important food sources, akin to the power bars of the food web.

With such diversity and adaptability, some say that invertebrates will inherit the Earth, while entomologists emphatically state that they already own it.

While there remains a dearth of information and understanding of the value of land snails in Belize, a building body of evidence suggests that snails are even more important than we ever imagined. Snails are a significant food source to small mammals like rats, mice, shrews, bats, otters, American and Morelet's crocodiles, freshwater turtles, a variety of birds such as oscillated turkeys (suspected), great curassows (suspected), crested guan (suspected), egrets and herons, snail and hooked billed kites, snakes (*Sibon* species), lizards (suspected), frogs (suspected), salamanders, carnivorous land snails and invertebrates. Snails are part of the smaller majority of organisms that are considered building blocks of ecosystems.

The following observations and information further supports the importance of land snails to the health of ecosystems in Belize and documents some behaviors not previously recorded. Hears what we now know.

## Mammals
The river otter along with its usual diet of fish also consumes large quantities of freshwater mussels and aquatic snails. Piles of mussel shells found on gravel bars indicates the feedings of a river otter. Opossums and coatis, known to be consumed by the harpy eagle, include snails as part of their complex diets. Small mammals such as moles, shrews, mice and rats also eat land snails as do several species of primates.

## Bats
In an intriguing relationship between gastropod and predator, a remarkable bat feeding behavior on gastropods has been documented deep within the Maya Mountains. Along massive outcrops of limestone under scattered overhanging rock-ledges, multiple shell fragments containing puncture marks from bat canines along with bat guano containing fragments of snail shells were found. At first glance, the discovery of these shell fragments were attributed to rodents that inhabit much of the same habitat. However, small rodents feed on land snails by biting a hole around the apex of the shell (leaving the shell essentially intact) through which snail flesh is then extracted and consumed (Dourson 2007). The shells found under overhanging rock in the Bladen Nature Reserve were in multiple fragments, illustrating a very different feeding strategy than found in rodents. Furthermore, shell fragments were also found in bat guano collected from these overhangs, making a direct link between bats and a diet of land snails.

31

The distance between canine puncture marks on several shells suggests the marks may belong to a fringe-lipped bat, an opportunistic feeder known to consume a variety of foods including frogs, lizards and even small rodents (pers. comm. Price Sewell 2012). But puncture marks have varied considerably in spacing on other snail shells recovered, so it can be assumed that a variety of bat species may be using the overhangs as feeding sites. Snail shells found also indicate that bats seem to consume snail flesh only, not the entire shell that has been reported in other studies (Bonato *et al.* 2004; Thabah *et al.* 2007). Bats also are reported to use the same night roost for consuming food caught during night feeding events. This interesting feeding behavior is believed to be the first reported observation of its kind and continues to be investigated.

Fringed-lipped bat with a Milky Cone, *Orthalicus livens,* Belize

Carol Foster

## Birds

Bats are not the only winged predators feeding on land snails. There are birds in Belize that hunt the protein rich flesh of gastropods but many of the exact culprits remain unknown. The evidence: hundreds of small stones covered in fragmented shells found throughout the jungle. One possible predator are wood thrushes, a common winter resident in the dense jungles of the Maya Mountains. In Great Britain, song thrushes can be a major predator on adult wood snails, *Cepaea nemoralis*, crushing the shells on stones to get at the soft snail within (Whitson 2005). Some shells found at these "breaking stone stations" (pictured below) are quite large and structurally solid; leading to the speculation that larger bird species like tinamou, great curassow, crested guan and oscillated turkey are likely consumers of land snails.

A typical "breaking stone station" in the Bladen Nature Reserve, Belize

While it is common knowledge that the snail kite's diet is aquatic apple snails, the diet of the hook-billed kite is less well documented. Recent observations by Ryan Phillips and Jan Meerman of the Belize Raptor Research Institute (BRRI) have made some rather remarkable discoveries about the hook-billed kite in Belize. Their research of a few nesting pairs in the Maya Mountains suggests that hook-billed kites feed almost exclusively on Cross Cones, *Orthalicus princeps crossei* (99%) and *Euglandina ghiesbreghti* at only (1%) (pers. comm. Jan Meerman, 2016). The picture on next page shows broken Cross Cones at the

Jan Meerman

Broken Cross Cones harvested and eaten by a Hooked-billed Kite, Belize

base of a tree were harvested by hook-billed kites from unknown locations in the jungle. Snails were brought to the roost tree, broken from the bottom side (above image figure, a), at which time snail flesh was extracted for consumption or fed to the young kites. The feeding tree was located about 30 meters from the actual nest tree (pers. comm. Jan Meerman, 2016). Where the shells are found and how the shells are broken can reveal a great deal about the snail-predator. It should be noted that shells are broken or accessed very differently than seen in bats, small mammals, and other birds.

Even the largest raptor in the Americas, the Harpy Eagle, indirectly depends on land snails for its survival. For example, harpy eagles feed on crested guans, monkeys, opossums and coatis, all which are reported to include snails as part of their varied diet.

Bat falcons feed on bats which feed on land snails. In fact, many birds feed on snails directly for the protein rich meats for the calcium carbonate found in the shells or feed on something that feeds on snails. These interesting discoveries in Belize clearly illustrate the importance of land snails to avian species and await further investigation.

### Snakes

Dipsadine snakes within the genus *Sibon* are generally referred to in the literature as feeding exclusively on slugs and snails (Peters 1960). However, there is little or no information available regarding which species of gastropods these snakes eat. Furthermore, there is some suspicion that the diets' of several snakes in the genus *Sibon* occasionally include non-molluscan prey

Above image of a Speckled Snail Sucker shows little interest in a Cross Cone, even moving away to avoid a confrontation. Bottom picture of the same snake feeding on a Banded Cone, its preferred food. Note the lower jaw of the serpent rests inside the snail shell where a paralyzing saliva is introduced into the gastropod by lacerations made by its teeth. The saliva will begin to immobilize the snail until it ceases to struggle and the snake begins extraction of the gastropod flesh. Bladen Nature Reserve, Belize

A Speckled Snail Sucker stalking and consuming a slug, Maya Mountains, Bladen Nature Reserve, Belize.

(Montgomery *et al.* 2007). During several studies conducted recently, various native foods were offered to eight individuals of the Speckled Snail Sucker, *Sibon nebulata* (Dourson *et al.* 2012). Feeding habitats were observed and recorded for each snake. Although the behavior witnessed was based on a meager 25 feeding events, certain characteristics remained surprisingly consistent, especially when the snakes fed on land snails in the genus *Drymaeus*. Before attempting to extract the snail flesh, *S. nebulata* made deliberate and calculated assessments of prey size, movement and position. When the snail moved, the snake used its tongue to delicately touch the snail, stopping all movement forward (pers. comm. Richard Foster 2012). Occasionally, the snail withdrew into its protective shell but the snake simply waited with great patience until the snail re-emerged. At this point, the snake began to hover over the snail, turning its head in contorted angles as it searched for an ideal strike angle, being careful not to make premature contact with the gastropod.

Next the snake carefully aligned its lower jaw with the lower opening of the snail aperture. A strike then followed with the precision of a thread passing through the eye of a needle. Only the lower maxilla of the snake entered the aperture while the snake's upper mandible came to rest on the outer surface of the last body whorl of the shell. Without delay, the snake made a few lower and upper maxilla adjustments before seizing down on the live snail flesh, holding the snail securely until it ceased to struggle. The snake began extracting snail flesh only when all snail movement ended, which lasted upwards of an hour or more for *Drymaeus* species. The hesitation to eat the snail immediately may be a direct response to the effectiveness of the snake's saliva. Actual feeding and extraction of the gastropod flesh took around five minutes or less and 80-90% of the snail was usually retrieved by the snake. The shell remained intact and undamaged.

When eating land snails such as the larger *E. ghiesbreghti, S. nebulata* seem to execute the usual strikes as seen in other snake species, grabbing, yet sometimes missing whatever portions of the snail remain outside the shell. Upon securing the snail's body, the snake holds its prey without moving, as long as 24 hours or more until the snail ceases to struggle before it is eaten. Slugs eaten by *S. nebulata* are eaten without the hesitation seen in shelled species with whole feeding events usually lasting only a few minutes. *Sibon nebulata* was offered and fed on live snail species: *Bulimulus unicolor, Euglandina ghiesbreghti, Drymaeus emeus, Drymaeus sulphureus, Leidyula floridana,* and *Leidyula moreleti* (both slugs).

Other land snail species offered but rejected included *Orthalicus princeps, Helicina rostrata, Helicina amoena, Halotudora kuesteri, Halotudora gruneri* and *Neocyclotus dysoni.* Five of the above six species offered had operculum structures that covered the aperture like a trap door. Operculum structures that prevent desiccation of the snail in drought conditions may also prevent the snake's lower jaw from entering the aperture of the shell. Non-mollusk organisms offered the snakes included frogs, lizards, small snakes, earthworms, beetle larva and other invertebrates, all which were declined. All foods presented were found locally and native to the Maya Mountains with the exception of *V. floridana* (a native slug to Florida, USA and Cuba).

These feeding behaviors suggest that *S. nebulata* is able to distinguish shelled and non-shelled gastropods (slugs) and differentiate shelled species based on size and aperture shapes, perhaps through visual or olfactory senses. Clearly, *S. nebulata* uses different feeding strategies for dissimilar snail species. Further, the sequence of events strongly suggests that the snakes use immobilizing saliva to relax or even kill shelled snails before consumption. This is thought to make extraction of the muscular and slippery snail flesh easier for the taking.

*Sibons* are reported to have a greater number of teeth in the lower right jaw than the lower left jaw, an evolutionary response to land snails in this region that possess right-sided apertures (openings). This allows the snake to reach deeper into the shell, resulting in removal of a higher percentage of snail flesh.

**Fish**
Few studies have focused on the diet of freshwater fish in Belize. One study however by Cochran (2007) found that 62% of the diet of multiple fish species (i.e. cichlids, bay snook and mountain mullet) from the Bladen River came from three aquatic snails, *Pachychilus corvinus*, *Pachychilus largillierti*, and the small, non-native *Melanoides tuberculata* or Asian Thorn.

**Crocodiles**
Among the largest freshwater snails in the world, the apple snails, *Pomacea* species are bulky gastropods about the size of a lime found throughout Belize in stillwater lagoons, oxbows and slow moving rivers. A study by Platt *et. al* (2006) found that apple snails comprised as much as 70% of the diet of adult Morelet's crocodiles. Apple snails were present in the stomach contents of 22.1% of all the 450 individuals studied. Surprisingly, consumption of snails increased with an increase in body size of the crocodile. The smaller-sized crocodiles fed primarily on insects and arachnids while the medium-sized crocodiles broadened their diet to include apple snails, fish and other vertebrates, and larger crocodiles fed primarily on apple snails, fish, and crustaceans. In fact, the largest adult crocodiles studied were found to have 70.8 % of their total diet to consist of apple snails (Platt *et. al* 2006). The loss of aquatic gastropods such as apple snails due to water pollution, dredging, or other forms of habitat degradation would clearly threaten a host of wildlife species including the highly endangered Morelet and American crocodiles.

**Invertebrates**
Drill-holes found in Belizean land snails remain a mystery and are waiting further investigation. The larvae of drilid beetles (Coleoptera:Drilidae) are reported to prey on land snails (Lawrence 1991) and are mainly present in Austria, Belgium, Luxembourg, France, Italy, Spain, Germany and Switzerland,

but not reported from the Americas. According Baalbergen *et al.*, aestivating land snails in western Turkey protect themselves from predators with hardened mucus or a rock face, blocking entry through the shell's opening. This forces the drilid beetle larvae to drill a hole through the side of the shell in order to extract the snail flesh. Örstan also reported finding similar drill holes in *Cerion* species from Caribbean islands although the exact culprit remains anonymous. In Belize, most land snails aestivate in the dry season months of March through May, protecting the entrance in much the same way while some species possess an operculum, yet another defense strategy. This added protection may further induce beetle predators in Belize to employ the same strategy. It is not uncommon to find single or multiple drill-holes through the shells of snails in Belize. Even though the precise intruder remains a mystery, a form of beetle larva is a suspected predator, perhaps like those below although this remains to be more fully explored.

*Brachypodella speluncae*

**Belize snails**

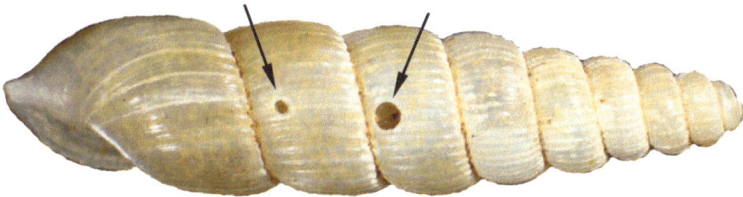

*Mayaxis martensiana*

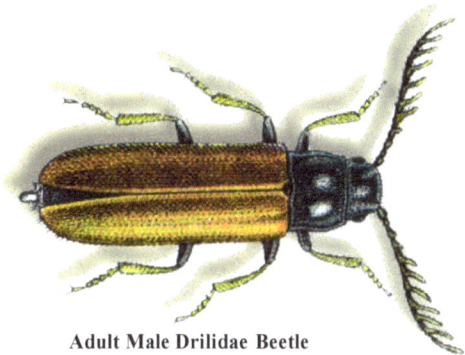

**Adult Male Drilidae Beetle**

**Adult Female Drilidae Beetle**

Images from Creative Commons, Worldwide Web

39

# 4 Collecting and Identifying Land Snails in Belize

## Site Selection
Collecting and identifying land snails in Belize begins by selecting an area to survey. When looking for land snails, searching specific habitats will give the collector the best opportunity to find land snails.

## Habitats
- Under leaf litter (the layers of leaves and first layer of soil)
- Underside of palm leaves
- Epiphytic orchids
- Rocky outcrops/Rock talus
- At the base of limestone cliffs
- Boulder fields
- Cave entrances/Caves
- Elfin forests of the volcanic regions of the Maya Mountains
- Sinkholes
- Rock crevices
- Under moss mats.
- Under bark of standing and/or downed dead trees
- Hollow trees or damaged trees oozing sap
- Under and on top of caps of mushrooms
- Seeps along streams
- Muddy lagoons and bajos
- Man-made features like roadsides, steep banks, retaining walls, cement structures, discarded bottles

## Field Equipment
- Ziploc bags (quart or gallon size)
- Leaf litter bags (cloth soil sample bags work best)
- Permanent marker (Sharpie)
- GPS unit or cell phone with compass (iPhone or other)
- Field notebook (preferably Rite-in-Rain or placed in Ziploc bag)
- Hand lens (4X power, if possible)
- Hand rake or forked stick for raking leaf litter

## Collecting Methods
Once a site has been selected, use the hand rake or forked stick to scrape leaves and loose soil to search for snail shells. Samples of larger (macro) snails 5mm and greater should be collected and placed in Ziploc bags with date, site num-

ber, GPS coordinates, and collector name written on the outside with permanent marker. Do not put paper labels in bags with live snails as they will eat the paper.

Samples of smaller (micro) specimens less than 5mm are best collected from leaf/soil collections. Collecting leaf litter can get unmanageable very quickly if you randomly scoop up leaf and soil material. It is best to determine micro-snail presence by collecting a handful soil/leaf litter and scanning the litter with a hand lens for evidence of micro specimens. If any snails are observed, a quart -sized cotton drying bag (or paper bag if drying bag is not available) is filled with the material from the site, labeled with the date, site number, collector's name and GPS coordinates. Leaf/soil samples are dried for approximately 2 weeks. In the rainy season, it may be necessary to place the samples in front of a fan to dry.

Dried samples should be sifted through a series of sieves ranging from 4.76 mm down to 500 micrometers. If sieves are not available, then material should be placed on a tray or pan and manually searched with the aid of a jeweler's or watch repair magnifier. In some cases, it may be necessary to use a dissecting microscope to determine the species of the small snails which have microscopic ornamentation.

## Identifying Land Snails
Successful identification of land snails often depends on your ability to notice shell detail. To identify a land snail, record observations of the following characteristics

**Shape:** Determine the general shape of the shell (see page 42). Shapes of same-species shells will sometimes vary slightly from site to site.

**Greater Diameter or Height:** Measure the greater diameter for heliciform shells and the height for cone or pupa-shaped shells (see page 42). Sizes will sometimes vary slightly.

**Whorls:** Count the number of whorls. (see page 45)

**Umbilicus:** Is the umbilicus imperforate, perforate, umbilicate, or rimate? (see page 44)

**Teeth:** Make note of any presence or absence of teeth in the aperture , their size and location on the aperture. (see page 43)

**Lip**: Is the lip simple or reflected? (see page 43)

**Micro-Features of Shell:** Determine the presence of any micro ornamentation (spiral striae, transverse striae, hairs, bumps, or wrinkles) on the shell by using strong magnification or a dissecting microscope. (see page 45).

Remember, the shell's shape (form), size and micro-features are the most important diagnostic features. In old shells that are bleached, slightly wetting their surface will sometimes bring out the micro-features.

# Basic Shell Shapes of Terrestrial Land Snails

Depressed heliciform

Heliciform

The shape of this shell lingers between depressed heliciform and heliciform so consider both shell shapes.

Roundish or Dome-shape

Tuba-shape

Succiniform

Oval-shape

Cone or Conical shape

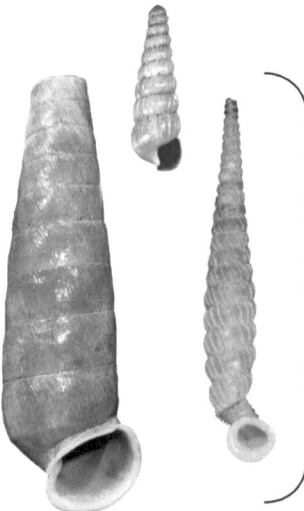

Cylinder-shape

Pupa-shape

42

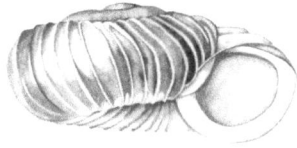

## Shell Morphology

Pictured (right) are three standard shell views used by malacologists to identify land snails. Always observe snails in these views. The *frontal view* shows the shell's general form and aperture shape. The *bottom view* shows one of the most important diagnostic features of any land shell, the umbilicus region. The *top view* shows the apex or embryonic whorl, the number of whorls, and the width of the whorls. In general, the frontal and bottom views are the most important diagnostic views of land snail shells. While the top view has limited value for separating snails within the same genus, it is a reliable view for separating snails of different families. Short black arrows or lines indicate key features of shells. All measurements used in this book are for greater diameter or height of adult shells.

**Bottom view**

**Top view**

## Terminology of the Shell

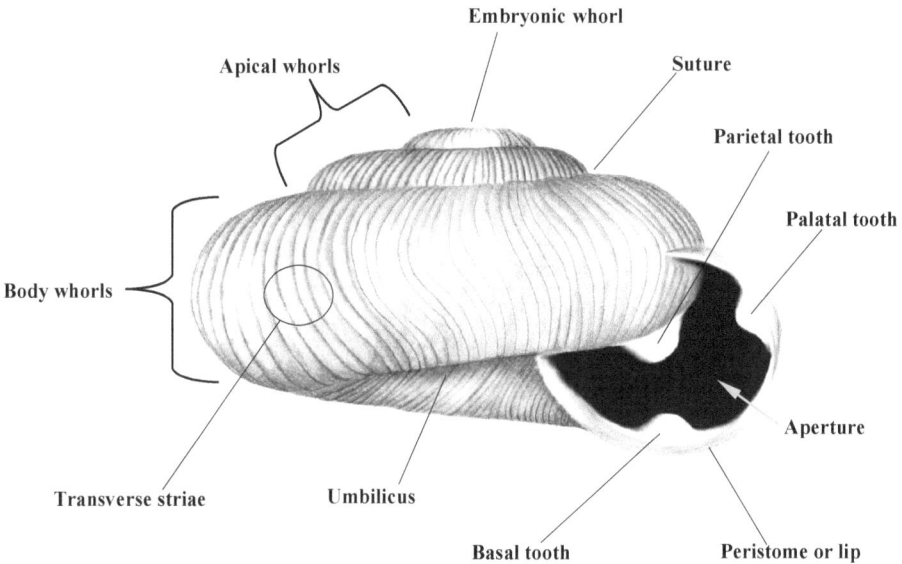

Embryonic whorl

Apical whorls

Suture

Parietal tooth

Palatal tooth

Body whorls

Aperture

Transverse striae

Umbilicus

Basal tooth

Peristome or lip

0 cm  1  2  3  4  5  6  7  8  9  10  11  12

# Umbilicus of the Shell (Burch 1962)

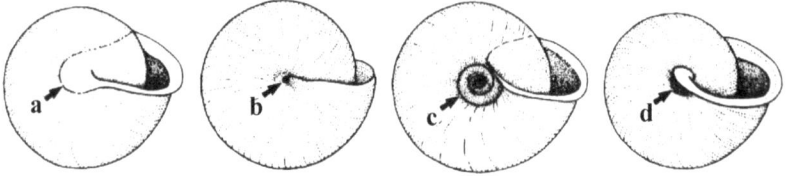

Figure A: **Imperforate** (closed umbilicus); Figure B: **Perforate** (small umbilical opening); Figure C: **Umbilicate** (wide umbilical opening); Figure D: **Rimate** (umbilical opening partially covered by aperture lip)

## Shell Measurements

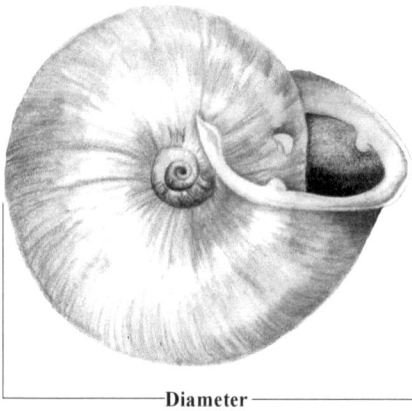

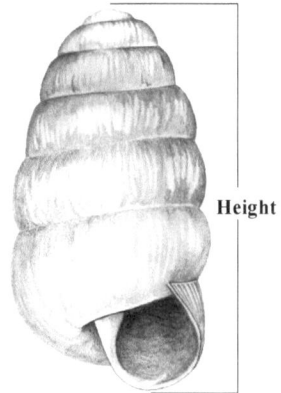

Height

—Diameter—

## Shell Periphery

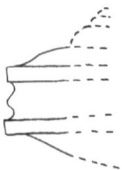

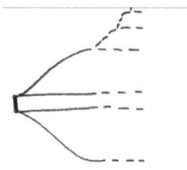

Doubly carinate          Carinate

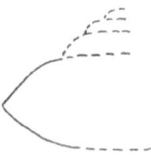

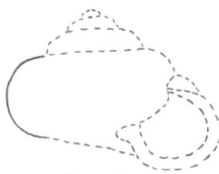

Angular          Round

## Peristome or Lip

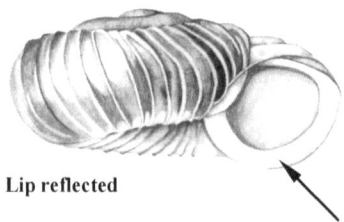

Lip reflected

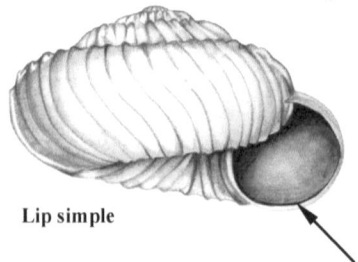

Lip simple

44

# Counting Whorls

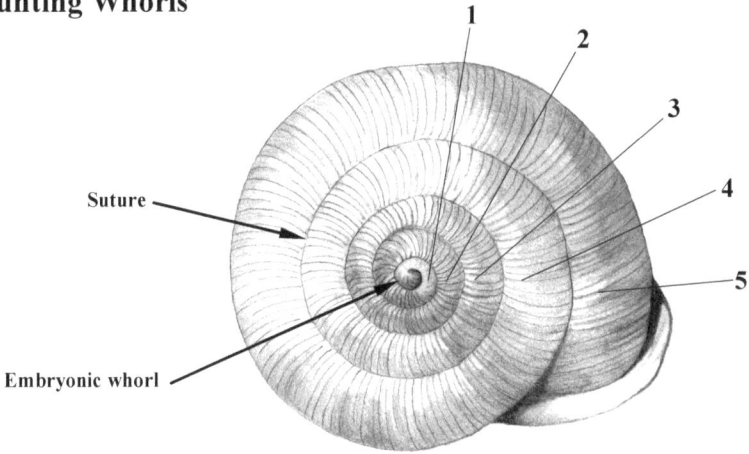

# Micro-features of the Shell Surface

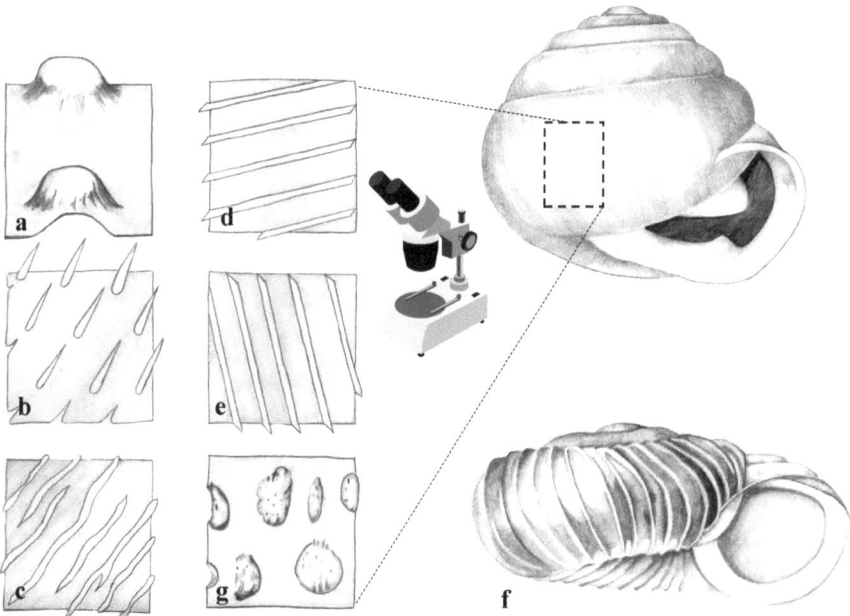

Papillae are raised bumps (a); hairs (b); wrinkles (c); spiral striae (can be in-dented or raised) run with the shell spire (d); transverse striae (can be indented or raised) are generally a micro-feature (e), but in some shells the raised striae are more rib-like (f); pits in the shell (g).

# Basic Land Snail Anatomy

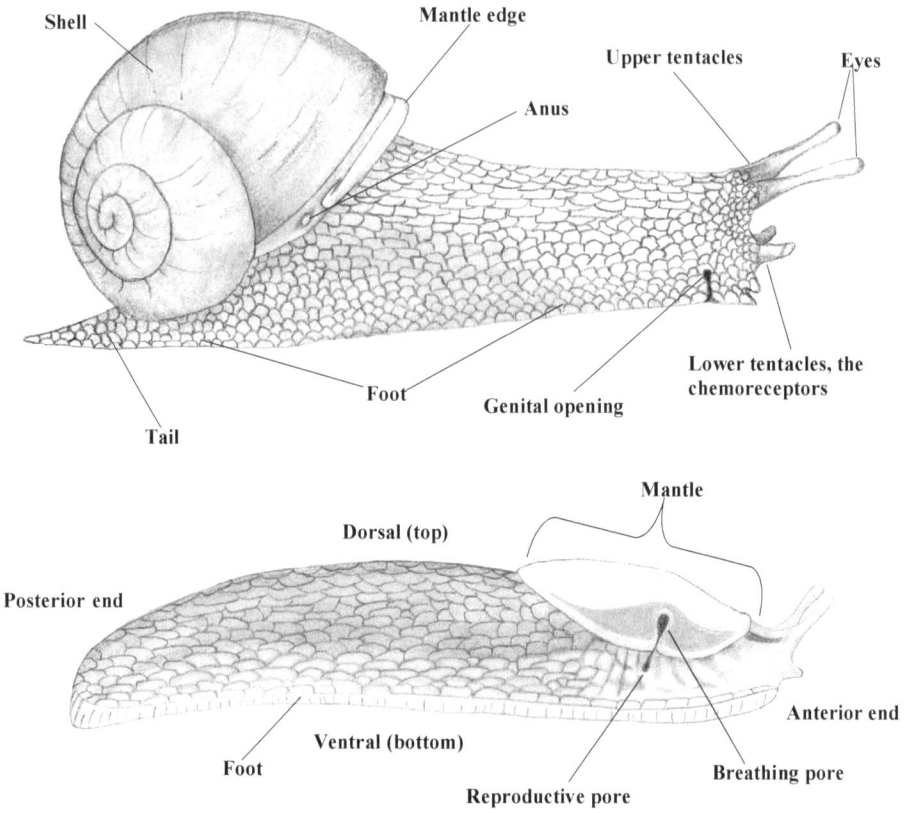

Shell

Mantle edge

Anus

Upper tentacles

Eyes

Lower tentacles, the chemoreceptors

Genital opening

Foot

Tail

Mantle

Dorsal (top)

Posterior end

Ventral (bottom)

Anterior end

Foot

Reproductive pore

Breathing pore

# Internal Anatomy

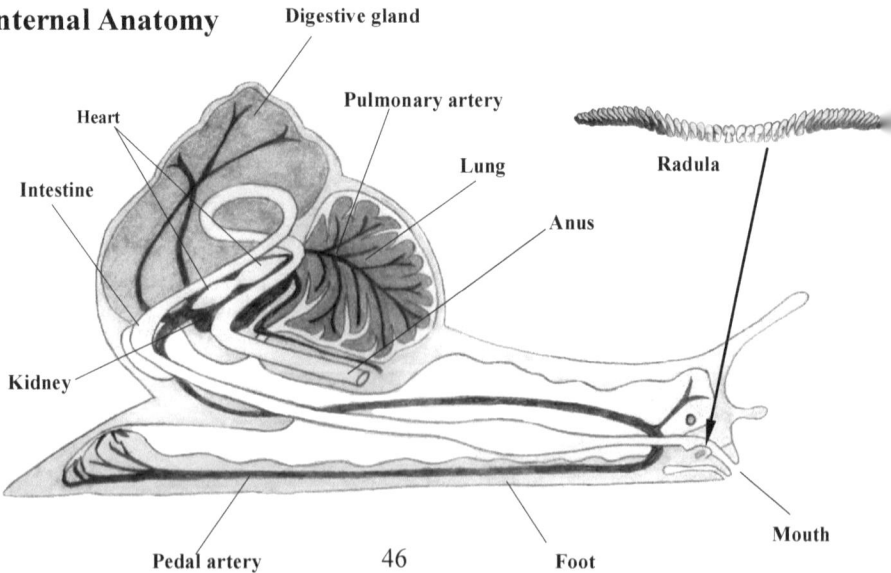

Digestive gland

Heart

Pulmonary artery

Intestine

Lung

Radula

Anus

Kidney

Pedal artery

Foot

Mouth

# How The Book Is Organized

The land snail species in this book are organized by shell form first then families within the forms. Species are classified into broad categories based on width or height first, followed by shape. It should be noted that some species of the same genus may not be found together due to the differences in size.

All gastropods here are accompanied with measurements in millimeters as a reference for their actual size. These measurements are for either the height (tall) or for the shell diameter (width). A millimeter ruler is included on page 43.

For **disc-shaped snails**, three standard views are pictured: the *frontal* showing aperture shape, the *bottom* showing umbilicus region and the *top* view showing the whorls. The *frontal* and *bottom* views are considered to be the best diagnostics for identifying most disc-shaped land snails in Belize. Sometimes as many as four views are shown to exemplify particular features of the shell.

For **conical-shaped snails,** a *frontal* view and *side* view are pictured for most species with additional views showing unusual or interesting features when necessary.

Letters used to depict certain shell features are consecutive on the same pages only (i.e. a, b). Black arrows point to important diagnostic features. Some minute snails have been enlarged to highlight important diagnostic features and sometimes shown to scale (life-size) on a small circle. These shells are not proportionate to the larger snail species that may occur opposite or on the same pages.

The following chapters illustrate all land snails known to occur within the boundaries of Belize including all native and exotic snails. Each species account includes the following information: **Common Name, Scientific Name, Description, Similar Species, Habitat, Status in Belize** and **Specimen** (providing the location of the particular specimen photographed). Species that are known only to occur in Belize are designated in blue and bold with the words: **ENDEMIC to BELIZE.**

Seventeen snails new-to-science are described within the text and are designated as **new species** after the name. In addition, land snails found during our surveys that could not be assigned to species needing additional specimens to make a final deliberation are also included and labeled as **undetermined.**

The text also includes land snails not yet reported but known to occur close to Belize and therefore, have a reasonable probability of occurring in the country. Land snails not yet reported will have no dots on the map to signify location within Belize. Other species included are erroneous taxa reported from Belize, several of which were misidentified specimens at museums (not an uncommon occurrence). Others were unlikely occurrences due to distance from known locations or lack of distinctive habitats not found in Belize (i.e. elevations above 2000 meters).

## Maps
Each species account includes a map showing approximate known locations designated by yellow dots. If locations are not known, as is the case for several historical records, the maps will display a circle where the species is predicted to occur based on our knowledge of the habitat requirements of that species.

## Abbreviations:
**UF or FLMNH**: Florida Museum of National History
**MCZ:** Museum of Comparative Zoology Harvard
**ANSP**: Academy of Natural Sciences Philadelphia
**NHMUK:** The Natural History Museum London
**MHNN:** Museo National de Historia Natural Mexico City
**BNR:** Bladen Nature Reserve
**ZMB:** Museum für Naturkunde, Humboldt Universität, Berlin, Germany

This symbol indicates a dissecting-scope is necessary or useful to view micro-features of the shell.

A scale bar is included to represent actual size of the shell.

Key features of the shell.

# 5 Shells Wider Than Tall, Greater Than 5 mm in Diameter

This section generally includes adult land snails wider than tall and greater than 5 mm in diameter, although a few illustrated here are a bit smaller and several groups such as *Helicina* are as tall, as they are wide. These gastropods are some of the most conspicuous and observed species in Belize and can be found from coastal plains to the highest mountaintops. In terms of color, these are the gems of the snail world. Many species boast a wide range of color morphs or variations (i.e. *Helicina oweniana*). Below one representative species from each family is illustrated to the right.

## Families and Genera Included:

**HELICINIDAE**
**NEOCYCLOTIDAE**
**XANTHONICHIDAE**
**POLYGRIDAE**
**THYSANOPHORIDAE**

# Remarkable Predators of Land Snails

Above image of a Hairy-tail Mole gripping a Magnolia threetooth, *Triodopsis tridentata* which will be consumed later in its underground burrow (Carvers Gap, Roan Mountain, Mitchell County, North Carolina, USA.) Another fascinating predator of land snails are ants. As with small mammals, ants may also be hunting land snails in the leaf litter but this remains entirely unstudied in Belize. Below is a picture of African weaver ants carrying a land snail to their hive for consumption.

© Piotr Naskrecki

# Family HELICINIDAE Ferussac, 1822

**DISTRIBUTION:** The Neotropical Realm, the Pacific Islands, Australia, Indian Ocean islands and southeast Asia.

**TAXONOMY:** The HELICINIDAE represent a family of over 500 described species. This group of gastropods were the first to evolve to terrestrial life independently from any other known group of land snails, the Pulmonates, their closest relatives being marine and freshwater snails. The HELICINIDAE make up a significant portion of the molluscan fauna of the tropics both in diversity and abundance. (Richling 2004)

Four genera including eleven recognized species and two species that remain undetermined are reported in Belize.

Land snails belonging to the family HELICINIDAE are generally dome-shaped or roundish in form, having either slightly to widely reflected lips and most species are without protective barriers such as teeth or lamellae. Shells can be cryptic, blending well with their natural surroundings (one species, *Pyrgodomus microdinus* covers its shell with clay) but most members of this diverse family have evolved some rather spectacular color and pattern configurations like *Helicina oweniana*. Most HELICINIDAE are small (<15 mm) and live in elevated environments such as trees, vegetation or on the side of rocky outcrops. Like most gastropods found in the tropics, life history for the family HELICINIDAE is largely unknown. As interests are cultivated in the study of land snails throughout Central America including Belize, it is expected that additional species new to science will be added to the assemblage.

# Genera Included:
(in order of appearance in text)

*Helicina*
*Pyrgodomus*
*Lucidella*
*Schasicheila*

## Non HELICINIDAE species

*Hyalosagda* (included for comparison)

## Astonishing Shell & Color Evolution in HELICINIDAE

HELICINIDAE are distinguished primarily by the presence of a calcareous operculum and one pair of antennae with eyes being at the base of these antennae. They are fairly proportionately distributed across the tropical and subtropical zones of the New World as well as Indopacific and Pacific islands including margins of the Asian and Australian continents (Richling 2004). The islands of the Pacific and the Caribbean Sea are the epicenter of this family with Cuba reporting more than eighty species and Jamaica around thirty-five species. The regions of Central America including Belize that drain into the Caribbean Sea are more diverse in species than those of the Pacific slope, not only because they are more extensive but also because they are more favored by a moist tropical climate (Von Martens 1890).

These are some of the most strangely-ornate and colorful gastropods on Earth. One need only study the intricate shell ornamentation of *Priotrochatella constellata* (pictured on opposite page) to fully appreciate one of nature's greatest achievements in calcium-architecture. These island gems from Cuba are stunning examples of convoluted-evolution.

*Emoda sagraiana percrassa,* **juvenile (Agayo and Jaume, 1954)  Cuba, 10 mm**

*Helicina adspersa* **(Pfeiffer, 1839) Cuba, 17 mm**

*Priotrochatella constellata* (Morelet, 1847) Cuba , 13 mm

Orange-lip Dome, *Helicina oweniana*, BNR

Toothed Dome, *Helicina rostrata*, BNR

# Orange-lip Dome        HELICINIDAE

*Helicina oweniana* **Pfeiffer, 1849**

**Diameter**: 8-10 mm, Height: 8 mm

**Description**: Dome-shape; lip reflected, orange; aperture roundish with an operculum; shell with 4.5 whorls; imperforate; color of shell can vary greatly from buff to grayish-blue or even yellow; with or without color bands (below images showing the remarkable shell color variation in *H. oweniana*), the orange lip appears to be a constant and reliable feature; periphery well rounded.

**Similar Species**: Other *Helicina* species in Belize are less colorful and usually without an orange lip.

**Habitat**: A species of limestone regions, becoming less common at higher elevation acidic soils; often on undersides of *Heliconia* leaves.

**Status**: Common, a widespread and frequent snail of the Maya Mountains.

**Specimen**: Belize, Toledo District, all specimens illustrated below were within 30 meters of each other on the undersides of *Heliconia* leaves next to Blue Creek Cave (Dourson collection).

**Type Locality**: Chiapas, Mexico.

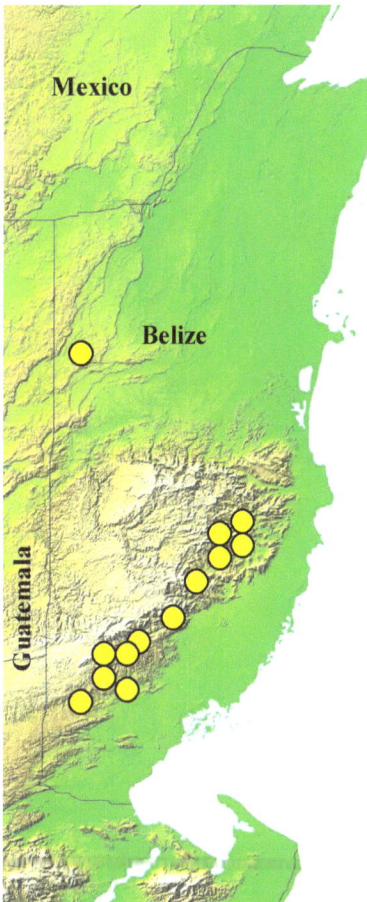

# Jungle Dome                    HELICINIDAE

*Helicina flavida* **Menke, 1828**

**Diameter**: 5-7 mm, Height: 8 mm

**Description**: Dome-shape; lip reflected (thick); <u>aperture roundish and proportionately smaller than other *Helicina* species</u> with an operculum; shell with 4.5 -5 whorls; imperforate; spiral striation strong; <u>usually with or without a single periphery band (a)</u> but many shells are without this feature; base color of shell white or yellow but in Belize the shell color varies greatly; live animal of this species is speckled and can be seen through the thin, juvenile shell (opposite page, bottom image), a feature thought to aid in crypsis; periphery rounded.

**Similar Species**: *H. arenicola* is larger, has a proportionately larger aperture, is less elevated and has a more engraved shell surface.

**Habitat**: Shells found under leaf litter on limestone hills and around talus, live animals found on the undersides of leaves on elevated live vegetation.

**Status**: Common, one of the most widespread *Helicina* species in Belize.

**Specimen**: All specimens below from Belize, Toledo District, base of Forest Hill, Bladen Nature Reserve (Dourson collection).

**Type Locality**: Misatlanta, Mirador, Mexico.

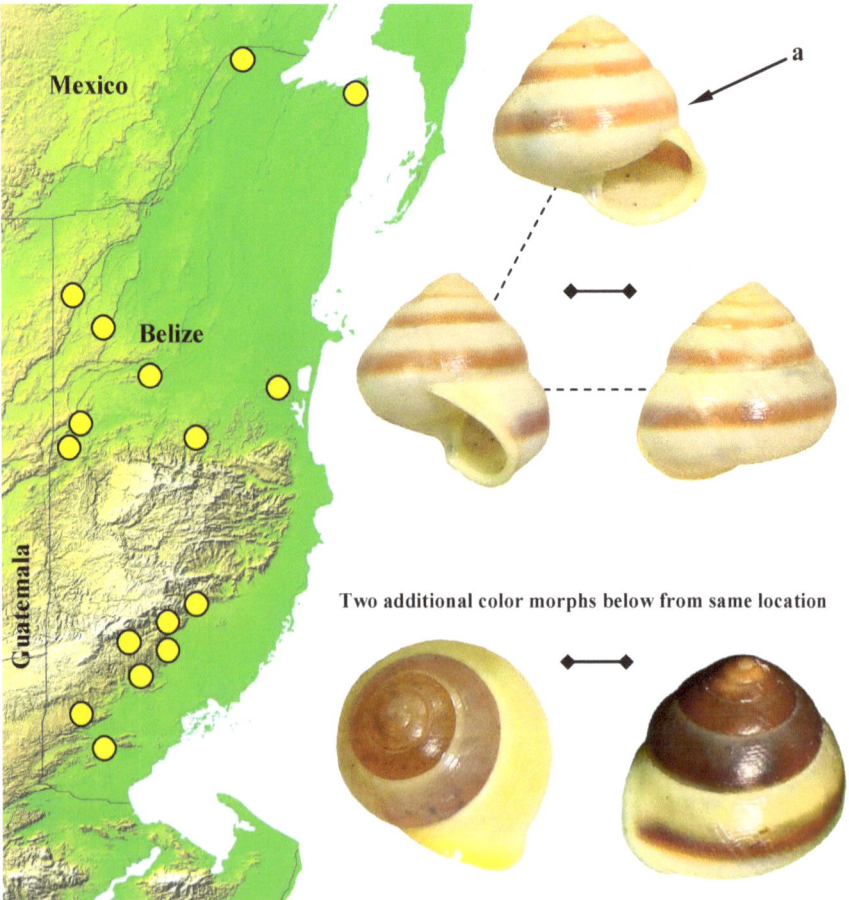

Two additional color morphs below from same location

Jungle Dome, BNR, Belize

The translucent shell of an immature Jungle Dome, Bladen Nature Reserve, Belize

57

# Engraved Dome                    HELICINIDAE

*Helicina arenicola* **Morelet, 1849**

**Diameter**: 6-10 mm, Height: 5-6 mm

**Description**: Dome-shape; lip reflected (thin); aperture roundish, with an operculum; shell with 4.5-5 whorls; imperforate; base color of shell usually yellowish-white (dark forms are not uncommon), with patches of brown that form numerous lines; <u>multiple color bands of varying widths</u> when present varying greatly; inside aperture is light yellow; spiral striae well developed.

**Similar Species**: *H. flavida* is similar to *H. arenicola* in form but is smaller by 3 to 4 mm and is usually with a single band.

**Habitat**: Found on steep talus slopes of karst hills.

**Status**: Common where it is found in the Peccary Hills, around Gracie Rock both in the Belize District, along the Coastal Highway, the Petén Karst Plateau of northern Belize and Shipstern Nature Preserve in Corozal District.

**Specimen**: Figure (a) from Belize, Belize District, Rockville Quarry, Gracie Rock (UF 207309) and figure (b) Belize District, Runaway Creek (Dourson collection).

**Type Locality**: Sisal, Yucatan, Mexico.

# Fragile Dome                    HELICINIDAE

*Helicina fragilis* **Morelet, 1851**

**Diameter**:5-6 mm wide, 5.5 mm tall

**Description**: Dome-shape and tall; lip reflected, aperture roundish, with an operculum; shell with 4.5-5 whorls; imperforate; base color of shell yellowish–horny, especially on the first whorls; without the usual color bands of other *Helicina*; shell surface nearly smooth or under a strong lens with weak transverse and spiral striae; shell thin for a *Helicina*; periphery broadly rounded.

**Similar Species**: *Helicina oweniana* has the same shell build but is larger and has an orange lip; *H. flavida* has single color band; *H. amoena* is larger and more compressed having an angular periphery.

**Habitat**: A species of karst hills and rocky outcrops; shells can be located on the ground under leaf litter, live individuals are found on elevated vegetation.

**Status**: Rare; in Belize known only from extreme southern Belize near the Mayan town of Dolores.

**Specimen**: Figure (a) from near Dolores, Toledo District, Belize (Dourson collection) and figure (b) from Guatemala, Petén, (NHMUK 1893.2.4.809, Lectotype, images by Ira Richling, Stuttgart©).

**Type Locality**: Petén, Guatemala.

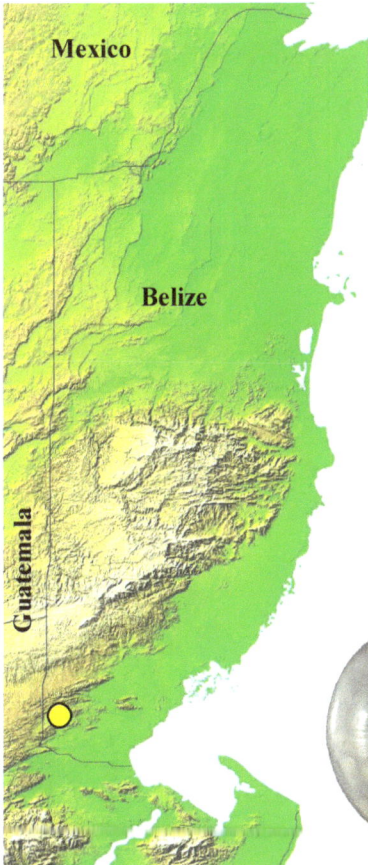

Mexico

Belize

Guatemala

a

b

59

# Toothed Dome                                 HELICINIDAE

*Helicina rostrata* **Morelet, 1849**

**Diameter**: 14-16 mm, Height: 10-12 mm

**Description**: Dome-shape; lip reflected (thick) with a small outward tooth projection, aperture roundish and containing an operculum; shell with 4.5-5 whorls; imperforate; overall color of shell is bone white, some specimens having a yellow tinge; usually with 2 to 3 bands but occasionally shells are found without; when crawling, the snail positions its head under the tooth, possibly protecting it from predators such as birds but this remains speculative.

**Similar Species**: No other *Helicina* species in Belize has the outer projecting tooth; *H. amoena* has an angular periphery.

**Habitat**: An arboreal snail living on trees and other elevated structure in lower elevation limestone hills.

**Status**: Common, a frequent species in the Maya Mountains, Rio Bravo Conservation Area and the Petén Karst Plateau of northern Belize

**Specimen**: Belize, Toledo District, base of Forest Hill, Bladen Nature Reserve (Dourson collection).

**Type Locality**: San Augustin Lanquin, Guatemala.

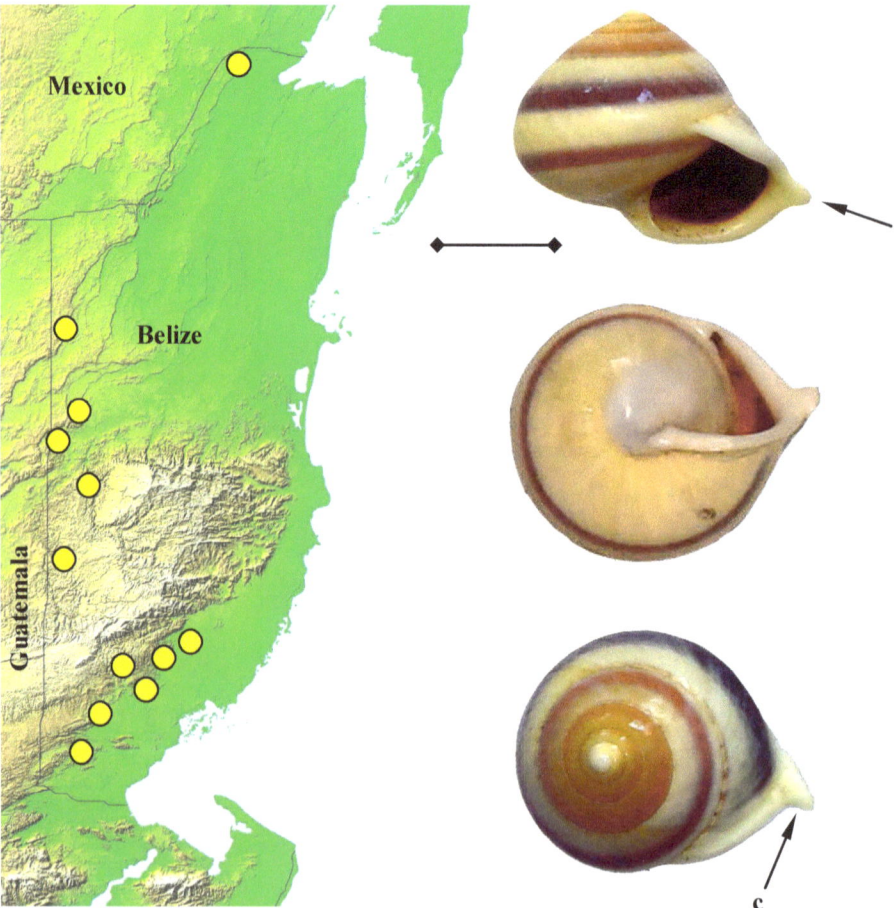

60

# Coastal Dome                    HELICINIDAE

### *Helicina bocourti* Crosse & Fischer, 1869

**Diameter**: 5 mm, Height: 3 mm

**Description**: Dome-shape; lip reflected, aperture oval with an operculum; shell with 5 whorls; imperforate, with or without (most without) a light band around the periphery; periphery bluntly angular (a).

**Similar Species**: Other *Helicina* species except for *H. amoena* (which is considerably larger) have rounded peripheries; *H. bocourti* from Honduras have a slightly lower shell profile and bolder color bands.

**Habitat**: The species appears restricted to the coastal plains of the Caribbean Sea, associated with palm species and oak forests where it is usually found crawling or resting on vegetation above the ground; also found in rather degraded habitats like the town of Punta Gorda.

**Status**: Common along the coastline of Belize

**Specimen**: Toledo District, Punta Gorda, Gomier's Restaurant (Dourson collection).

**Type Locality**: Utila Island, Honduras.

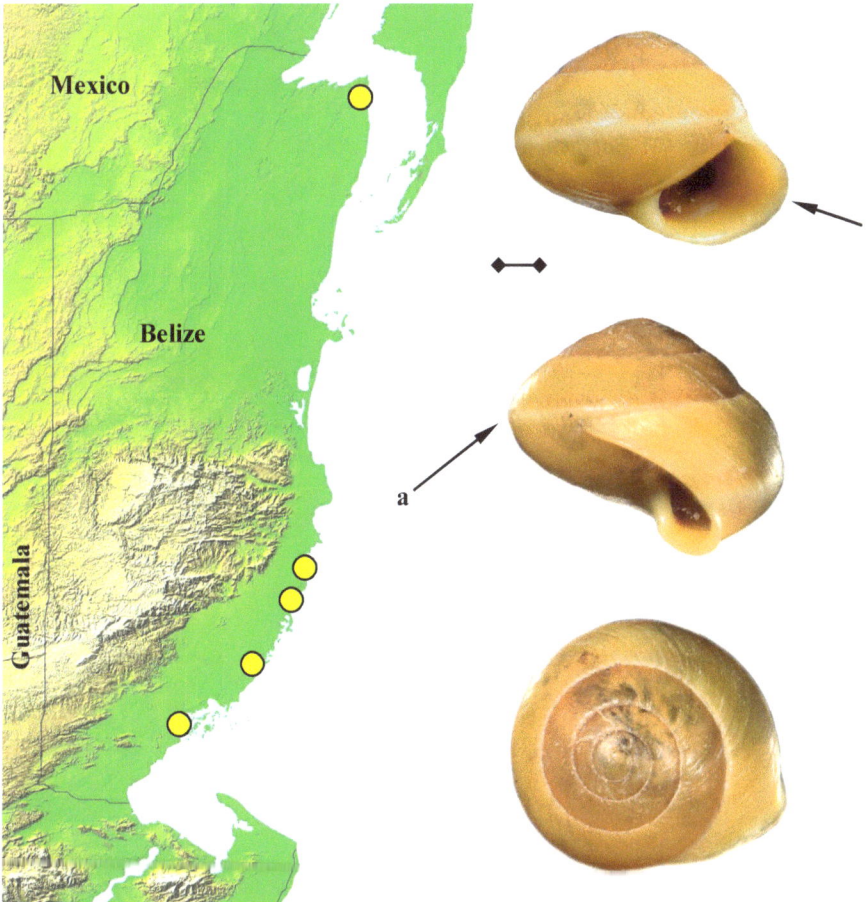

# Angled Dome                                        HELICINIDAE

### *Helicina amoena* Pfeiffer, 1849

**Diameter**: 13 mm, Height: 8-10 mm

**Description**: Dome-shape; lip reflected, aperture oval with an operculum; shell with 5 whorls; imperforate, usually with multiple light-colored broken bands, but some shells, especially older ones may be without; young shells with 3 whorls or less are fringed; periphery of shell angular (a).

**Similar Species**: Other *Helicina* species except for *H. bocourti* (a species of the coastline) have rounded peripheries.

**Habitat**: Generally found in floodplains under forest litter but also a species of higher elevation forests; like most *Helicina* species, *H. amoena* is found climbing on low vegetation during wet weather.

**Status**: Common; this is one of the most common and widespread *Helicina* species in Belize.

**Specimen**: Belize, Toledo District, around Richardson Creek, Bladen Nature Reserve (Dourson collection).

**Type Locality**: Honduras.

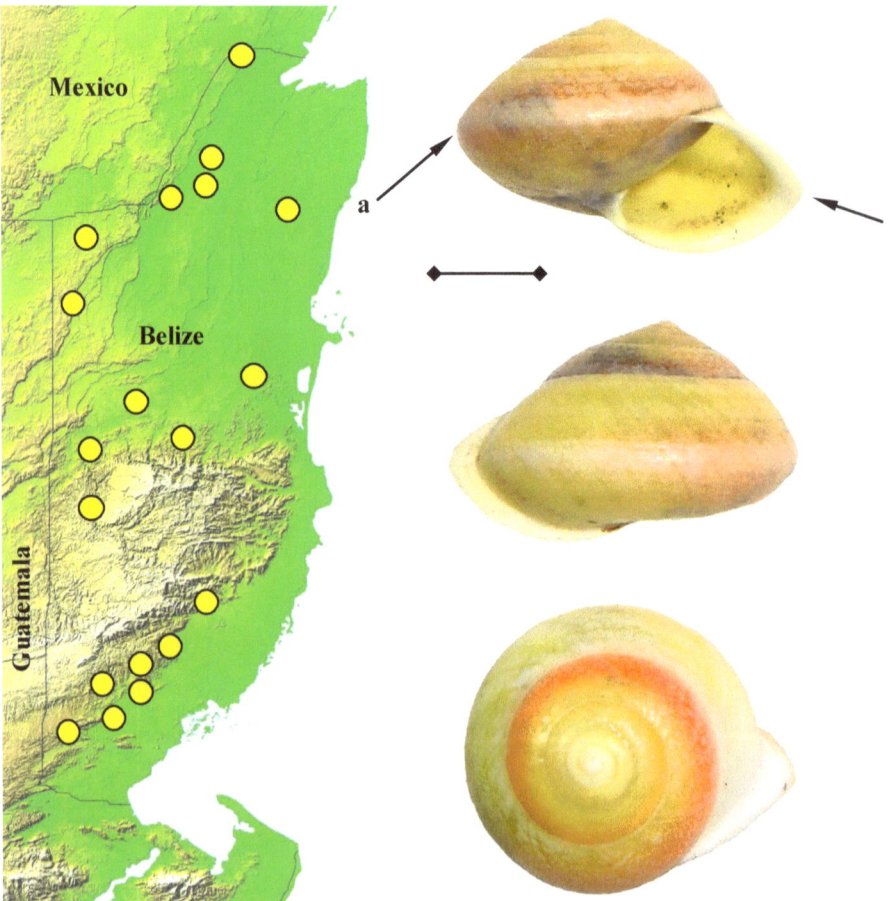

# Shaggy Dome                                    HELICINIDAE
## *Schasicheila* species (undetermined)
**Diameter**: 9-10 mm, Height: 8 mm

**Description**: Dome-shape; lip reflected at maturity with an operculum; shell with 4.5 whorls; imperforate, without color bands; shell with <u>closely spaced spiral fringes,</u> the longest fringes growing from the periphery, exceeding 2mm in length; <u>periphery of shell bluntly angular.</u>

**Similar Species**: The Shaggy Dome does not refer to any other *Schasicheila* species found in neighboring countries and likely represents a new species.

**Habitat**: Found on a karst hill in talus formations around Pass Valley near AC Camp, Bladen Nature Reserve.

**Status:** Rare, **ENDEMIC to BELIZE (?)** and possibly the Bladen Nature Reserve (?). <u>Additional specimens are needed to make a final determination.</u>

**Specimen**: Belize, Toledo District, Pass Valley near AC Camp, Bladen Nature Reserve; adult shells have not been observed, the specimen below is of a nearly mature individual (without the reflected lip), to date the only shell found of this exceptionally rare species (Dourson collection).

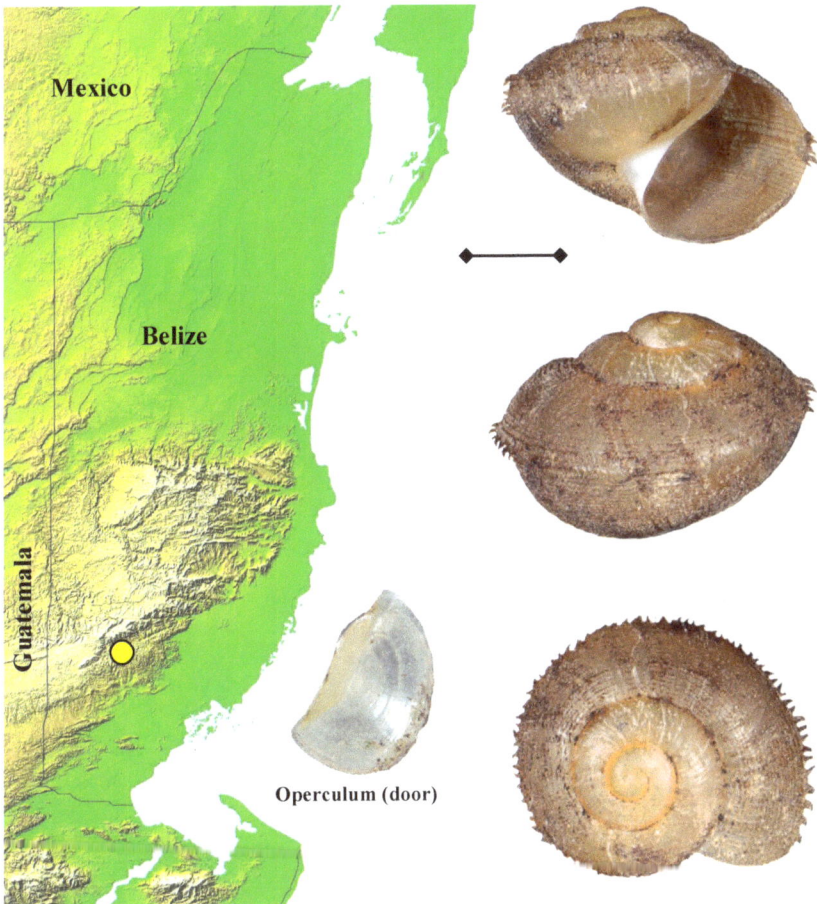

Mexico

Belize

Guatemala

Operculum (door)

63

# Fringed Lucidella                    **HELICINIDAE**

*Lucidella lirata* (Pfeiffer, 1847)

**Diameter**: 4 mm, Height :2-3 mm

**Description**: Dome-shape or  heliciform; lip reflected, outer lip edge is wavy; aperture oval with an operculum; shell with 4 whorls; imperforate; without color bands; small basal tooth present (a); shell surface has distinctly raised spiral fringes; periphery bluntly angular.

**Similar Species**: Other HELICINIDAE are without raised fringes (except for juvenile shells), usually with color bands and notably larger in size; differs from *Lucidella caldwelli* by being smaller, having a higher shell profile, a smaller basal tooth and fringes instead of hairs; the two species are occasionally found together and are easily separated by size alone.

**Habitat**: Generally found under leaf litter and among rocky places, sometimes found in large numbers around the base of limestone cliffs and rocky outcrops.

**Status**: Common; the species is generally frequent where it is found.

**Specimen**: Belize, Toledo District, base of Forest Hill, Bladen Nature Reserve (Dourson collection).

**Type Locality**: Isla de Carmen, Mexico.

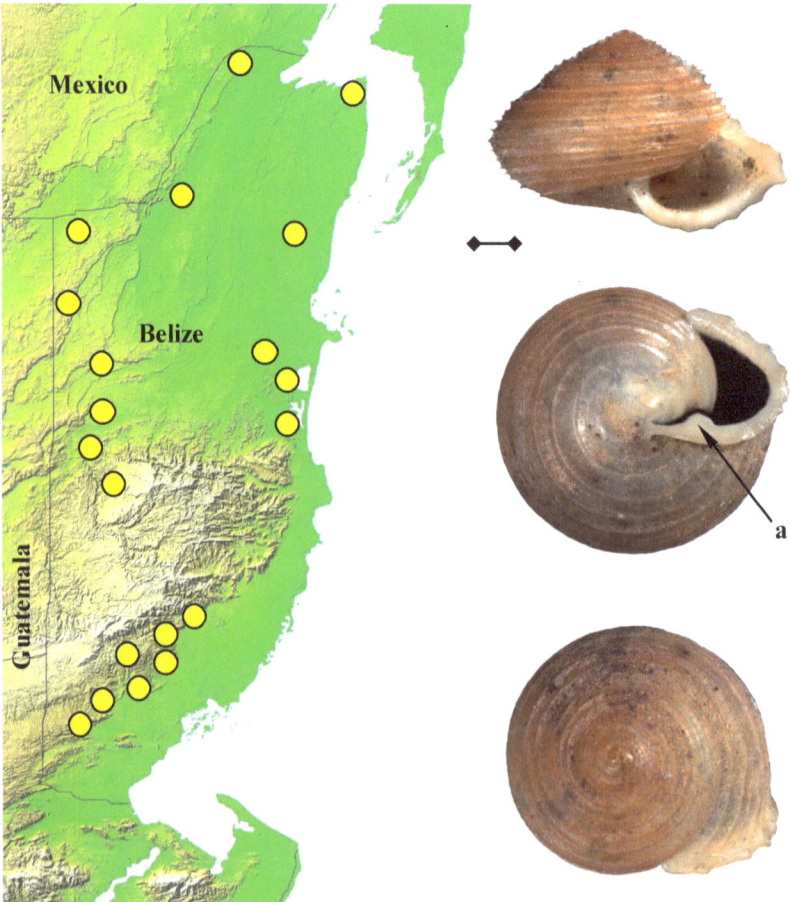

64

# Hairy Lucidella                          HELICINIDAE

*Lucidella caldwelli* (new species)

**Diameter**: 5-6 mm, Height: 3-4 mm

**Description**: Heliciform; lip reflected, outer lip edge is wavy; aperture oval, with an operculum; shell with 4 whorls; imperforate, without color bands; medium basal tooth present (b); shell surface has distinctly raised spiral hairs growing from the sutures; periphery roundish.

**Similar Species**: This species differs from *L. lirata* (page 64) by being larger, having a lower shell profile, a larger basal tooth and hairs instead of fringes.

**Habitat**: Only recently discovered in 2012 under leaf litter, among rocky limestone outcrops and talus slopes.

**Status:** Uncommon, ENDEMIC to BELIZE known from scattered locations in central Belize.

**Specimen**: Belize, Stann Creek District, Runaway Creek Nature Preserve, (Dourson collection). Holotype UF 505429 Runaway Creek, Belize and Paratypes UF 505430 from same location, not pictured.

**Type Locality**: Runaway Creek, Belize (17-10-15 N, 88-22-57 W).

**Etymology**: Named in recognition of Dr. Ronald Caldwell's important contributions in land snail work in Belize.

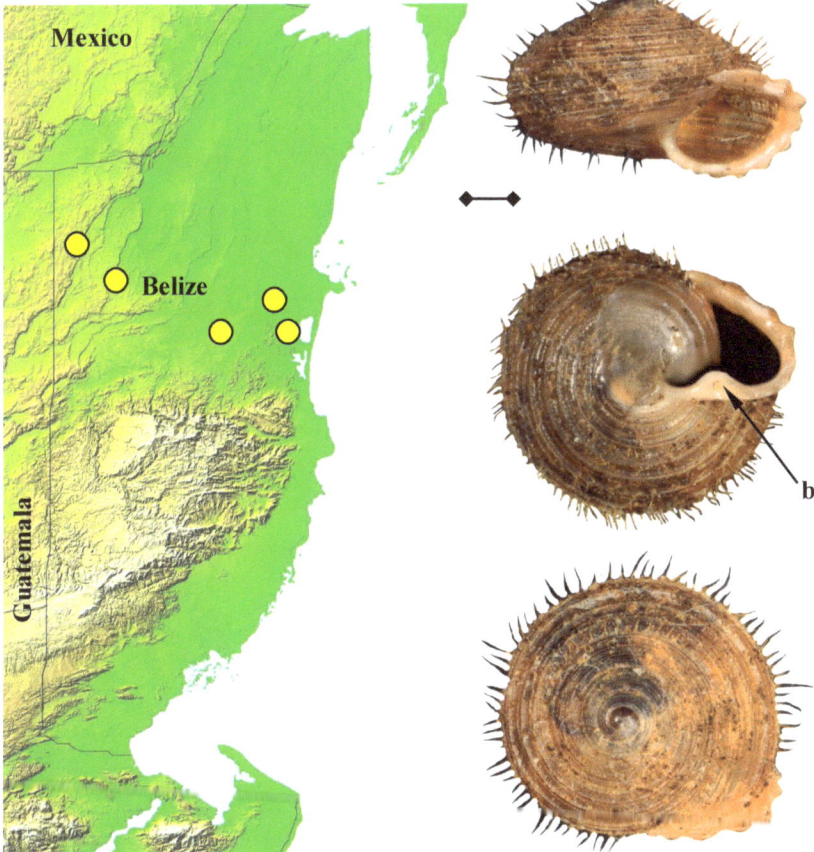

65

# Muddy Temple                    HELICINIDAE

## *Pyrgodomus microdinus* (Morelet, 1851)

**Diameter**: 3-4 mm, Height: 3.8 mm (size variable)

**Description**: Dome-shape or <u>pyramid–shape</u>; lip simple, outer lip edge not wavy as in *Lucidella* species; aperture roundish, with an operculum; shell with 4 whorls; imperforate, yellowish, without color bands; no teeth; <u>fresh shells with well developed, raised spiral striae but in older shells sometimes a wanting feature; periphery bluntly angular.</u>

**Similar Species**: Similar in form to *Pyrgodomus simpsoni* but notably larger in size, differing in color and the apex whorl is smaller.

**Habitat**: Common around rocky limestone outcrops and among talus slopes, live specimens often found on cliff faces; this species will cover its shell with a clay residue (typical of the genus *Pyrgodomus*), presumably acting as a mask, either visual or olfactory (pers. comm. Fred Thompson, 2010).

**Status:** Common, found in leaf litter collections throughout Belize.

**Specimen**: Belize, Toledo District, base of Forest Hill, Bladen Nature Reserve (Dourson collection).

**Type Locality**: Vera Paz, Guatemala.

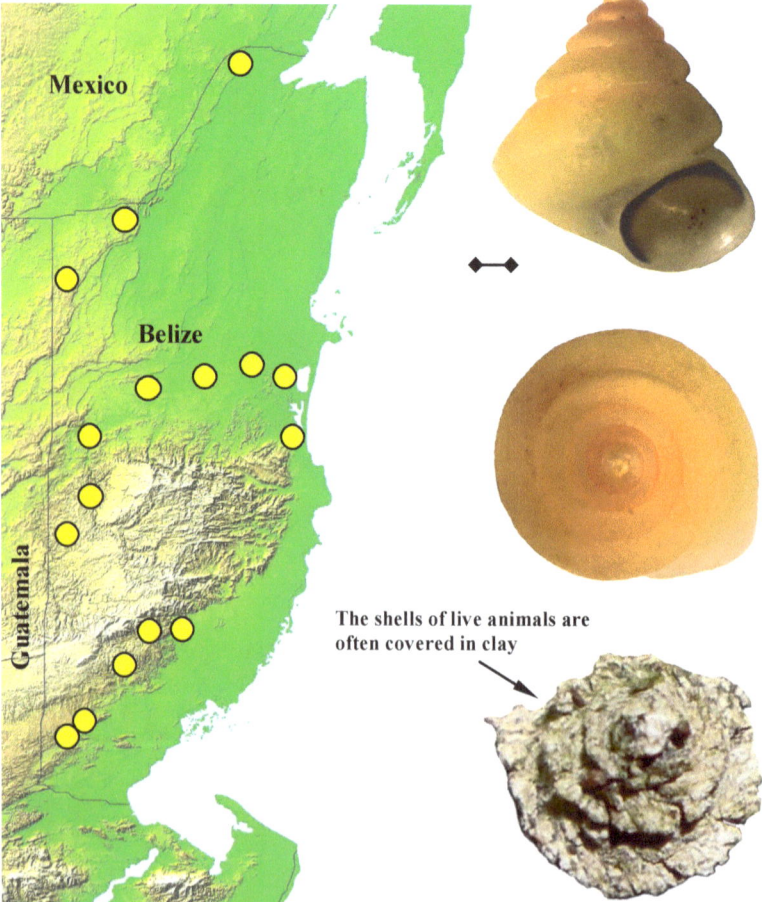

The shells of live animals are often covered in clay

# Petite Temple                    HELICINIDAE

## *Pyrgodomus simpsoni* (Ancey, 1886)

**Diameter**: 2.6 mm, Height: 2.6 mm (size variable)

**Description**: Dome-shape or pyramid–shape; lip simple, outer lip edge not wavy as in *Lucidella* species; aperture round to oval with an operculum; shell with 4 whorls; imperforate, without color bands; no teeth; fresh shells with well developed, closely-spaced, raised spiral striae, but in older shells sometimes a wanting feature; periphery bluntly angular.

**Similar Species**: Similar in form to *Pyrgodomus microdinus,* but notably smaller in size, having a different color and occurring much less frequently.

**Habitat**: Rocky limestone outcrops and among talus slopes covered in broad leaf rainforest.

**Status:** Rare, new for Belize, this recently discovered species of *Pyrgodomus* is known only from the Peccary Hills region.

**Specimen**: Belize, Belize District, Peccary Hills, at base of limestone outcrops (Dourson collection).

**Type Locality**: Utila Islands, Honduras.

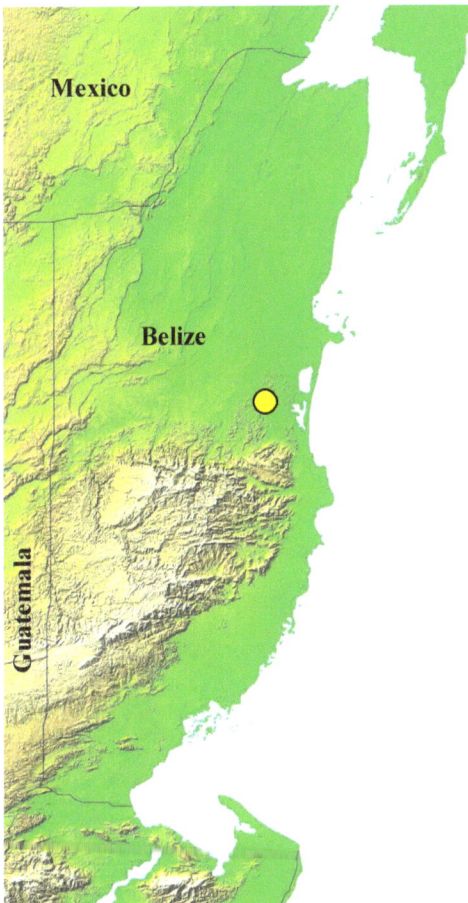

**Hairy Lucidella,** *Lucidella caldwelli* **(new species)**
**This interesting** *Lucidella* **species is endemic to Belize.**

# Belize **HELICINIDAE** & **SAGDIDAE** shells compared and proportionate
(all images on this page taken from Belize material, for location details see text)

10 mm

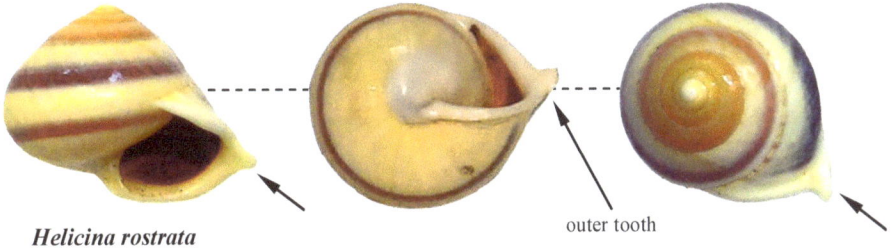

*Helicina rostrata*

outer tooth

*Helicina oweniana*

orange lip, multiple colors

*Helicina fragilis*

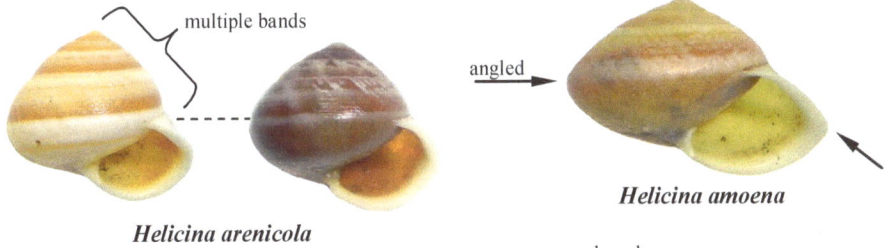

multiple bands

*Helicina arenicola*

angled

*Helicina amoena*

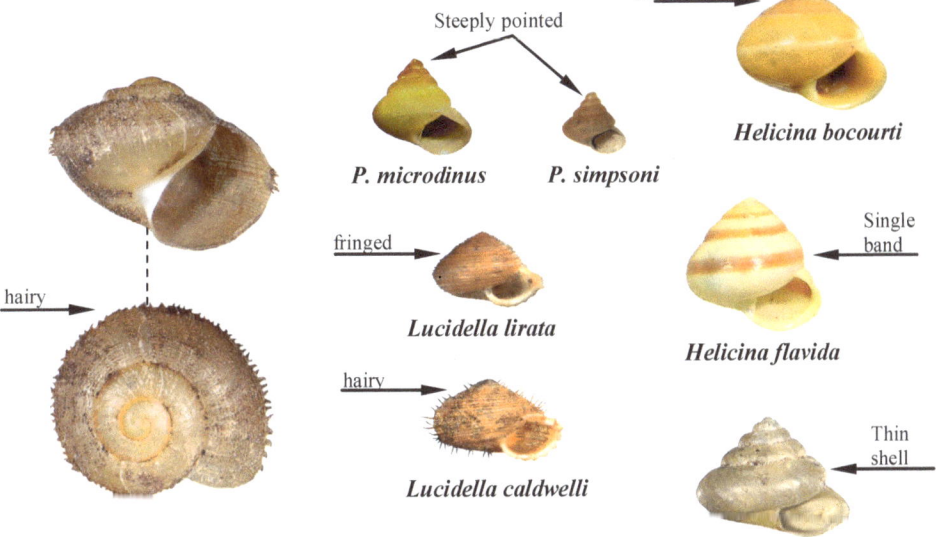

low dome

*Helicina bocourti*

Steeply pointed

*P. microdinus*　　*P. simpsoni*

hairy

fringed

*Lucidella lirata*

Single band

*Helicina flavida*

hairy

*Lucidella caldwelli*

*Schasicheila* **species**

Thin shell

*Hyalosagda turbonella*
see page 264

69

**HELICINIDAE** species reported from nearby areas of Guatemala & Mexico (possibly occurring in Belize but not yet documented)

*Helicina pterophora* Sykes, 1901, Guatemala (NHMUK 20140075, Type Specimen, images by Harry Taylor ©)

*Helicina ghiesbreghti* Pfeiffer, 1856, Mexico (UF 214176).

*Helicina tenuis* Pfeiffer, 1849, Guatemala (UF 190327).

*Helicina notata* Pfeiffer, 1856; Mexico, Cordova (MIZ 9003; likely holotype), image ©Ira Richling, Stuttgart

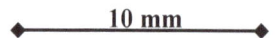

10 mm

*Helicina durangoana* Mousson, 1883; Mexico, Jalisco (UNAM 1876), image ©Ira Richling, Stuttgart

*Schasicheila hinkleyi* Pilsbry, 1919, Guatemala (UF 189975).

## Family NEOCYCLOTIDAE
## Kobelt & Moellendor, 1897

**DISTRIBUTION:** American tropics, South Pacific islands.

**TAXONOMY:** NEOCYCLOTIDAE is a family of land snails that possess an operculum or trap door that closes to ward off predators and prevent desiccation (drying) when weather conditions are dry. While they are not true pulmonates, they lack a pneumostome, instead having developed a vascularized lung from the mantle cavity with a much larger opening to the cavity that is usually closed by the operculum when the animal retreats into its shell. A separate breathing tube or notch in the shell, sometimes present, allows air in and out.

Three genera which includes four recognized species and four subspecies are reported from Belize. Members of this diverse family are extremely varied in terms of size and shell form. They are among the most common and largest terrestrial gastropods in Central America. Included are the tall and slender aquatic-looking *Tomocyclus* and one of the largest land snails in Central America *Amphicyclotus* which can grow to more than 50 mm in diameter. These are the giants of ground dwelling gastropods generally occupying calcareous soils, limestone outcrops and occasionally, the entrances of caves. One species, *Neocyclotus dysoni*, actually is a complex of species that includes four described subspecies. Unfortunately the variation within the genus *Neocyclotus* is not completely understood due to the lack of molecular information. Future genetic studies will more than likely render some of these subspecies obsolete. Little is known of the natural history of this family especially diet and reproduction but many are suspected to be detritivores consuming decaying leaf matter and other organic material.

# Genera Included:
(in order of appearance in text)

*Amphicyclotus*
*Neocyclotus*
*Tomocyclus*

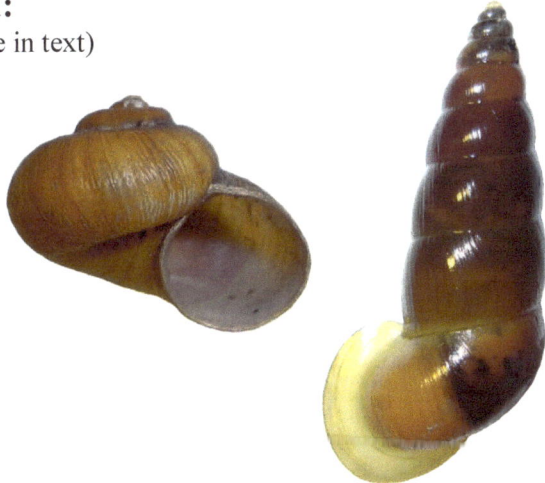

**Key to the reported <u>*Neocyclotus dysoni*</u> complex in the region of Belize** (adapted from Torres *et. al*, 1942)

A) Shell diameter less than 33 mm

    <u>**Axial sculpture with strongly wavy riblets (see figure A below)**</u>    **Page**

        1) Transverse striae not closely spaced. (3-4 riblets per mm).
           whorls somewhat flattened apically
              shell 20-27 mm
                  base not typically bicolor………………....…………*dysoni*   73
        2) Transverse striae closely spaced (4-5 riblets per mm).
           shell 18-29 mm
                  base bicolor………………........................................*hinkleyi*   74

    <u>**Axial sculpture without strongly wavy riblets (figure B below).**</u>

        1) Axial transverse striae of penultimate whorl straight
           sculpture not rough or course.
              color bands present……………………………………*berendti*   75
        2) Axial transverse striae of penultimate whorl not straight.
           base not bicolor.
              color bands absent…………………………………….*cookei*   76

B) Shell diameter greater than 33 mm, base bicolor...………..……………..*dyeri*   77

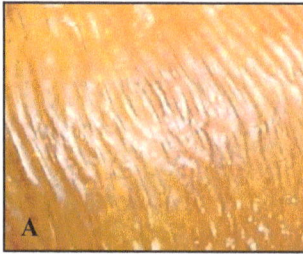

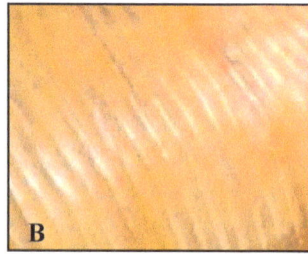

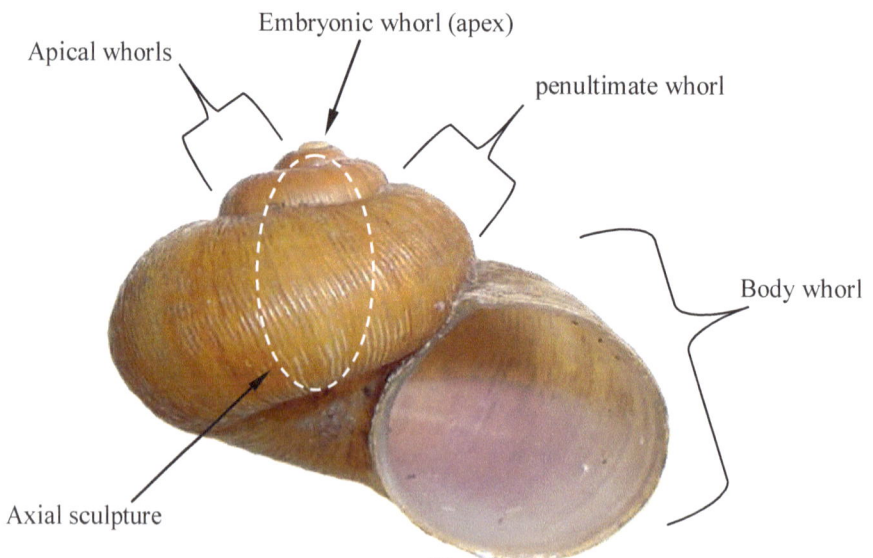

Embryonic whorl (apex)

Apical whorls

penultimate whorl

Body whorl

Axial sculpture

# Common Crater        NEOCYCLOTIDAE

## *Neocyclotus dysoni dysoni* (Pfeiffer, 1851)

**Diameter**: 20-27 mm wide, Height: 20 mm

**Description**: Heliciform; lip simple; aperture oval with an operculum (a); shell with 4 whorls; umbilicate; bronze; the periostracum is not worn as in *Amphicyclotus;* shell has a dull glossy surface; <u>base not typically bicolor</u>; transverse striae wavy; periphery well rounded; a detritivore.

**Similar Species**: *Neocyclotus dysoni cookei* is a form that appears little different in shape and habitat, but is generally lighter in color, more compact in form, has a roundish aperture and is smaller by around 5 mm in diameter; similar to *Amphicyclotus* but smaller with a more narrow umbilicus.

**Habitat**: Found in nearly every habitat in Belize including volcanic valleys, limestone hills and floodplains.

**Status**: The most common and widespread land snail in Belize

**Specimen**: Belize, Toledo District, top of Forest Hill, Bladen Nature Reserve (Dourson collection).

**Type Locality**: Stated only as Honduras.

# Hinkley's Crater                    NEOCYCLOTIDAE

*Neocyclotus dysoni hinkleyi* (**Bartsch & Morrison, 1942**)

**Diameter**: 18-29 mm wide, Height: 20 mm

**Description**: Heliciform; lip simple, aperture roundish, with an operculum; shell with 4.5 whorls; umbilicate; shell rather large; interior of aperture bluish-white or white; covered with a brownish olivaceous periostracum; a distinctive bicolor base of the last whorl; irregularly wavy riblets which are as wide as the spaces that separate them; these riblets are rather fine compared with most of the genus and closely spaced; periphery is well rounded; a detritivore.

**Similar Species**: *Neocyclotus d. hinkleyi* appears little different from *Neocyclotus d. dysoni* but may be distinguished by its bicolor base and closer spaced traverse striae, these two features however not always distinct between the two subspecies.

**Habitat**: Limestone foothills under leaf litter covered in jungle.

**Status**: Although not yet reported from Belize, suitable habitat exists throughout the county.

**Specimen**: Guatemala, Dept. Izabel Las Escobas, Rio Escobas (UF 190580).

**Type Locality**: Stated only as Guatemala.

# Banded Crater NEOCYCLOTIDAE

## *Neocyclotus dysoni berendti* (Pfeiffer, 1861)

**Diameter**: 17-21 mm, Height: 15 mm

**Description**: Heliciform; lip simple, aperture oval to roundish, with an operculum; shell with 4 whorls; umbilicate; olivaceous, straw colored with <u>multiple light bands of various widths</u> (a); shell has a dull gloss; periphery is well rounded; like most species of *Neocyclotus,* probably a detritivore, feeding on decaying vegetation.

**Similar Species**: *Neocyclotus dysoni* has a higher shell profile and is without the revolving bands; similar to *Amphicyclotus* but smaller with a much thinner shell and a more narrow umbilicus.

**Habitat**: Limestone hills and rocky outcrops.

**Status**: Uncommon; found scattered throughout Belize; this is the only subspecies in the "<u>*dysoni* complex</u>" found in Belize that is easily separated from *Neocyclotus d. dysoni,* and in the authors' opinion probably merits full species status.

**Specimen**: Mexico, Yucatan, 1.3 km NE Bechanchen (UF 19170).

**Type Locality**: Chitzen Itza, Mexico.

a

## *Neocyclotus dysoni cookei* (Bartsch & Morrison, 1942)

**Diameter**: 15-19 mm, Height: 12-14 mm

**Description**: Heliciform; lip simple, aperture oval to roundish, with an operculum (seen in frontal view); shell with 4 whorls; umbilicate; the first whorl pale flesh-colored, succeeding whorls rosy red with the last whorl pale chestnut brown or olive; striae straight (not wavy); overall color beige, the top of shell a pale rose; dull glossy; periphery well rounded; a detritivore that feeds on decaying vegetation.

**Similar Species**: *Neocyclotus dysoni cookei* is a form that appears little different in architecture to *N. dysoni dysoni* but the inside of the aperture is rose-red not bronze as in *N. d. dysoni* and is smaller by around 5 mm.

**Habitat**: Found on limestone hillsides and tropical savannah throughout the northern half of Belize with one site near Punta Gorda, Toledo District, Belize.

**Status**: Relatively common in Belize; unfortunately we don't understand the variation within the genus *Neocyclotus* completely due to the lack of molecular work and future genetic studies will more than likely render some of these subspecies obsolete.

**Specimen**: Belize, Cayo District, Xunantunich Maya Ruin (UF 135081).

**Type Locality**: Uaxachtun, Guatemala.

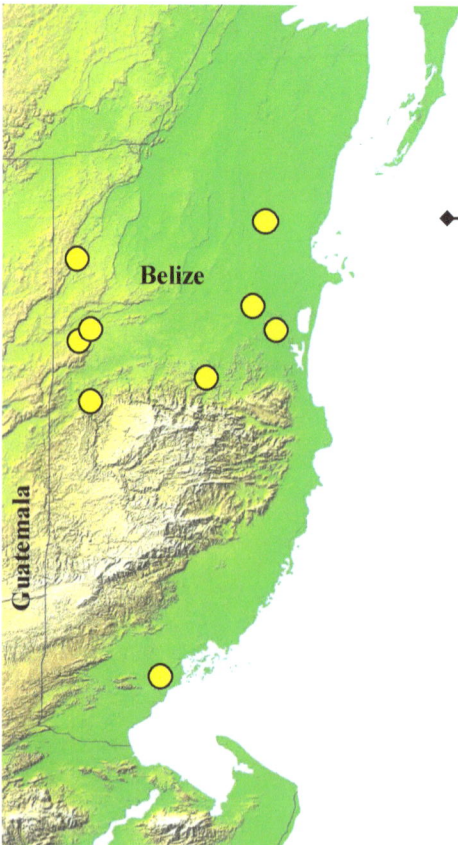

# Broad-banded Crater             NEOCYCLOTIDAE

*Neocyclotus dysoni dyeri* **(Bartsch and Morrison, 1942)**

**Diameter**: 30-40 mm, Height: 24 mm

**Description**: Heliciform; lip simple, aperture roundish, with an operculum (a); shell with 4.7 whorls; umbilicate; shell very large for the genus, the shell is an olivaceous-buff color, the bottom of the shell lighter in color giving the shell a distinctive bicolor base on the last whorl; early whorls when denuded are rosy; periphery is well rounded; a detritivore.

**Similar Species**: The largest of the *N. dysoni* complex reaching at least 5-10 mm larger than its counterparts; similar to *Amphicyclotus* but smaller with a much thinner shell and a more narrow umbilicus.

**Habitat**: Found on limestone foothills of the Maya Mountains and at the entrances of limestone caves.

**Status**: Rare in Belize; unfortunately we don't understand the variation within the genus *Neocyclotus* completely due to the lack of molecular studies therefore, the status of this subspecies in Belize remains questionable.

**Specimen**: Honduras, Olancho Dept., Agalta National Park, limestone ridge, 6 km NW of Catacamas (UF 221469).

**Type Locality**: La Ceiba, Honduras.

Common Crater
*Neocyclotus dysoni dysoni*

Close up of a Common Crater and its eye located at the base of the antenna. Most snails have eyes positioned at the tips of the top two antenna.

# Mayan Crater                    NEOCYCLOTIDAE

## *Amphicyclotus ponderosus* (Pfeiffer, 1851)

**Diameter**: 35-53 mm, Height: 25-27 mm

**Description**: Depressed heliciform; lip simple, aperture oval, with an operculum; shell with 4 whorls; umbilicate; <u>shell thick and solid</u>; the periostracum is often worn especially on the bottom of shell; shell has a dull surface and a crisscross pattern (a); periphery well rounded.

**Similar Species**: Similar to *N. dysoni* but much larger, with a thicker shell and a wider umbilicus; it often occurs with *N. dysoni* but is always the less common of the two species.

**Habitat**: Lives on higher elevation wet limestone hillsides; occasionally found at cave entrances; shells also can be found deeper just beyond the light zone.

**Status**: Uncommon to Rare, this large land snail is currently known only from southern portions of the Maya Mountains including the upper Bladen River watershed.

**Specimen**: Belize, Toledo District, Pass Valley near AC camp, Bladen Nature Reserve (Dourson collection).

**Type Locality**: Senahu, Guatemala.

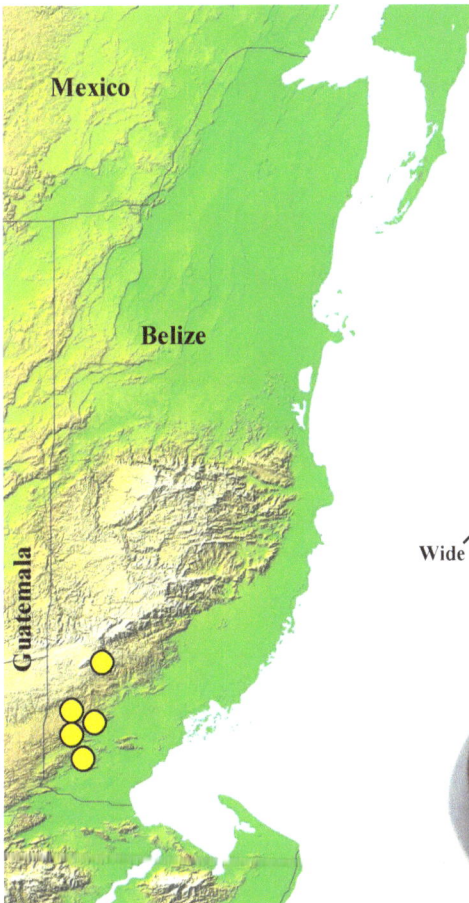

Mexico

Belize

Guatemala

Wide

# Topless Horn                    NEOCYCLOTIDAE

## *Tomocyclus simulacrum* (Morelet, 1849)

**Height**: 35-40 mm, Diameter: 10-12 mm

**Description**: Conical-shape; lip reflected; aperture roundish containing an operculum; whole shells will carry around 8-10 whorls but most shells referred to as "decollate" are without the last 4 or 5 whorls (a & b) with the apex segment or top usually lost in juvenile animals; without apex segment shells have around 5-6 whorls; perforate; the pseudosiphon is a small and rounded indentation (c); juvenile shells (d) are common where adults shells are found and may be mistaken for other species of conically-shaped land snails.

**Similar Species**: See opposite page for *Tomocyclus fistularus*.

**Habitat**: This aquatic–looking land snail is found under leaf litter on wooded limestone hillsides of lower elevation tropical rainforests.

**Status**: Uncommon, the species becomes increasingly common farther south in Belize especially in karst hills near San Jose, Toledo District.

**Specimen**: All specimens below from Belize, Toledo District, Bladen Nature Reserve (Dourson collection).

**Type Locality**: Petén, Guatemala.

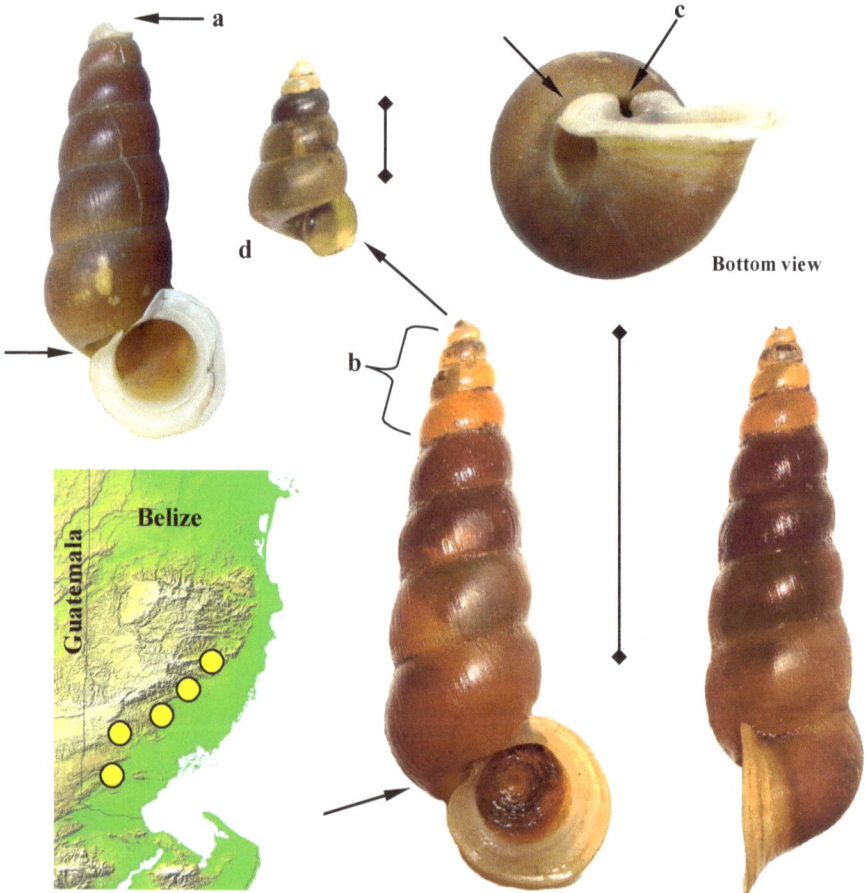

a

c

d

b

Bottom view

Belize

Guatemala

# Cayo Horn                                    NEOCYCLOTIDAE

*Tomocyclus fistularus* Thompson, 1963

**Height**: 28.5-33 mm, Diameter: 11.2 mm

**Description**: Conical-shape; lip reflected; with an operculum; whole shells have around 8-10 whorls; perforate; decollate shells (a) have 5-6 whorls.

**Similar Species**: *Tomocyclus fistularus* is separated from *T. simulacrum* on the basis of five characters: 1) the diameter of the fourth from the last whorl is about 0.60 times the diameter of the last whorl; 2) the aperture is 7.45-9.50 mm in diameter; 3) the collar is relatively narrow, 0.10-.024 times the diameter of the aperture 4) the pseudosiphon is rectangular (not rounded) with a larger indentation (b), 5) the basal carina is larger in circumference, passing in front of the reflected outer lip (c & d) (Thompson 1963).

**Habitat**: Limestone hillsides of upper elevation tropical rainforests.

**Status**: Uncommon; in Belize known only from the Maya Mountains.

**Specimen**: Figure (e) Belize, Cayo District, 50 M SW of San Ignacio, Valentine Camp (UMNZ 194095, Holotype, images by Taehwan Lee) and figure (f) from Belize, Toledo District, Blue Creek Cave area (Dourson collection).

**Type Locality**: Valentine Camp, Belize.

Bottom view

81

Above live Cayo Horn, *Tomocyclus fistularus* and its scat. Below the same animal with its operculum, Blue Creek Cave area, Toledo District, Belize.

# Family XANTHONICHIDAE Strebel & Pfeiffer, 1880

**DISTRIBUTION**: A member of the Pulmonata subclass, XANTHONICHIDAE is a Neotropical family that is widely distributed from Panama north to Tamaulipas, Mexico on the east coast, and to Sonora on the west coast. (Thompson 2011).

**TAXONOMY**: In general, the family is poorly known and there remains a number of undescribed species (per. comm. Thompson 2009). *Leptarionta* and *Trichodiscina* species that are members of this family are some of the most brightly and multicolored gastropods found in the region.

Other members of this family, the Giant *Lysinoe* species, are the largest heliciform land snails in Central America and some of the largest in the world, reaching shell diameters of nearly 80 mm. Endemic to southern Mexico, Guatemala and western Honduras (Belize?). These jungle-giants (see next page) live in steamy, montane cloud forests. Although snails in the genus *Lysinoe* are not yet reported from Belize, there are small unsampled patches of suitable habitat found at the highest elevations that could harbor these giants.

## The Genera Included in this Section:
(in order of appearance in text)

*Leptarionta*
*Trichodiscina*

## Land Snails of Size and Color Variability

In the mountainous regions of northern Central America live some rather re-markable land snails in the family, XANTHONICHIDAE (see opposite page). Snails in the genus *Lysinoe* are perhaps the most interesting in terms of their enormous size. These are the giants of cloud forests, several species reaching shell diameters of nearly 80 mm (below image showing an undescribed species from Honduras). Amusingly, when the live animal of *Lysinoe ghiesbreghti* (next page, bottom figure) is agitated, it jerks its enormous shell from side to side in systematic movements for reasons that remain somewhat of a mystery. It lives in lofty forests and gradually becomes more rare near settlements that regularly burn the lands to reduce forest cover. In former times, it was used as Lenten-food by the Indians of the department of Vera Paz, Guatemala (Von Martens 1890). It is not known if these large land snails were eaten by earlier cultures such as the Maya but such a sizeable food source must have been sought.

Other XANTHONICHIDAE species like the Belize Globe, *Leptarionta trigonostoma,* although smaller in size, have evolved some rather interesting color configurations. No doubt numerous species living in remote and inaccessible mountain ranges of northern Central America including Belize await discovery and scientific description.

**A giant undetermined species of *Lysinoe*, highlands of Honduras**

# ANTHONICHIDAE

40 mm

*Leptarionta aff. trigonostoma*
BNR, Belize

*Lysinoe eximia* (Pfeiffer, 1844), Guatemala

*Lysinoe leai,* Honduras, Thompson 1996

An undetermined species of
*Leptarionta* from Honduras

*Lysinoe ghiesbreghti* (Nyst, 1841) Guatemala

   All specimens from FLMNH (UF)

# Belize Globe                    XANTHONICHIDAE

*Leptarionta aff. trigonostoma* (Pfeiffer, 1844)

**Diameter**: 25-30 mm, Height: 20-27 mm

**Description**: Heliciform; lip reflected, aperture oval and without an operculum; shell with 4-5 whorls; imperforate; shell surface glossy, thick with conspicuous (bold) black bands of varying width; shell color bone white, some specimens having a yellowish tinge; periphery rounded to sub-angular.

**Similar Species**: Some specimens labeled as *Leptarionta trigonostoma* from Florida Museum of Natural History (UF) are strikingly different in shape and color (opposite page).

**Habitat**: Found sparsely in tropical rainforests along wet limestone in the lower Bladen River valley and Macal River Gorge, becoming more frequent in southern portions of the Maya Mountains.

**Status**: Uncommon; the species appears sparsely but widespread in southern Belize and is never very abundant where it is found; what appears to be a variable species likely represents multiple taxa awaiting further study.

**Specimen**: Belize, Toledo District, karst hills, 5 miles west of San Jose (Dourson collection).

**Type Locality**: Alta Vera Paz, Guatemala.

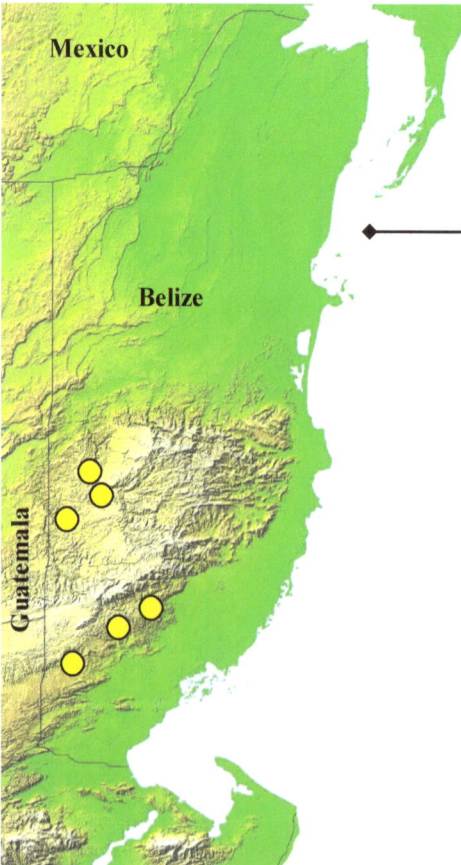

*Leptarionta* aff. *trigonostoma,* Belize, Toledo District, Columbia River Forest Reserve, Quartz Ridge, Camp 2 (UF 268085)

*Leptarionta trigonostoma,* Guatemala, 30 km south of Puerto Barrios (UF 190595)

*Leptarionta trigonostoma,* Guatemala, Chalhuitz (UF 103145)

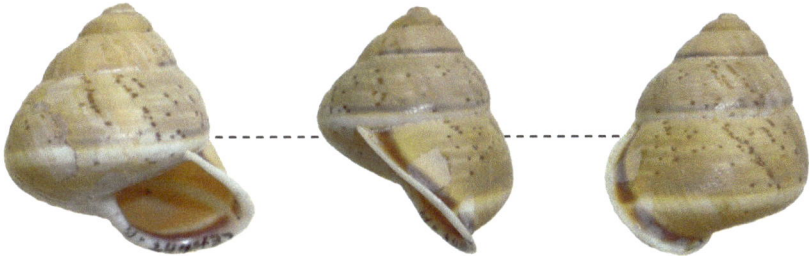

*Leptarionta trigonostoma stolliana,* Guatemala, Dept. Izabal, Sierra de Caral, Morales, San Miguelito (UF 244437)

Live *Leptarionta trigonostoma* from Fischer and Crosse (1892)

# Flat Button                    XANTHONICHIDAE

## *Trichodiscina coactiliata* (Férussac, 1838)

**Diameter**: 10-12 mm, Height: 5 mm

**Description**: depressed heliciform, discoidal (meaning flat or disc-like); lip slightly reflected, aperture turning downward, aperture oval and without an operculum; shell with 4 whorls; widely umbilicate; shell surface a dull glossy, thin and very fragile with 4 or 5 light color bands becoming nearly undetectable in old shells, (wetting shells with water or saliva will usually bring out the bands); young shells covered in fine hairs; periphery roundish.

**Similar Species**: Other *Trichodiscina* species color bands are either absent or much less conspicuous; *Trichodiscina hinkleyi* has lighter color bands, long hairs and a downward facing aperture.

**Habitat**: *Trichodiscina coactiliata* appears to prefer living in epiphytes of trees and other elevated situations.

**Status**: Common throughout Belize but shells do not weather well and deteriorate quickly.

**Specimen**: Belize, Toledo District, Forest Hill; BNR (Dourson collection).

**Type Locality**: Realjo, Nicaragua

# Hairy Button                    XANTHONICHIDAE

## *Trichodiscina hinkleyi* Pilsbry, 1919

**Diameter**: 12-14 mm, Height: 6-8 mm

**Description**: Depressed heliciform, discoidal; lip slightly reflected, aperture turning abruptly downward (a); shell with 4 whorls; widely umbilicate; shell surface dull-glossy, thin and fragile with one faint color band; juvenile and mature shells are covered in long hairs (see page 91), these hairs disappear as snail ages; periphery roundish and shouldered.

**Similar Species**: *Trichodiscina coactiliata* is more compressed, has bolder color bands and lacks the distinctive hairs.

**Habitat**: Karst hills living in elevated habitat such as trees.

**Status**: Uncommon, only known from three locations within the Bladen Nature Reserve and one site from Gallon Jug; in general, the genus is poorly known and there remains a number of undescribed species (per. comm. Thompson 2009).

**Specimen**: Belize, Toledo District, Bladen Nature Reserve (Dourson collection).

**Type Locality**: Livingston, Guatemala.

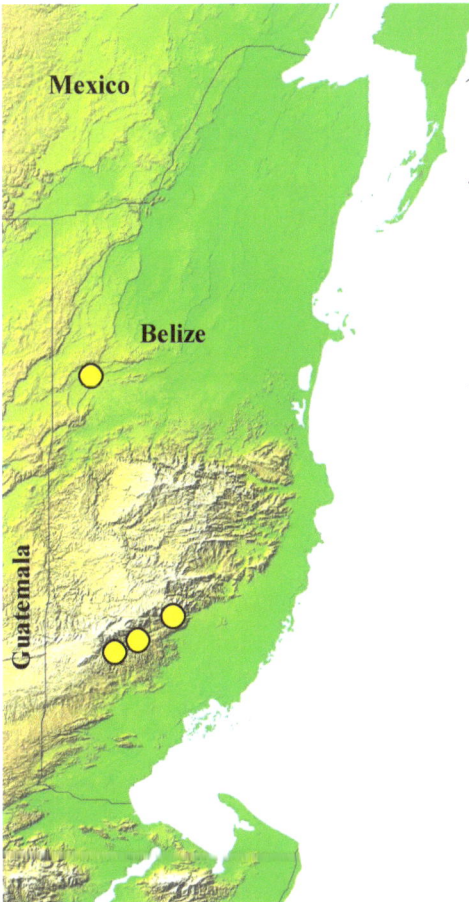

# Common Button
# XANTHONICHIDAE

*Trichodiscina suturalis* **(Pfeiffer, 1846)**

**Diameter**: 9-12 mm, Height: 7 mm

**Description**: Depressed heliciform; lip only slightly reflected, aperture roundish and without an operculum; shell with 4 whorls; sutures deeply placed (a); widely umbilicate; shell surface dull-glossy, thin and fragile without any notable color bands; young shells are covered in fine hairs; periphery roundish.

**Similar Species**: *Trichodiscina coactiliata* is more compressed and has notable color bands in fresh shells; *Trichodiscina hinkleyi* has long hairs that cover the entire shell and a downward facing aperture.

**Habitat**: *Trichodiscina suturalis* is a species of karst hills where it is found under leaf litter.

**Status**: The most common *Trichodiscina* species in Belize.

**Specimen**: Belize, Toledo District, base of Forest Hill, Bladen Nature Reserve (Dourson collection).

**Type Locality**: Listed as only Honduras.

Hairy Button, *Trichodiscina hinkleyi*

Hairy Button

Night lizards are also diurnal (active during the daytime), living in rotting logs and other moist places and in Belize at the entrances of caves. They are reported to eat a variety of insect and plant material but are highly suspected predators on land snails especially those species that live in and around outcropping limestone and caves. Night lizards have strong jaws and would have little problem crushing the fragile shells of snails like the Pearly Tuba, *Halotudora kuesteri* and the juvenile Common Crater, *Neocyclotus dysoni*. This however remains to be more fully investigated. Below image of a Night lizard stalking a land snail, Rio Frio Cave, Cayo District, Belize.

# Family POLYGRIDAE Pilsbry, 1895

**DISTRIBUTION:** POLYGYRIDAE is a family of air-breathing terrestrial mollusks occurring from temperate North America south to northern Central America. One species, *Praticolella griseola* has been widely introduced to other tropical and subtropical regions (Thompson 2011).

**TAXONOMY:** Emberton (1995) lists twelve genera belonging to this subfamily (Thompson 2011). While POLYGYRIDAE make up a significant portion of the land snail fauna of eastern North America, only two genera and three species are recognized and reported from Belize.

## The Genera Included in this Section:
(in order of appearance in text)

*Polygyra*
*Praticolella*

Does my helix make me look fat?

93

# Flat Three-tooth Disk                    POLYGRIDAE

*Polygyra dysoni* (Shuttleworth, 1852)

**Diameter**: 4.5 mm, Height: 2.5 mm

**Description**: Depressed heliciform; lip widely reflected; shell with 5 whorls; widely umbilicate; shell surface dull-glossy; wide parietal tooth, the basal and palatal teeth are 1/3 as large and more narrow in form; well developed traverse striae that are rib-like, strongest on the top and sides of the shell, weaker on the base of the shell; upper periphery shouldered (a).

**Similar Species**: *Polygyra yucatanea* is smaller, has a wider umbilicus and a larger, U-shape parietal tooth.

**Habitat**: *Polygyra* species are usually found living where outcrops of limestone occur and where soils are rich in calcium.

**Status**: Rare, **ENDEMIC to BELIZE** this apparently rare species was collected in only two locations along the gravel road to Vaca Dam; little literature exist for this endemic gastropod.

**Specimen**: Belize, Cayo District, along Arenal Road to Vaca Dam, cleared and burned milpa farm, 17°1'27"N, 89°6'55"W (Dourson collection).

**Type Locality**: Listed only as "Honduras", when in parentheses means Belize.

# U-tooth Disk                    POLYGRIDAE

*Polygyra yucatanea* **(Morelet, 1849)**

**Diameter**: 8.5-10 mm, Height: 5 mm

**Description**: Depressed heliciform; lip thickened and reflected; shell with 4-5 whorls; <u>widely umbilicate</u>; shell surface dull-glossy; large <u>U-shaped parietal tooth</u>, basal tooth also large but set deeper in the aperture of the shell; well developed traverse striae that are rib-like, strongest on the top and sides of the shell, weaker on the base of the shell; <u>upper periphery shouldered</u>.

**Similar Species**: *Polygyra yucatanea* refers to *P. dysoni* but is more compact, containing a larger U-shaped parietal tooth and has a notably wider umbilicus.

**Habitat**: *Polygyra* snails are usually found living where outcrops of limestone occur and where soils are rich in calcium; sometimes found in large numbers in relatively small places

**Status**: Uncommon, reported from areas around Pulltrouser Swamp, Orange Walk District (Turner and Harrison 1983), Gallon Jug, San Ignacio, Rockville Quarry, Gracie Rock, Xunantunich Archaeological Site and the Eastern Block-faulted Coastal Plain area of Belize.

**Specimen**: Belize, Belize District, Gracie Rock (UF 00135140).

**Type Locality**: "Yucatan".

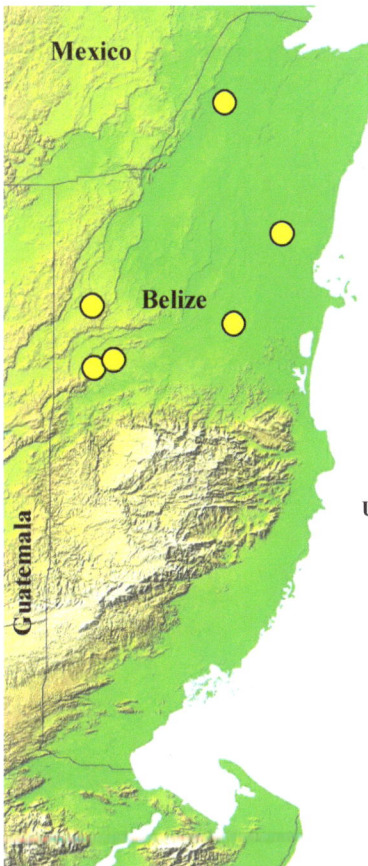

Mexico

Belize

Guatemala

Umbilicus wide

# Central American Scrubsnail     POLYGRIDAE

## *Praticolella griseola* (Pfeiffer, 1841)

**Diameter**: 8.5-10 mm, Height: 7 mm

**Description**: Heliciform; lip reflected; shell with 5-6 whorls; perforate; shell surface glossy with <u>one or multiple (black or brown) bands</u> (below two variations illustrated) on all whorls, although some specimens are without these bands; without any teeth; traverse striae weakly developed, no spiral striae; periphery rounded.

**Similar Species**: No other native species in size-range (8-10 mm) in Belize has the strong banding features of *P. griseola;* the exotic *Bradybaena similis* is slightly larger, flatter in build and, if present, lighter color bands.

**Habitat**: In general, found in limestone regions covered in at least some type of vegetation.

**Status**: Uncommon, known only from two sites in the northern half of Belize.

**Specimen:** Figure (a), 3 shells, Mexico Veracruz (UF 140091) and figure (b), 3 shells, Belize, Orange Walk District, 7.4 mi. SE Carmelita on the old Northern Highway (UF 135163).

**Type Locality**: Vera Cruz, Mexico.

Mexico

Belize

Guatemala

a

b

Slightly open

96

**DISTRIBUTION**: Middle America from Arizona and New Mexico south to Colombia and Venezuela; the Caribbean region and south Florida (Thompson 2011).

**TAXONOMY**: Five genera are recognized. Three genera containing six described and two undetermined species are reported from Belize. One species, Ghost Snail, *Itzamna sigoidius* is included here because of its larger size (over 5 mm). The remaining species in THYSANOPHORIDAE are less than 5 mm and illustrated in Chapter 11 on page 245.

## The Genera Included in this Section:
(in order of appearance in text)

*Itzamna*

# Ghost Snail                    THYSANOPHORIDAE

*Itzamna sigmoides* (Morelet, 1851)

**Diameter**: 10-15 mm, Height: 8 mm

**Description**: depressed heliciform, discoidal with flat embryonic whorl; lip reflected at the columellar insertion, aperture broad, nearly half the size of the shell in frontal view; 3-4 whorls; umbilicus rimate; shell translucent, thin and very fragile, breaking easily under the pressure of two fingers; entire shell covered in short diagonally-arranged hairs (a) that may act as an adaptive disruptive feature, allowing the snail to hide in the open; periphery rounded.

**Similar Species**: Unlike any other species found in Belize.

**Habitat**: Little is known of the species habitat and all shells in Belize have been located in limestone boulder talus and along cliffs.

**Status**: Rare. New record for Belize, the species appears to be restricted to southern Belize; previously known only from the type locality.

**Specimen**: Belize, Toledo District, rocky outcrops around Blue Creek Cave (Dourson collection).

**Type Locality**: Vera Paz, Guatemala.

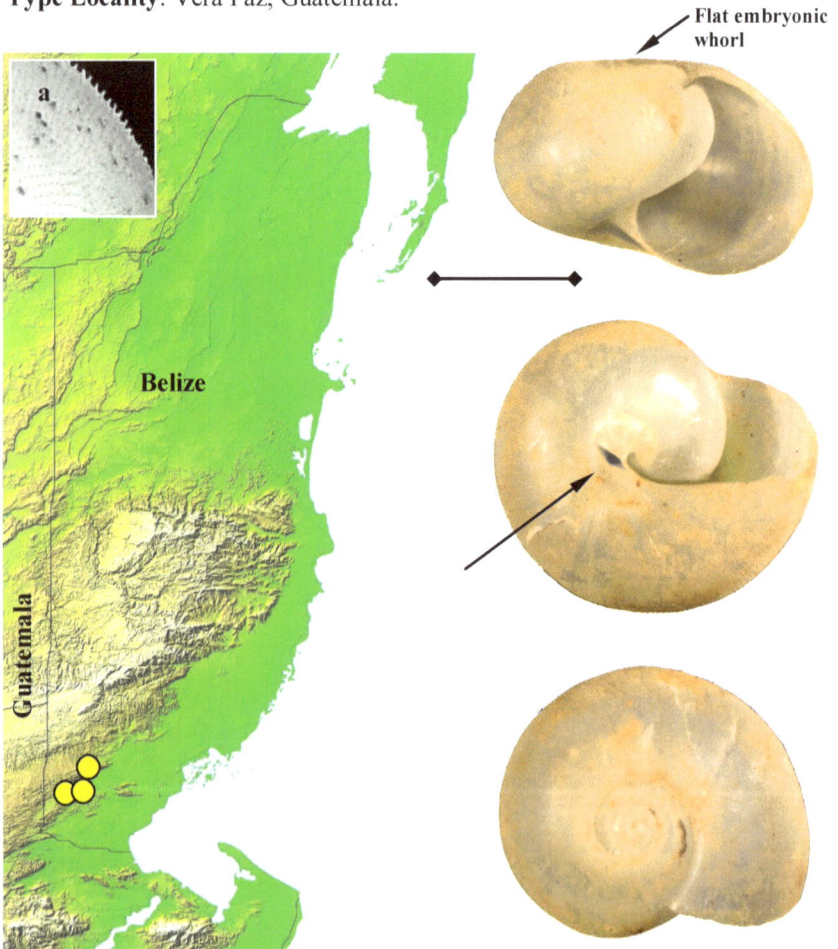

Flat embryonic whorl

a

Belize

Guatemala

# 6 Shells Taller Than Wide, Greater Than 5 mm in Height

This section includes adult land snails that are taller than wide, and in most cases, greater than 5 mm tall. These gastropods are among the largest and most colorful terrestrial snails in Central America, thriving from lowland plains to mountain peaks. This group also includes carnivorous snails that hunt and eat other gastropods. One representative species from each family is illustrated to the right below.

## The Families and Genera Included in Chapter 6

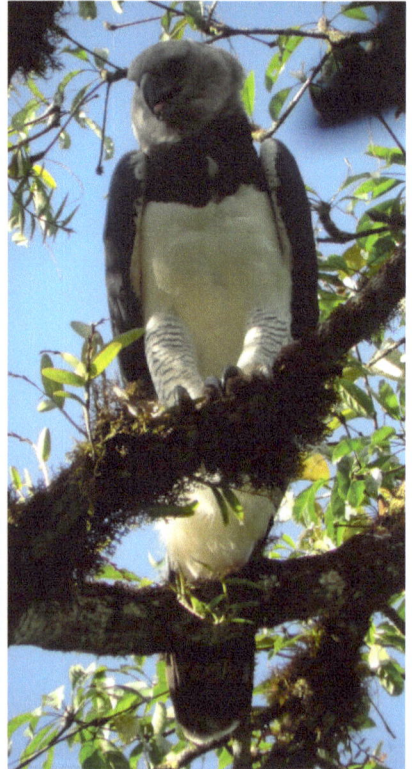

Above images from a recent 2016 National Geographic Waitt Foundation Grant Expedition in the Bladen Nature Reserve to study a bottom-up-relationship between land snails and harpy eagles. (Images by Kasia Biernacka, Dan Dourson and William Garcia, respectively).

**DISTRIBUTION:** Tropical South America, the West Indies, southern US to the tip of Florida, Central America and Mexico.

**TAXONOMY:** The nomenclature employed in this study follows Breure & Schouten (1985). Six genera are recognized in the subfamily Orthalicinae. Four occur in Central America and Mexico (excerpted from Thompson 2011). Three genera containing 14 recognized species, 2 subspecies and 4 undetermined species are currently known from Belize. Members of this large and diverse family include some of the most brilliantly colored and robust land snails on Earth. More than 1400 species have been described to date and there remains numerous undescribed species, especially in remote, unsampled regions of Central and South America. Most species live in elevated environments such as trees and shrubs from sea level to cloud forests. Adult shells of *Drymaeus* in good condition are rarely encountered, a result of the fragile shells. See next page for ORTHALICIDAE distribution and several examples.

## The Genera Included in this Section:
(in order of appearance in text)

*Orthalicus*
*Drymaeus*
*Bulimulus*

# Charmers of the Land Snail World

With nearly 1500 described species, snails in the family ORTHALICIDAE are some of the most diverse and colorful families in the Americas, if not the world. These arboreal snails possess ornately decorated shells sporting every color imaginable in a staggering number of pattern configurations, painted on an overwhelming number of shell forms. Most species are cone-shape and found living on elevated structures such as cliff faces, trees and shrubs from sea level to steamy cloud forests. Several early cultures traded these vivid shells for currency or used them as decorative neck wear. Species in the genus *Drymaeus* are the principal diet of the co-evolved snail-eating snakes in the genus *Sibon* of Central and South American rainforests.

*Sultans (Metorthalicus) vicaria*, Peru (UF)

*Plekocheilus gibbonius*, Colombia (UF)

*Sultans (Metorthalicus) deburghiae*, Ecuador (UF)

*Sultans (Metorthalicus)* species, Ecuador (UF)

*Sultans (Metorthalicus) labeo*, Peru (UF)

# Distribution of ORTHALICIDAE in the Americas

*Rabdotus dealbatus*, USA, Kentucky

*Liguus fasciatus* 3 color morphs illustrated, USA, Florida

*Liguus virgineus* Hispaniola

*Orthalicus princeps*, Belize

*Sultana sultana*, Peru

*Odontostomus pantagruelinus*, Brazil (UF)

*Zaplagius navicula*, Brazil (UF)

*Auris bilabiata*, Brazil (UF)

103

# Princess Cone                    ORTHALICIDAE

## *Orthalicus princeps princeps* (Broderip, 1833)

**Height**: 50-75 mm, Diameter: 30 mm

**Description**: Cone-shape; lip simple; shell with 6 whorls; imperforate; shell thick and solid, surface glossy; vertical color streaks (typically black or brown) on all whorls, many forming large chevrons; in some specimens, these color features are vivid while other shells will be less impressive; the last whorl without spiral bands and the 2 or 3 vertical stripes (a) are wide and forked above (b); base of young shells have continuous narrow bands.

**Similar Species**: *Orthalicus p. deceptor* has smaller and more faded chevrons with a few bold elongated streaks; *Drymaeus* species are much smaller in size.

**Habitat**: Found in a variety of elevated habitats such as trees and other aerial vegetation that provide food and cover from predators.

**Status**: Common, the most widespread large arboreal snail in Belize.

**Specimen**: Belize, Toledo District, karst hills around Blue Pool, Bladen Nature Reserve at an unexcavated ancient Maya archaeological site.

**Type Locality**: Conchagua, El Salvador.

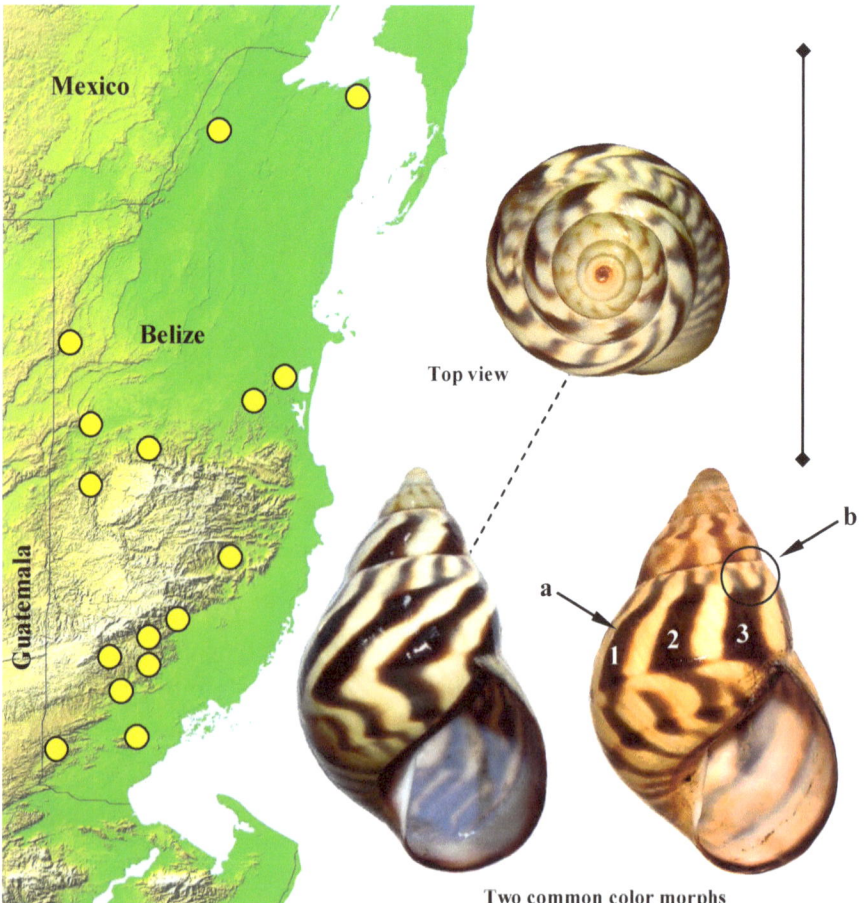

Top view

Two common color morphs

104

Better in Belize image

Two different color morphs of *Orthalicus princeps princeps* from Belize, above from Better in Belize Eco-Village, Macal River Gorge, Cayo District and below from around Blue Creek Cave in the Toledo District.

Above image of an adult Cross Cone, *Orthalicus princeps crossei,* feeding on algae and below its hard-shelled eggs and newly hatched snails feeding on a decomposing leaf. All images from the Bladen Nature Reserve, Belize.

Eggs

10 mm

Eggs found under leaves at base of tree

# Cross Cone                                    ORTHALICIDAE

*Orthalicus princeps crossei* **(Von Martens, 1893)**

**Height**: 50-55 mm, Diameter: 30 mm

**Description**: Cone-shape; lip simple; shell with 6 whorls; imperforate; shell rather thick, surface glossy; narrow, vertical color streaks with chevrons are intensified on this subspecies of *O. princeps*, with a more vertical arrangement on the shell and are far more numerous; there are 3 angulated-narrow and continuous but broken bands on the last whorl (a); base color is tawny, streaks dusky purple to brown; periphery broadly rounded. The preferred diet of hook-billed kite (Meerman and Phillips, pers. comm. 2016).

**Similar Species**: *Orthalicus p. deceptor* has smaller and more faded chevrons with an occasional bold thin streak.

**Habitat**: Same as previous species.

**Status**: Uncommon in Belize; unfortunately the variation within the genus *Orthalicus* is not completely understood, due to the lack of molecular studies and future genetic studies will more than likely render some of these subspecies obsolete.

**Specimen**: Belize, Toledo District, Bladen Nature Reserve (Dourson collection).

**Type Locality**: Belize.

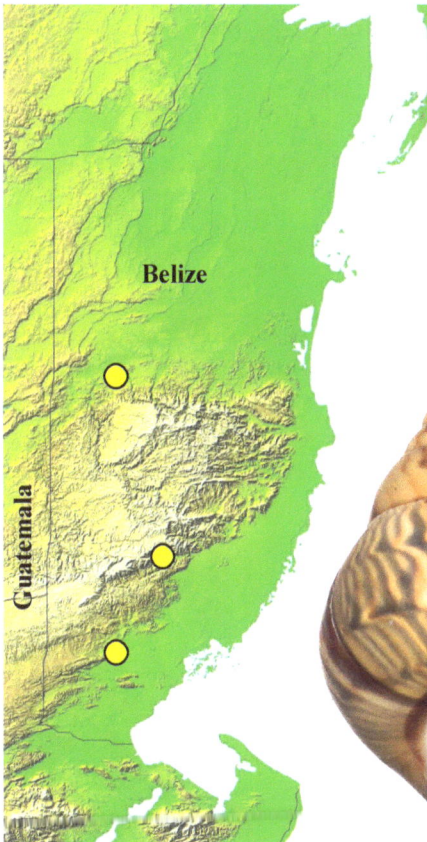

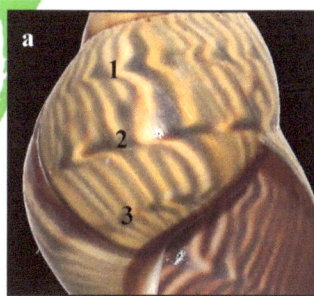

# Deceptor Cone                    ORTHALICIDAE

*Orthalicus princeps deceptor* (Pilsbry, 1899)
**Height**: 49-67 mm, Diameter: 30-40 mm
**Description**: Cone-shape; lip simple; shell with 6 whorls; imperforate; shell thick and solid, surface glossy; last body whorl with a broken peripheral band (a) of oblong blackish spots alternating with buff and mottled; long random dark streaks which are growth stoppages in the shell (b); on the spire (c) there are 1-2 narrow bands; overall color features (except for the black streaks) can be quite faded; the base color of the shell is generally a light yellow; periphery broadly rounded.
**Similar Species**: Similar to *O. princeps crossei* but appears morphologically different only by its bands located on the spire and its much lighter color streaks; *O. princeps crossei* and *O. princeps princeps* are without bands.
**Habitat**: Same as other *Orthalicus* species.
**Status**: Uncommon in Belize; unfortunately the variation within the genus *Orthalicus* is not completely understood, due to the lack of molecular studies; the status of this subspecies remains highly questionable.
**Specimen**: Figure (d) *O. princeps deceptor* from Nicaragua, Syntype ANSP 5099; figure (e) *O. princeps deceptor* from Guatemala, UF 109559.
**Type Locality**: Polvon, Nicaragua.

# Milky Cone                                    ORTHALICIDAE

## *Orthalicus cf. livens* Shuttleworth, 1856

**Height**: 55-67 mm, Diameter: 30 mm

**Description**: Cone-shape; lip simple; shell elongated with 6 whorls; imperforate; shell thick and solid, surface glossy with dark vertical color streaks that vary in width on a dull-buff background; aperture rather small for *Orthalicus* species; last whorl narrowly and inconspicuously two broken banded (a); periphery broadly rounded .

**Similar Species**: Similar to *Orthalicus princeps princeps* but having 2 obscure broken bands below the periphery (a) and by having a more slender build.

**Habitat**: Found in a variety of elevated habitats such as trees and shrubs.

**Status**: Rare, in Belize, reported from Xunantunich, an ancient Maya archaeological site (Kavountzis 2009); although the specimen below lacks the purple-black apex, it shares most other characteristics of *Orthalicus cf. livens*.

**Specimen**: Figure (b) *O. livens* from Vera Cruz, Mexico Syntype MHNN, and figure (c) *O. cf. livens* from Belize, Cayo District, forests around Xunantunich, an ancient Maya archaeological site, UF 135080.

**Type Locality**: Near Vera Cruz, Mexico.

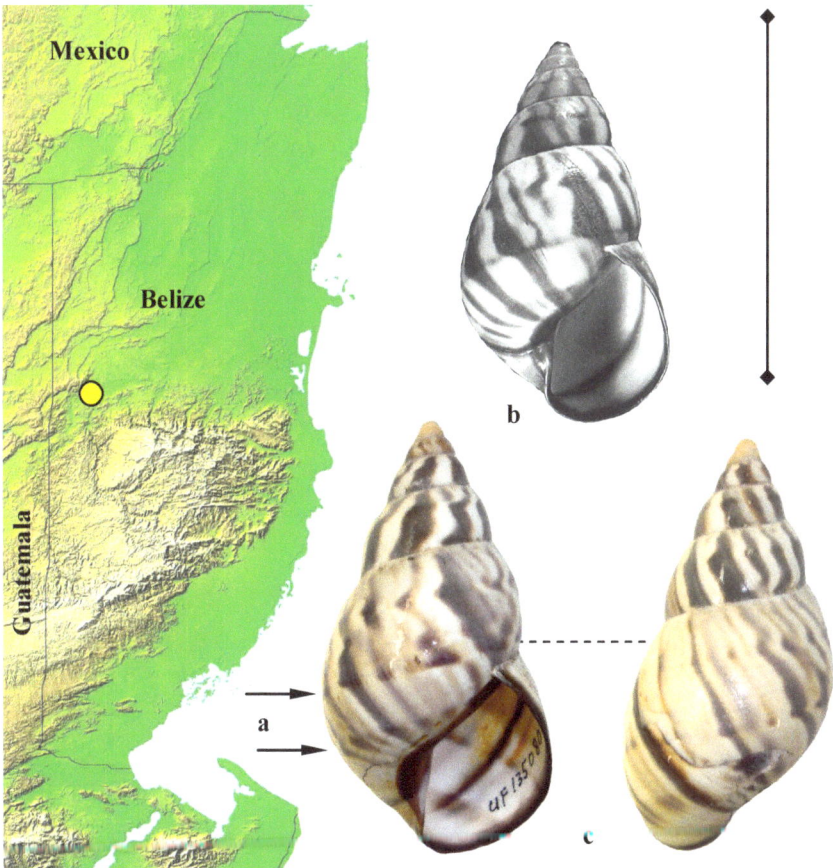

109

## The *Drymaeus* Complex

The genus *Drymaeus* is quite large, consisting of nearly 600 named species. In Central America, this includes 65 species and 50 subspecies. Numerous intermediate species and division into smaller units has eluded authors since its inception. Many species have colorful shells and local varieties have frequently been named, adding to the confusion when working with such a large group. Although these varieties have little basis for taxonomic recognition, authors have given them formal names (Thompson 2011). All forms known to occur within Belize as well as those that are close to its arbitrary borders are included in the text. Future genetic studies will more than likely render some of these varieties obsolete.

Two subgenera recognized within the study area *Drymaeus* and *Mesembrinus*, are distinguished by the development of the peristome (lip) and some anatomical differences (Breure & Eskens 1981). The shell in the subgenus *Drymaeus* has an expanded to broadly reflected outer lip (a), whereas the outer lip is simple to narrowly expanded (b) in the subgenus *Mesembrinus* (all *Drymaeus* in the text except for *D. serperastrus*). This distinction is often arbitrary (Thompson 2011) and has proved unreliable for several species found in Belize. While the two examples illustrated below show the distinctions between *Drymaeus* and *Mesembrinus* groups splendidly, the peristome of *Drymaeus attenuatus* (page 118), in the "*Drymaeus* group" lingers somewhere between the two, making its separation difficult based solely on the peristome.

Drymaeus Group-*Drymaeus lattrei lattrei*          Mesembrinus Group-*Drymaeus shattucki*

*Drymaeus* is further divided into species groups on the basis of other similar shell traits such as shape but even this task has proved to be challenging due to the wide range of shell and color variations from different regions, even within same populations. The distinctions in this text include previous authors' observations and our own from the examination of numerous museum specimens as well as field collections.

# Other *Drymaeus* Characters

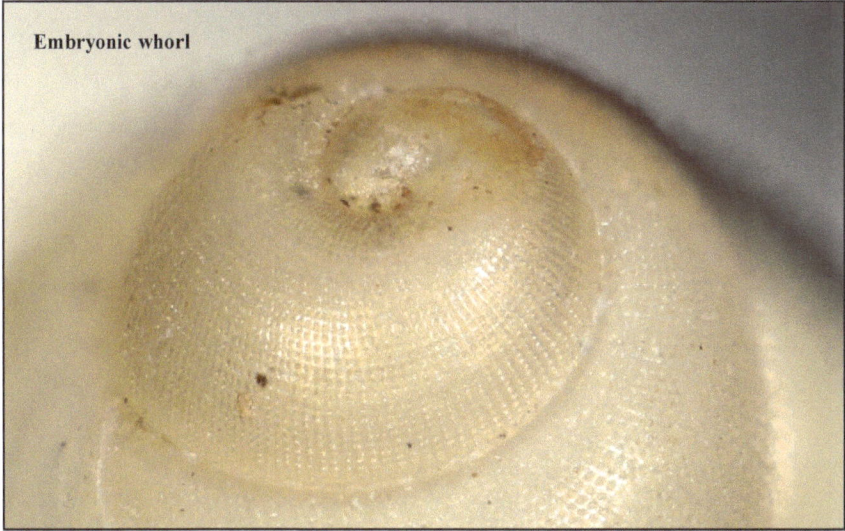

Embryonic whorl

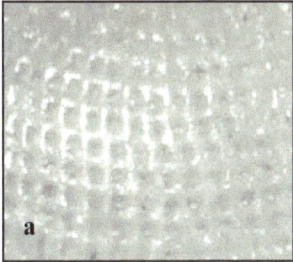

a

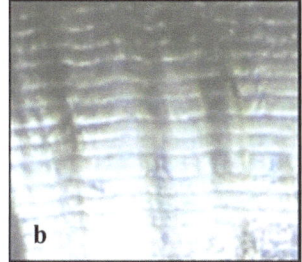

b

Above images show the diagnostic spiral papillae (a) found on the embryonic whorl or protoconch of all *Drymaeus* species (a feature only seen under a strong lens). The embryonic whorls of *Bulimulus* are wrinkled and *Orthalicus* are mostly smooth, making separation of the three genera straightforward. Most *Drymaeus* are also ornamented with engraved spiral striae (b) but the transverse striae, typically well developed on *Bulimulus* shells, may be a wanting feature in most *Drymaeus*.

Other traits used in separating *Drymaeus* from *Bulimulus* include color bands and stripes; *Drymaeus* almost always with these features and *Bulimulus* in Belize are typically without.

*Drymaeus*

*Bulimulus*

# Banded Cone                    ORTHALICIDAE

## *Drymaeus emeus* (Say, 1829)

**Height**: 30-35 mm, Diameter: 14-16 mm

**Description**: Cone-shape, variable; lip very slightly reflected; shell with 6-7 whorls; perforate; columellar notably curved (a); shell thin and glossy; color either white or pale yellow; usually with 5 reddish-brown bands that may vary from continuous or interrupted into rows of spots; one wide, dark band emerges from the aperture (b), continuing to the outer lip; crowded spiral striae are exceedingly fine; spiral papillae on embryonic whorls; periphery broadly rounded.

**Similar Species**: *Drymaeus sulphureus* is without notable color features; *D. tropicalis* is sinistral; *D. serperastrus* usually has 2 dark bands emerging from its aperture and has a yellow tinge; not yet positively identified from Belize.

**Habitat**: An arboreal species on limestone hills of the Maya Mountains.

**Status**: The most common *Drymaeus* in Belize.

**Specimen**: Figure (c & d) from Belize, Toledo District, Bladen Nature Reserve (Dourson collection) and figure (e)

**Type Locality**: Jalapa, Mexico.

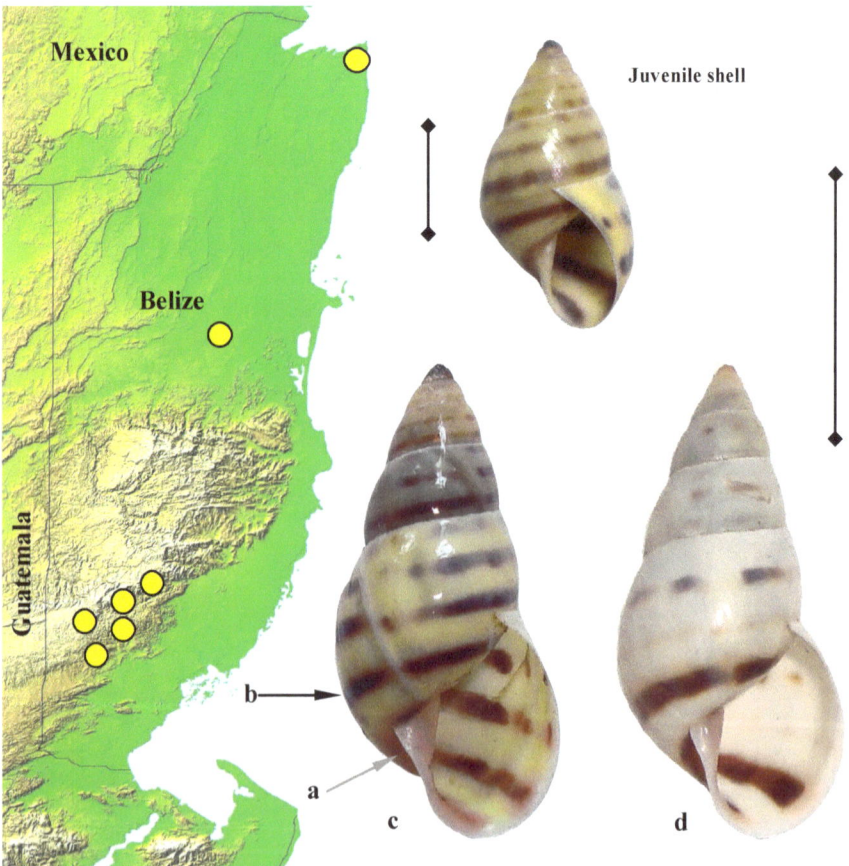

Juvenile shell

112

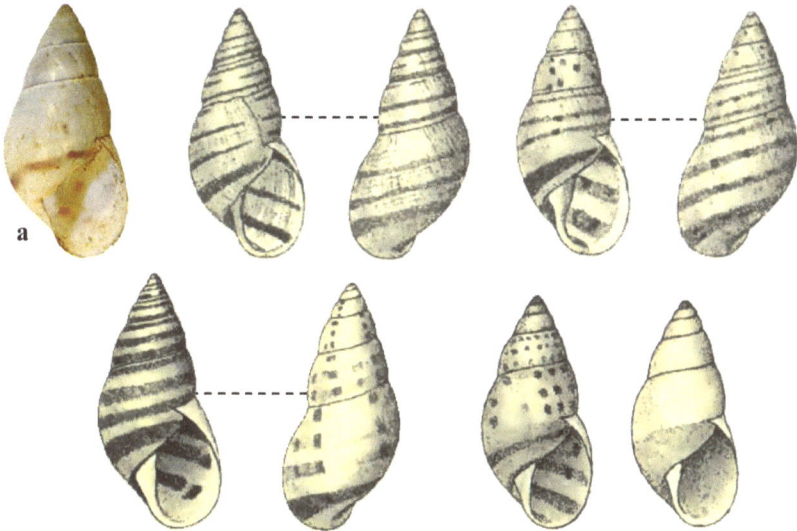

Figure (a) *Drymaeus emeus* from Mexico, Vera Cruz, Jalapa, Syntype ZMB 109.947a; all other figures of *Drymaeus emeus* illustrations from Manual of Conchology, Pilsbry 1899.

# Sulfur Cone                    ORTHALICIDAE
## *Drymaeus sulphureus* (Pfeiffer, 1857)

**Height**: 24-29 mm, Diameter: 9-15 mm

**Description**: Cone-shape; lip simple or slightly reflected; shell with 5 whorls; perforate; shell thin, surface very glossy; without color bands, but occasionally with faded vertical streaks; base color of shell can be bone white to bright yellow; as in all *Drymaeus* species the first few whorls display spiral papillae (opposite page), while later whorls have faint spiral striae; live animals bluish-white with yellowish tentacles; periphery broadly rounded; *Drymaeus* species are the preferred food of snail-eating *Sibon* snakes (Dourson *et al.* 2011).

**Similar Species**: *Orthalicus* species are much larger and are streaked with bold color patterns; most other *Drymaeus* are with some variety of color configuration.

**Habitat**: An arboreal species of limestone foothills, in wet weather usually found crawling on the leaves of the cohune palm.

**Status**: Found only in the southern sections of the Maya Mountains, it is common where it occurs.

**Specimen**: All figures from Belize, Toledo District, Blue Creek Cave area (Dourson collection).

**Type Locality**: Cordoba, Mexico.

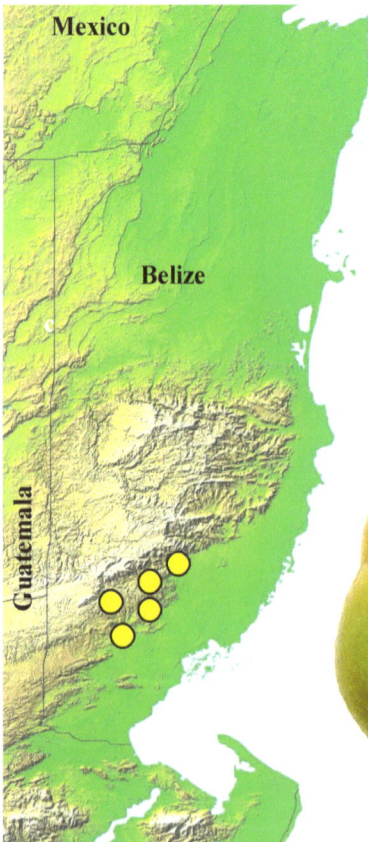

The beautiful Sulfur Cone, *Drymaeus sulphureus*, photographed on a palm leaf around Blue Creek Cave, Toledo District in southern Belize.

## Sulfur Cone

Above a juvenile Sulfur Cone, *Drymaeus sulphureus,* on a leaf. Below the same species illustrating two different color morphs. All figures from Blue Creek Cave area, Toledo District, Belize.

Yellow morph

White morph

# The Sulfur Cone: A Favorite Food of Snail-Eating Snakes

Above a Ringed Snail Sucker, *Sibon sartorii,* stalking a Sulfur Cone, *Drymaeus sulphureus,* in a tree, BNR, Belize. Below a Speckled Snail Sucker, *Sibon nebulata,* feeding on a Sulfur Cone, Forest Hill, BNR, Belize.

# Central American Cone    ORTHALICIDAE
## *Drymaeus attenuatus* (Pfeiffer, 1851)

**Height**: 28-35 mm, Diameter:  12 mm

**Description**: Cone-shape; lip simple with a slight reflection; shell with 6 whorls; imperforate; shell thin, surface glossy; 4-5 variable, vertical color stripes with disruptions in their uniformity; these stripes are observed on all whorls, the strongest ones on the final whorl; spiral papillae on embryonic whorls and later whorls with spiral striae throughout; periphery rounded.

**Similar Species**: *Orthalicus* species are much larger; *Drymaeus sulphureus* is without notable color features; *D. tropicalis* is sinistral; *D. emeus* has revolving bands not vertical stripes.

**Habitat**: Likely an arboreal species found in trees and other elevated habitats.

**Status**: Rare; in Belize, currently known only from around the entrance of Rio Frio Cave.

**Specimen**: Figure (a) from Costa Rica, Osa Peninsula, UF 176771, figure (b) from Belize, Cayo District, mouth of Rio Frio Cave, representing the only shell from Belize.

**Type Locality**: Vera Cruz, Mexico.

118

# Rio Frio Cone             ORTHALICIDAE

*Drymaeus* species (undetermined)

**Height**: 19-26 mm,  Diameter: 12 mm

**Description**: Cone-shape; lip simple; shell with 6 whorls; perforate; shell thin and glossy; shell with or without 3 faint bands above the periphery, <u>marked with irregular intervals by nearly square brown spots, arranged in vertical rows</u>; below the periphery, one continuous or broken band on the base and another, often obsolete, around the umbilicus; spiral papillae on embryonic whorls; faint spiral striae throughout; periphery broadly rounded.

**Similar Species**: *Drymaeus sulphureus* is without notable color features; *D. tropicalis* is sinistral; *D. emeus* has bands not spots; *D. tzubi* has no lower body band.

**Habitat**: An arboreal species of karst hills around Rio Frio Cave, the species appears to be most frequent in jungles with rocky soils.

**Status**: Rare, **ENDEMIC to BELIZE** currently known only around Rio Frio Cave.

**Specimen**: All figures below from Belize, Cayo District, limestone hills around Rio Frio Cave (Dourson collection).

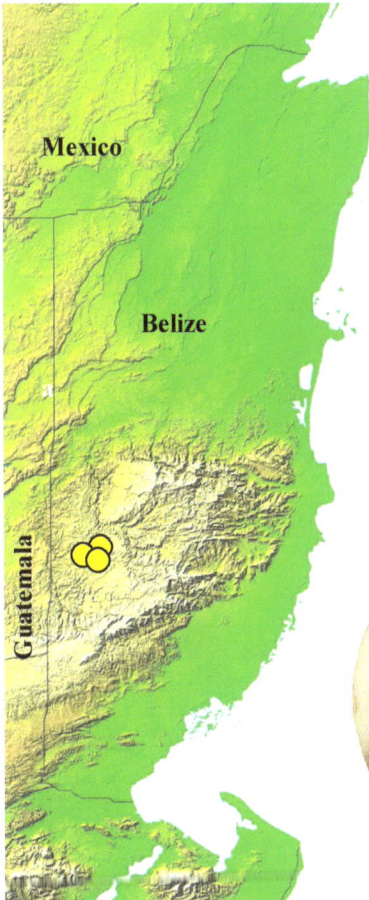

Mexico

Belize

Guatemala

Juvenile shell

119

# Striped Cone                                ORTHALICIDAE
## *Drymaeus translucens alternans* (Beck, 1837)
**Height**: 18-20 mm, Diameter: 10-12 mm

**Description**: Cone-shape; lip simple with a slight reflection near the umbilicus; shell with 5 whorls, perforate; shell white or with a faint yellow tint, glossy; 5 largely continuous brown bands on the last whorl (a), the lower one forming a wide umbilical dark-patch (b); 3 continuous bands on the 2 apical whorls (c) located above the body whorl; spiral papillae on embryonic whorls.

**Similar Species**: *Drymaeus sulphureus* is without notable color features; *D. tropicalis* is sinistral; *D. emeus* is larger, generally with discontinuous bands on the first 3-4 whorls.

**Habitat**: Found under the leaves of cohune plum trees.

**Status**: Rare, in Belize known only from around San Jose in southern Belize.

**Specimen**: Two figured; figure (d) Belize, Toledo District, around San Jose and figures (e) from Guatemala, Guatemala Dept. Guatemala City, District 10, UF 161206.

**Type Locality**: Saboga, Panama.

120

# Belizean Cone                                    ORTHALICIDAE

## *Drymaeus cf. hondurasanus* (Pfeiffer, 1846)

**Height**: 18 mm, Diameter: 10 mm

**Description**: Cone-shape; lip simple shell with 6-7 whorls; <u>shell openly perforate</u>; shell thin, smooth and glossy, yellowish-white; <u>on the body whorl and penultimate whorl there are 3 bands each, 4 bands on the body whorl if you count the rose-brown umbilical patch</u>; spiral papillae on embryonic whorls; periphery broadly rounded.

**Similar Species**: *Drymaeus sulphureus* is without notable color features; *D. tropicalis* is sinistral; <u>*D. hondurasanus* may be inseparable from *D. emeus*</u>.

**Habitat**: An arboreal snail of vegetation in limestone hills.

**Status**: Rare, **ENDEMIC to BELIZE** little information exists for this species.

**Specimen**: Figure (a) from "Honduras" Lectotype NHMUK 1975265, figure (b) *Drymaeus cf. hondurasanus* from Belize, Belize District, Rockville Quarry, UF 135009, labeled as *D. alternans,* correction noted at museum.

**Type Locality**: "Honduras"; meaning Belize, the precise location unknown.

# Checkered Cone                    ORTHALICIDAE

## *Drymaeus tzubi* (new species)

**Height**: 18-22 mm, Diameter: 12 mm

**Description**: Cone-shape; lip simple with a slight reflection; shell with 5 whorls; perforate; shell thin, surface glossy; multiple rectangles arranged spirally and regularly on all whorls; spiral papillae on embryonic whorls; periphery broadly rounded..

**Similar Species**: Most similar to *Drymaeus* species (undetermined), page 119, but without the revolving band on the last whorl, instead having spots and a larger aperture; *D. sulphureus* is without notable color features; *D. tropicalis* has a sinistral aperture; *D. emeus* has bands not spiral spots on the shell.

**Habitat**: Found in karst hills of southern Belize.

**Status**: Exceedingly Rare, ENDEMIC to BELIZE currently known only from the type locality near the Maya village of San Jose.

**Specimen**: Belize, Toledo District, 3.25 miles west of San Jose; to date the specimen illustrated below represents the only shell collected, Holotype UF 505455.

**Type Locality**: San Jose, Toledo District, Belize (16°15'42"N, 89°8"28"W).

**Etymology:** Named in honor of Belizean Valentino Tzub, of San Jose, Toledo District, one of Belize's top biological field technicians. Valentino discovered many new land snail records and several new species to science including the recently described *Eucalodium belizensis* of southern Belize.

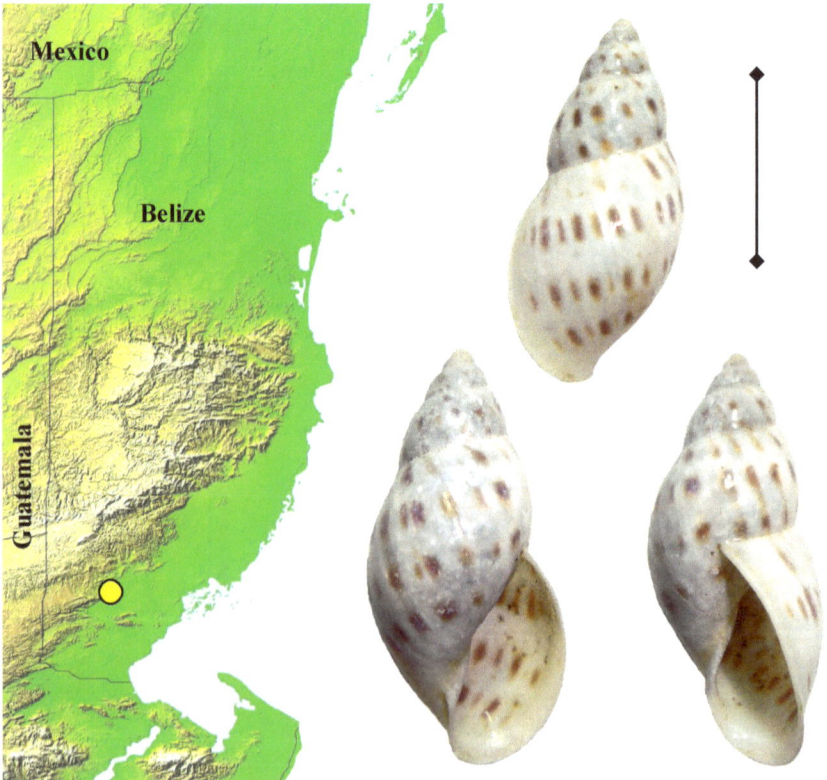

# Florida Cone                                    ORTHALICIDAE

## *Drymaeus dominicus* (Reeve, 1850)

**Height**: 15.5 mm, Diameter: 7-8 mm

**Description**: Cone-shape; lip simple with a slight reflection; shell with 5.5 whorls; perforate; shell thin, fragile and translucent; color whitish or yellowish; typically with 4 or 5 dark-brown bands, the upper three typically interrupted into small spots, the lower two continuous or nearly so, emerging from the aperture (a) as in *D. serperastrus*, contiguous and nearly midway between the axis and the periphery on the base; spiral papillae on embryonic whorls; periphery broadly rounded.

**Similar Species**: *Drymaeus sulphureus* is without notable color features; *D. tropicalis* has a sinistral aperture; *D. emeus* has bolder bands and is larger with a thicker shell; *D. tzubi* is larger and decorated with spots or rectangles instead of bands like *D. dominicus* and has a thicker shell.

**Habitat**: An arboreal species found in trees and other low vegetation; like most *Drymaeus* species it is most active after rains.

**Status**: Rare, reported by Fred Thompson from one location in the Toledo District of southern Belize.

**Specimen**: USA, Florida, Broward County, Ft. Lauderdale, UF 00109188.

**Type Locality**: Florida, USA.

# Reverse Cone          ORTHALICIDAE

## *Drymaeus cf. tropicalis* (Morelet, 1849)

**Height**: 20-27 mm, Diameter: 12 mm

**Description**: Cone-shape; lip simple; aperture is sinistral (an uncommon aperture position in Belize's land snails); shell with 6 whorls; perforate; shell thin, surface glossy; several light color band on each whorl, spiral papillae on embryonic whorls; periphery broadly rounded.

**Similar Species**: No other Orthalicidae snails from Belize are known to be sinistral.

**Habitat**: An arboreal species found in trees and other elevated habitats.

**Status**: Rare; In Belize, two specimens were found in a small block of jungle around the Spanish Lookout Mennonite Community, figure (a). It refers to *D. tropicalis,* but has a single light band on the final whorl only, lacking the multiple bands seen in typical *D. tropicalis.*

**Specimen**: Figure (a) *Drymaeus cf. tropicalis* from Belize, Cayo District, Spanish Lookout (Dourson collection) and figure (b) *D. tropicalis*, Mexico, Yucatan, Campeche, Lectotype NHMUK 1893.2.4.210.

**Type Locality**: Campeche, Mexico.

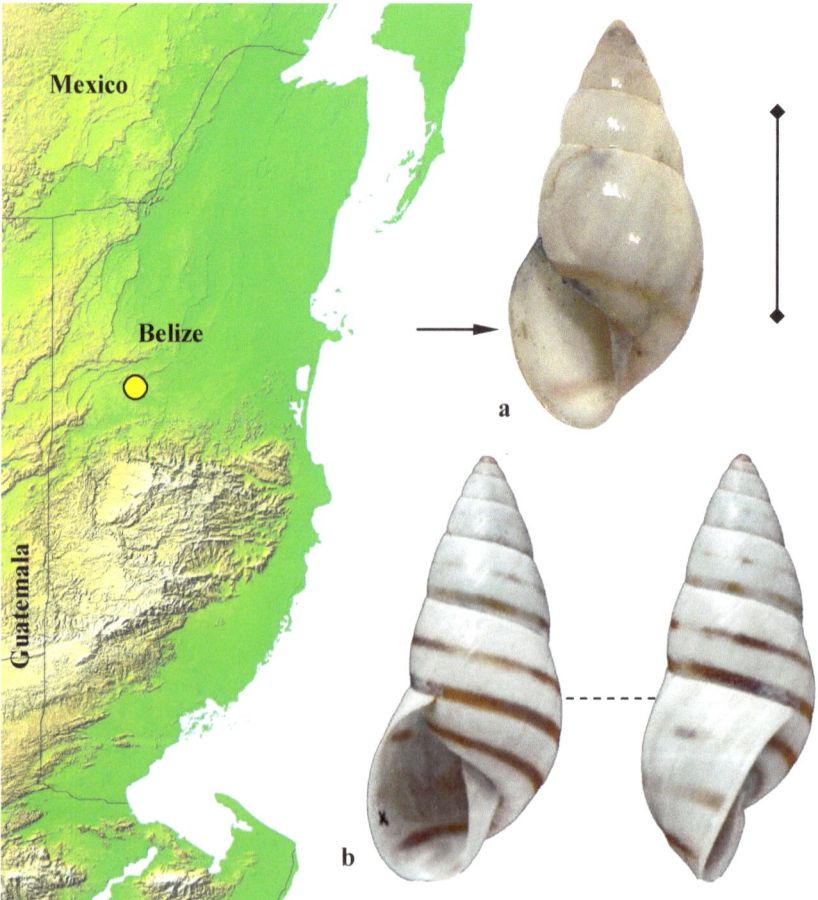

# Yucatan Cone                    ORTHALICIDAE

## *Drymaeus shattucki* Bequaert & Clench, 1931

**Height**: 17-21.5 mm, Diameter: 7-10 mm

**Description**: Cone-shape; lip simple with a slight reflection near umbilicus; aperture rather small for *Drymaeus* (a); shell with 5 whorls; perforate; shell thin, surface glossy in live and fresh dead shells; several light color bands on each whorl, the last and final whorl having one bold, usually unbroken band; spiral papillae on embryonic whorls; periphery broadly rounded.

**Similar Species**: *Drymaeus shattucki* is similar to *D. tropicalis*, possibly a sinistral race of that species and may not be separable from *D. hondurasanus* (Bequaert & Clench 1931); other *Drymaeus* have proportionately larger apertures.

**Habitat**: Adapted to human settlements; a camp-follower (Goodrich & Van der Schalie 1937).

**Status**: Not yet in Belize but known from Tikal Maya Archaeological Site, Guatemala.

**Specimen**: Figure (b) from Mexico, 3 km E of San Miguel, Holotype UF 00120445 and figure (c) from Mexico, Yucatan, Chichen Itza, Paratype MCZ 79396 .

**Type Locality**: Chichen Itza, Yucatan, Mexico.

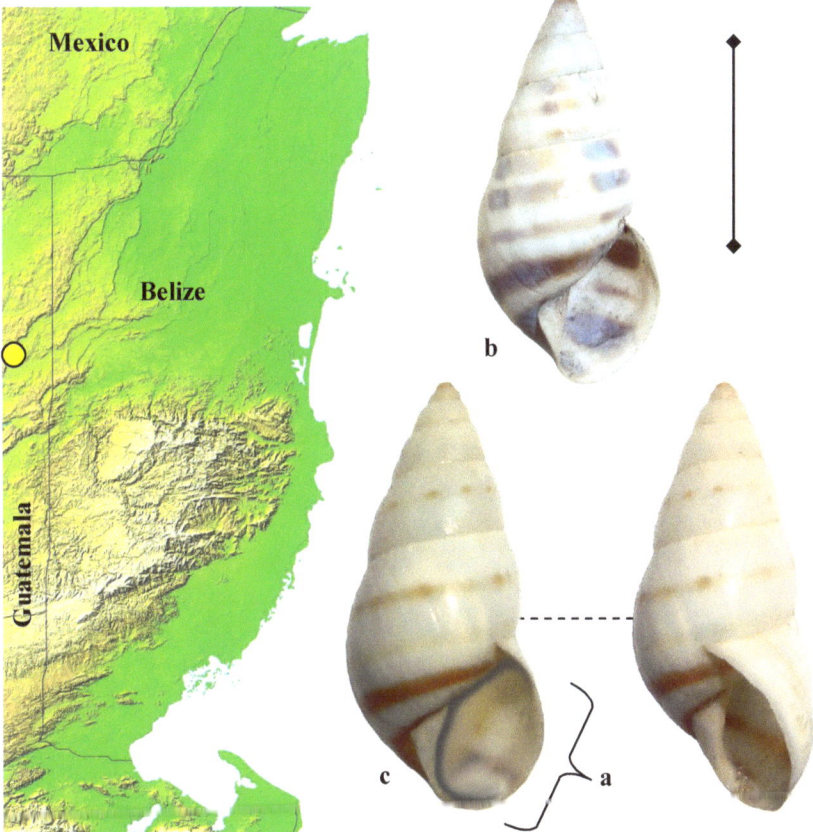

# Banded Cone ORTHALICIDAE

## *Drymaeus serperastrus* (Say, 1829)

**Height**: 25-38 mm, Diameter: 12-18 mm

**Description**: Cone-shape, elongated but variable; lip expanded; shell with 6 whorls; umbilicate; shell thin, shiny, nearly smooth or with faint spiral striae; the body whorl has 6 largely fragmented, blackish bands, the upper four nearly always irregularly interrupted into oblong spots, while 2 mostly continuous bands emerge from the aperture (a); shell white or ochre tinted.

**Similar Species**: *Drymaeus hondurasanus* has only 3 bands; *D. tropicalis* has a sinistral aperture; *D. dominicus* has two emerging bands from the aperture as in *D. serperastrus* but is only half the size and has a much thinner shell.

**Habitat**: An arboreal species of karst regions often found on a shrub called "huichin" a Spanish name in Mexico for *Berbesina persicifolia.*

**Status**: The residency of *D. serperastrus* in Belize seems doubtful; all specimen reported as *D. serperastrus* from Belize are likely *D. emeus*; no type specimens are known for *D. serperastrus* (pers. comm. Breure, 2015).

**Specimen**: Figure (b) from Mexico, 7.1 mi SW of Campeche, UF 00019335 and figure (c) from Mexico, Yucatan, 5 mi east of Rio Lagartos, UF 155817.

**Type Locality**: Vera Cruz, Mexico.

*Drymaeus serperastrus* **shells compared**

Figure (d) from Fisher &Cross, plate 24 fig 4; all other figures from Tryon and Pilsbry, Manual of Conchology(1899), vol. 12, plate 9, fig. 34-41, 4 band variations illustrated.

# River Cone                                    ORTHALICIDAE

*Bulimulus* **species (undetermined)**

**Height**: 13 mm, Diameter: 8 mm

**Description**: Cone-shape; lip simple; shell with 5-6 whorls; perforate; shell thin, surface dull-glossy; translucent white, multiple faint color streaks especially on the body whorl; protoconch without spiral papillae; transverse striae faint, strongest on last whorl.

**Similar Species**: *Bulimulus corneus* is corneous and a little more compact in build; *B. coriaceus* is around the same size but has a broader last whorl and has a distinctive band on the first few whorls.

**Habitat**: Found along large river floodplains in arboreal vegetation.

**Status**: Unknown, the specimen below was labeled as *Bulimulus umbraticus,* which has been reclassified in the genus *Drymaeus* (Breure & Ablett 2014); it does not match the Lectotype images and is without the characteristic embryonic papillae of *Drymaeus*. Additional specimens are needed for a final determination.

**Specimen**: The collection data reads "Honduras" Belize, Belize River, labeled as *Bulimulus umbraticus*, UF 00109715.

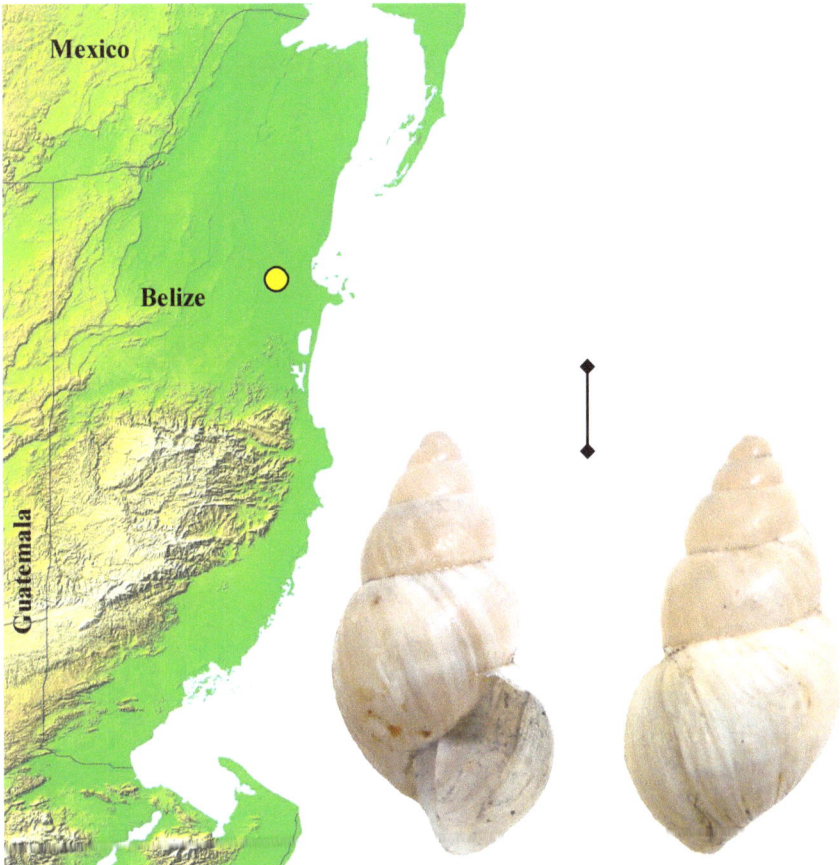

# Brown Cone                          ORTHALICIDAE

### *Bulimulus unicolor* (Sowerby,1833)

**Height**: 15-18 mm, Diameter: 12 mm

**Description**: Cone-shape, <u>wide but variable</u>; lip simple; shell with 5-5.5 whorls; perforate; shell thin, surface glossy; most shells brownish; with or without light vertical streaks; shell form and color vary considerably from population to population and even within same communities; no spiral papillae on embryonic whorls (a); last whorl not broadly rounded as seen in *Drymaeus*.

**Similar Species**: *Orthalicus* species are much larger; *Drymaeus* are larger, have spiral papilla on the first whorls and are generally with color streaks or bands; *Bulimulus corneus* is smaller and horn-colored.

**Habitat**: Found in a variety of habitats, karst foothills, ravines under leaf litter, and coastal plains.

**Status**: Common; found throughout Belize.

**Specimen**: Belize, Toledo District, Bladen Nature Reserve (Dourson collection).

**Type Locality**: Perico Island, Panama.

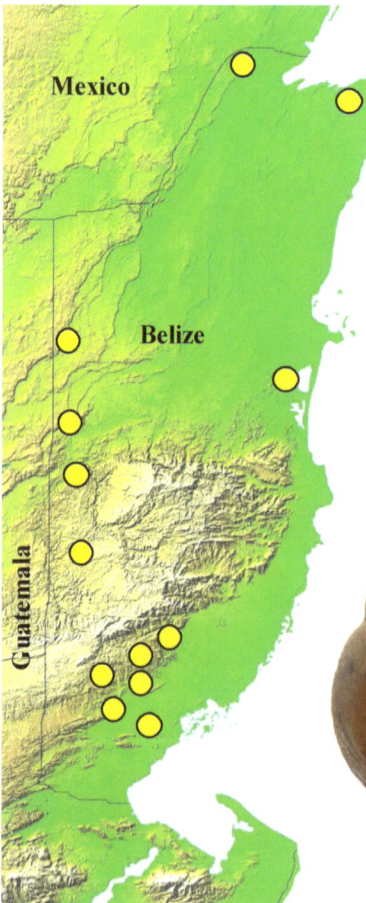

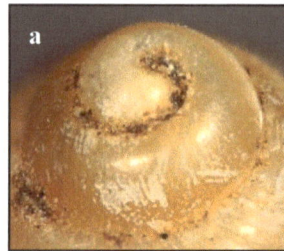

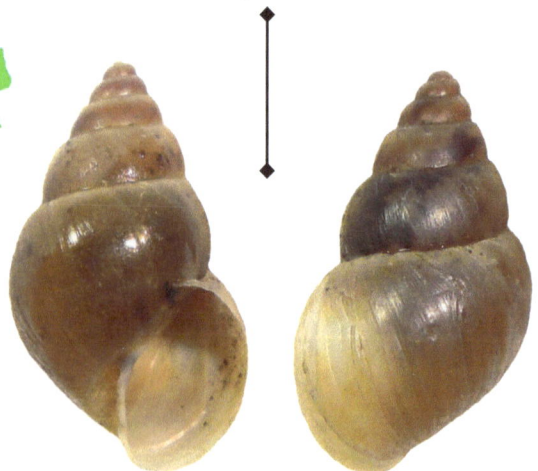

# Dyson's Cone                    ORTHALICIDAE
## *Bulimulus dysoni* (Pfeiffer, 1846)
**Height**: 17-20 mm, Diameter: 10 mm

**Description**: Cone-shape, <u>narrow but variable</u>; lip simple; shell with 5-5.5 whorls; perforate; shell thin, surface glossy; most shells brownish; shell form and color may vary slightly from population to population and even within same communities; no spiral papillae on embryonic whorls; last whorl not as broadly rounded as seen in *Drymaeus*.

**Similar Species**: *Orthalicus* species are much larger; *Drymaeus* are larger, have spiral papillae on the first whorls and are generally with color streaks or bands; <u>*Bulimulus unicolor* is slightly smaller and more obese in build</u>.

**Habitat**: Karst foothills under leaf litter.

**Status**: Uncommon; found at lower elevations in central Belize.

**Specimen**: Labeled only as "Honduras" Belize, Syntype NHMUK 1975453,n (images by Abraham Breure).

**Type Locality**: "Honduras" Belize (?).

# Ribbed Cone                                    ORTHALICIDAE

### *Bulimulus* species (undetermined)

**Height**: 18-22 mm, Diameter: 10 mm

**Description**: Cone-shape; lip simple; shell with 5-5.5 whorls; perforate; shell thick (solid); surface glossy; most shells corneous or horn color, sometimes with light vertical streaks; shell form and color may vary considerably from population to population and even within same communities; suture lines are wavy and uneven due to well developed transverse striae that are more rib-like (a); no spiral papillae on embryonic whorls; last whorl rounded.

**Similar Species**: *Orthalicus* species are much larger; *Drymaeus* species are larger, have spiral papillae on the first whorls and are generally with color streaks or bands.

**Habitat**: Karst hills around Caracol, an ancient Maya archaeological site; likely arboreal in habit, especially during wet weather.

**Status**: Little is known about the overall range and position of this species. Additional specimens are needed to make a final determination.

**Specimen**: Belize, Cayo District, Caracol Archaeological Site (Dourson collection).

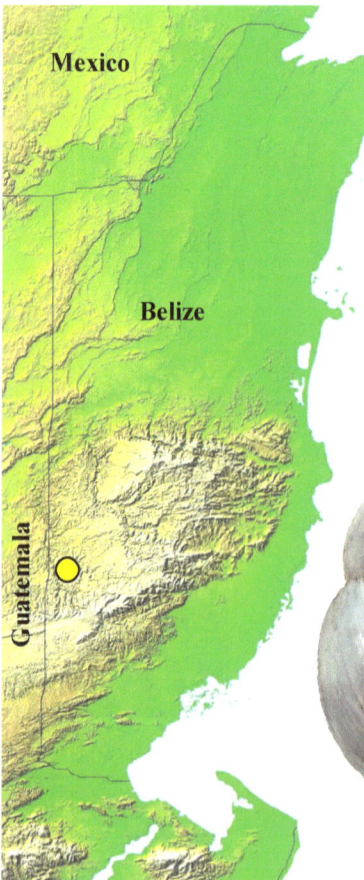

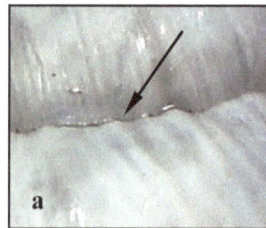

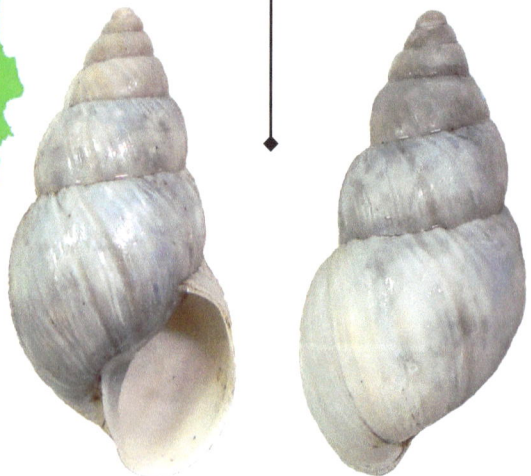

# Pale Cone                                    ORTHALICIDAE

## *Bulimulus corneus* (Sowerby, 1833)

**Height**: 15 mm, Diameter: 8 mm

**Description**: Cone-shape; lip simple; shell with 5-6 whorls; perforate; shell thin, surface glossy; <u>most shells corneous (horn color)</u>, sometimes with light vertical streaks; shell form and color may vary considerably from population to population and even within same communities; no spiral papillae on embryonic whorls; last whorl not broadly rounded as seen in *Drymaeus*.

**Similar Species**: *Orthalicus* species are much larger; *Drymaeus* are larger, have spiral papillae on the first whorls and are generally with color streaks or bands; most similar to *Bulimulus coriaceus* but a different color and lacks the light chestnut bands in early whorls.

**Habitat**: A resident of karst regions and likely arboreal in habit, especially during wet weather.

**Status**: Uncommon; found in three locations in the northern half of Belize.

**Specimen**: Mexico, town of Chetumal, UF 176362.

**Type Locality**: Polvon, Nicaragua.

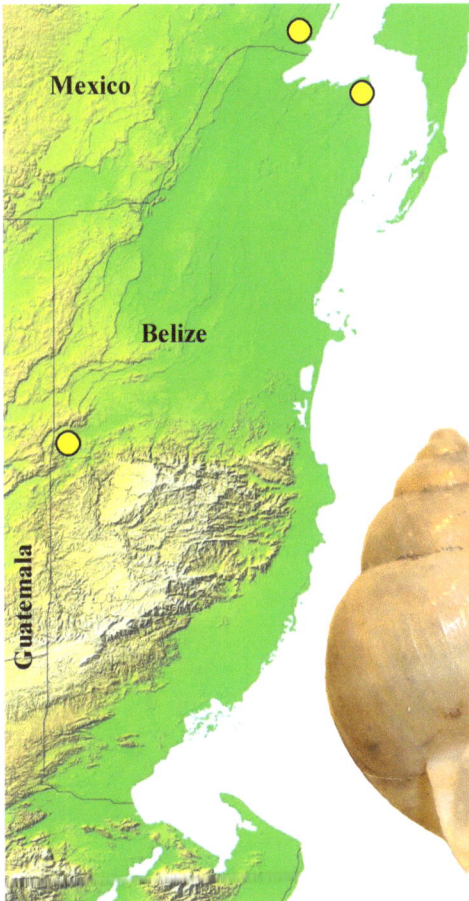

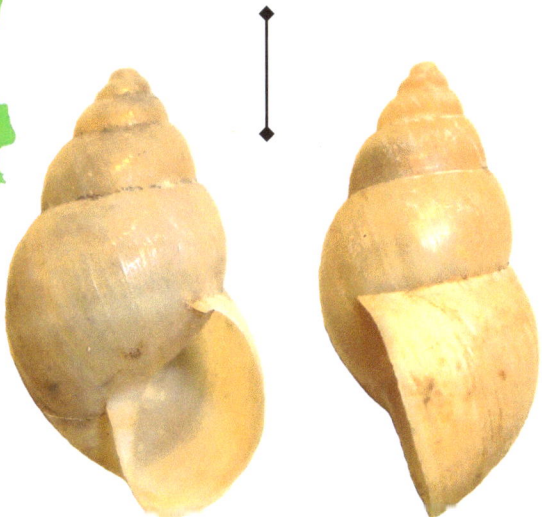

# Rosy Cone                                    ORTHALICIDAE

## *Bulimulus coriaceus* (Pfeiffer, 1856)

**Height**: 14-18 mm, Diameter: 10 mm

**Description**: Cone-shape; lip simple; shell with 5-6 whorls, last whorl ample in size; perforate; shell thin, surface dull-glossy; light brown, early whorls of the spire with a chestnut band above the sutures (a), fading on its upper margin; apex is light; sometimes with light vertical streaks; shell form and color may vary considerably from population to population and even within same communities; transverse striae or growth winkles are faint; very close spiral striae weak; last whorl not broadly rounded as seen in *Drymaeus*.

**Similar Species**: *Bulimulus corneus* is around the same size but lacks the distinctive band on apex whorls.

**Habitat**: A resident of karst regions and likely somewhat arboreal in habit, especially during wet weather.

**Status**: Rare. In Belize, reported from only one site, Belize District around Burrell Boom.

**Specimen**: Mexico, Veracruz limestone hill 3 km NE of Atoyac, UF 190898.

**Type Locality**: Vera Cruz, Mexico.

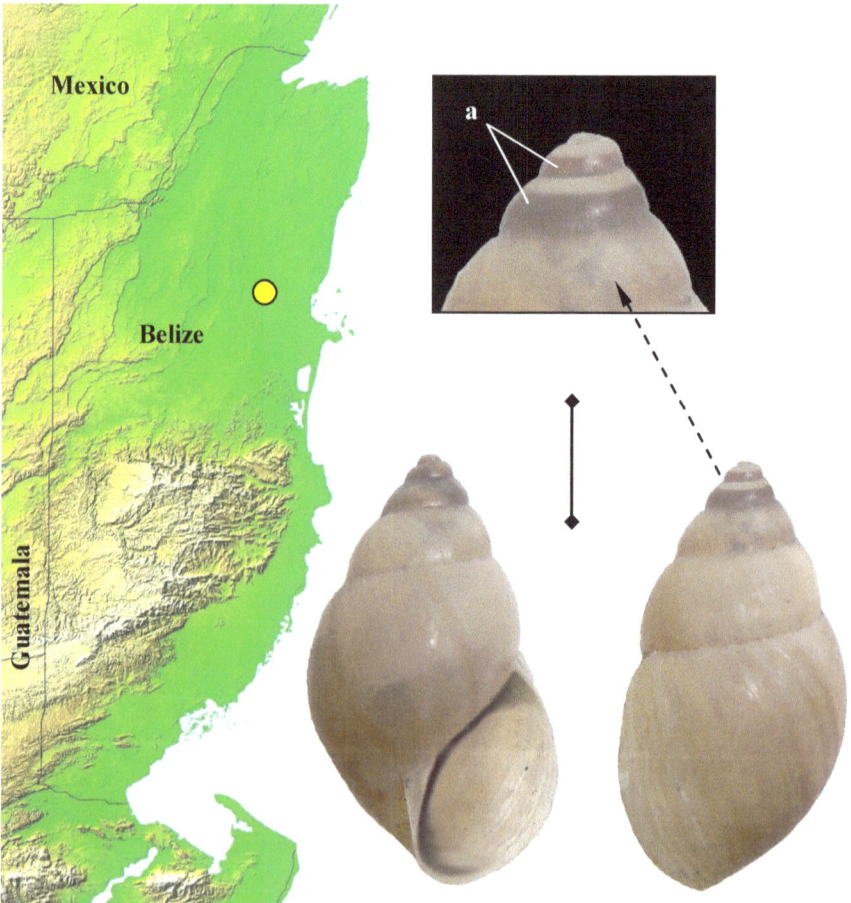

132

# Spotted Cone                    ORTHALICIDAE

*Bulimulus* species **(undetermined)**

**Height**: 16-17 mm, Diameter: 10 mm

**Description**: Cone-shape; lip simple; shell with 5-6 whorls, last whorl capacious; perforate; shell thin, surface dull-glossy; bi-color, the top and bottom of the shell darker than the center in frontal view; entire shell decorated with small tannish spots; transverse striae fairly well developed and closely spaced; without notable spiral striae; last whorl not broadly rounded as seen in *Drymaeus* and the much larger *Orthalicus.*

**Similar Species**: *Bulimulus coriaceus* is around the same size and build but without the scattered spots and has a smoother surface.

**Habitat**: A resident of limestone regions where it is likely arboreal in habit, especially during wet weather.

**Status**: Unknown; little is known on the overall range and taxonomic position of this interesting species. Additional specimens are needed to make a final determination.

**Specimen**: Belize, Belize District, Altun Ha Maya Archaeological Site UF 00134985.

# Breaking Stones

Pictured above is a "breaking stone station" used by birds to crack open snail shells to access the protein-rich flesh within; Bladen Nature Reserve, Belize. In Great Britain, song thrushes are a major predator on wood snails, *Cepaea nemoralis*, crushing the shells on stones to get at the soft flesh within (Whitson 2005). Numerous breaking stones have been documented across Belize.

*Amphicyclotus ponderosus*

*Euglandina ghiesbreghti*

*Orthalicus princeps*

*Neocyclotus dysoni*

*Leptarionta cf. trigonostoma*

# Family SPIRAXIDAE H.B. Baker, 1939

Three subfamilies representing 258 species and an additional 40 subspecies of the SPIRAXIDAE family are recognized in Mexico and Central America. Most species are very restricted in their distributions, reflecting a high degree of local endemism. The majority of the species are known from only a few regions of the study area. This is a reflection of its early colonial history with limited accessibility into many of the regions and the very small number of malacologists who have worked there. Large areas of the study area remain very poorly surveyed. Undoubtedly the number of SPIRAXIDAE species will be vastly increased with further biodiversity surveys of the many biotic provinces that comprise Mexico, Central and South America including Belize. Even areas that are relatively well known such as central Veracruz yield new taxa with each additional investigation. (Excerpted from Thompson 2011). This is the most speciose family in Belize containing 9 genera, 25 recognized species and 5 undetermined species representing taxa new to science. Shell shape and size varies greatly in the family which is primarily carnivorous, feeding on other snails.

## Genera Included:
(in order of appearance in text)
*Varicoglandina*
*Streptostyla*
*Salasiella*
*Mayaxis*
*Pseudosubulina*
*Volutaxis*
*Rectaxis*

## The Jaguars of Land Snails

Northern Central America is a gastropod "hotspot" containing some of the richest tropical land snail faunas in the world. In particular, land snails in the family SPIRAXIDAE have done especially well in the region. Many of the species found in this family are carnivorous, most often consuming other snails alive, earning them the nickname "the jaguars of land snails." The largest flesh -eating land snail in the world (about the size of a large avocado) is the Titan Marauder, *Euglandina titan* of Guatemala. A giant among its kind, the Titan Marauder eats salamanders and frogs for lunch, trapping them in bromeliads (pers. comm. Fred Thompson, 2009). Large areas of northern Central America remain poorly studied and undoubtedly, the number of species will be vastly increased as biodiversity surveys continue in the many biotic provinces that comprise southeastern Mexico, Guatemala, Honduras and Belize. Size and body form, not color, are the best characters used in separating *Euglandina* species.

Mayan Marauder, *Euglandina ghiesbreghti*

Sulfur Cone

Above image of a Mayan Marauder, *Euglandina ghiesbreghti*, searching for entry through the aperture of a Sulfur Cone. Having little chance of escape, the Sulfur Cone will be eaten alive leaving an empty, undamaged shell.

Gaining entry through the aperture (above), the Mayan Marauder goes in for the slow kill and begins to consume the flesh of the Sulfur Cone (below).

Cross Cone,
*Orthalicus princeps*
*crossei* (the prey)

a

Mayan Marauder,
*Euglandina ghiesbreghti* (the predator)

Above image illustrates a violent struggle between the Mayan Marauder, *Euglandina ghiesbreghti* and a Cross Cone, *Orthalicus princeps princeps*. The jaws of the Mayan Marauder are powerful and used like a wood-rasp to tear raw flesh from the side of the Cross Cone; flesh wounds (a) are clearly visible. Having little chance of escape, the Cross Cone is eaten alive. Although the hunt and consumption of snail flesh all happens at a snail's pace, it is no less theatrical than a jaguar taking down a brocket deer, just in slow motion!

138

Carol Foster

Above, a Mayan Marauder evaluating potential prey with its chemoreceptors. While the Mayan Marauder is a voracious predator, it is also prey, eaten by birds, lizards, small mammals and the snail-eating snakes. Below, a Speckled Snail Sucker feeding on a ill-fated Mayan Marauder.

# Mayan Marauder                                    SPIRAXIDAE

## *Euglandina ghiesbreghti* (Pfeiffer, 1856)

**Height**: 50-60 mm, Diameter: 20 mm

**Description**: Oval-shape, variable; lip simple; shell with 6-6.5 whorls; imperforate; shell thick; surface dull-glossy; rosy color, transverse and spiral striae are well developed and easily seen with a hand lens of 10X; protoconch whorls smooth; last whorl quite variable in shape, either broadly rounded or distinctly shouldered (a).

**Similar Species**: *Orthalicus* species are more capacious and carry vivid color patterns on their shells; most similar to *Euglandina ghiesbreghti* but with a more narrow build and less squared aperture.

**Habitat**: Found in numerous locations including under leaf litter, around cliffs, climbing trees and vegetation; a voracious predatory gastropod that hunts and eats other land snails.

**Status**: Common; occurs throughout Belize.

**Specimen**: Figure (b) from Mexico, Tabasco UF 190675 and figure (c) from Belize, Toledo District, Bladen Nature Reserve (Dourson collection).

**Type Locality**: Chiapas, Mexico.

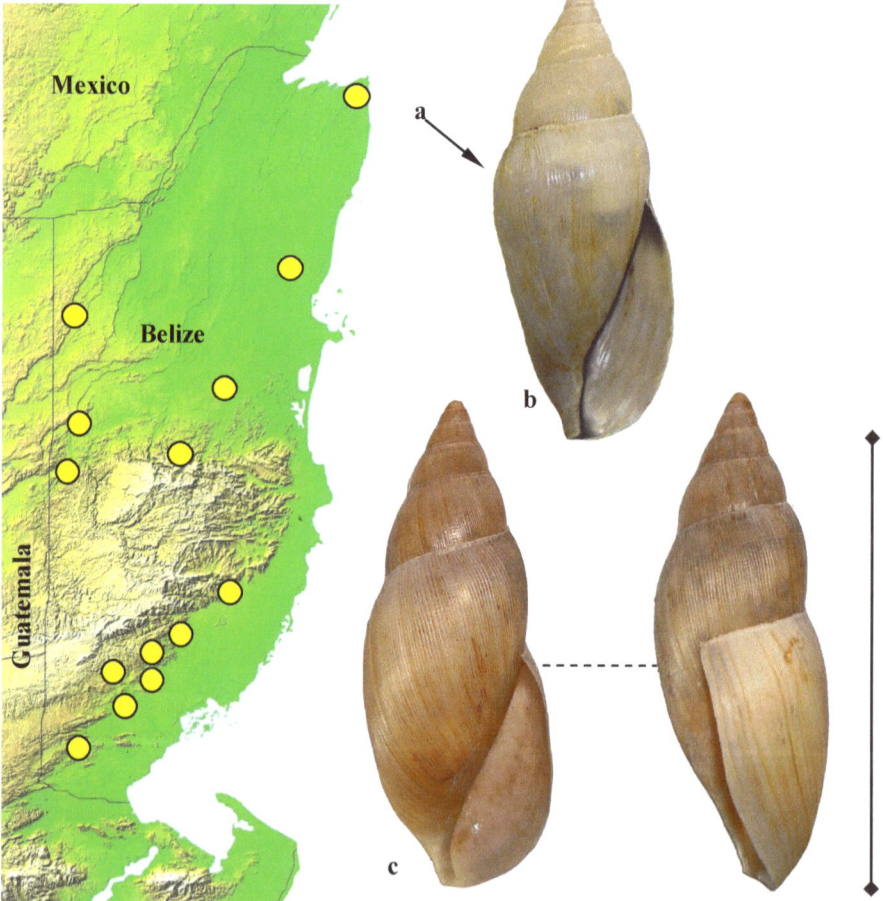

# Rio Frio Marauder                    SPIRAXIDAE

***Euglandina* species (undetermined)**

**Height**: 50 mm, Diameter: 20 mm

**Description**: Oval-shape; lip simple; shell with 6-7 whorls; imperforate; shell thick; surface dull-glossy; pale-rosy but color may be variable; transverse striae are crossed by spiral striae giving the shell surface a netted texture (a); sutures are thickened; the last whorl not broadly rounded as in other *Euglandina* but more squared and shouldered.

**Similar Species**: *Orthalicus* species are more capacious and carry vivid color patterns on their shells; *Euglandina ghiesbreghti* is more narrow in form and the outer edge of its aperture is more curving.

**Habitat**: Found alive under leaf litter at the base of a limestone cliff near the entrance of Rio Frio Cave.

**Status**: Rare; may be endemic to the Maya Mountains (?) below is the only specimen ever collected of this apparently restricted species. Additional specimens are needed to make a final determination.

**Specimen**: Belize, Cayo District, Rio Frio Cave, UF 463283.

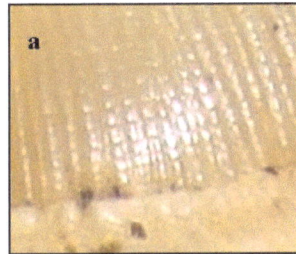

141

# Chetumal Marauder                     SPIRAXIDAE

*Euglandina cylindracea* **(Phillips, 1846)**

**Height**: 30-35 mm, Diameter: 15 mm

**Description**: Bullet-shape; lip simple; aperture proportionately smaller than seen in other *Euglandina* (a); shell with 6-7 whorls; imperforate; shell thick; surface dull-glossy; pale rosy color; spiral striae are well defined (hand lens of 10X or scope required to view this feature); suture lines thickened with calcium deposits (b); columellar plait forming an "S" shape.

**Similar Species**: Most similar to *Euglandina fosteri* but widest in the bottom third of the shell (*E. fosteri* is widest in the middle of the shell) and most importantly, has a proportionally smaller aperture; *E. ghiesbreghti* is larger with a longer aperture and more pointed apex.

**Habitat**: Found in leaf litter and terrestrial habitats surrounding swamps.

**Status**: Uncommon; Shipstern Nature Preserve and Pulltrouser Swamp, Orange Walk District (Turner and Harrison 1983).

**Specimen**: Mexico, Campeche, 6 km. S Hopelchen, UF 19281.

**Type Locality**: Yucatan, Mexico.

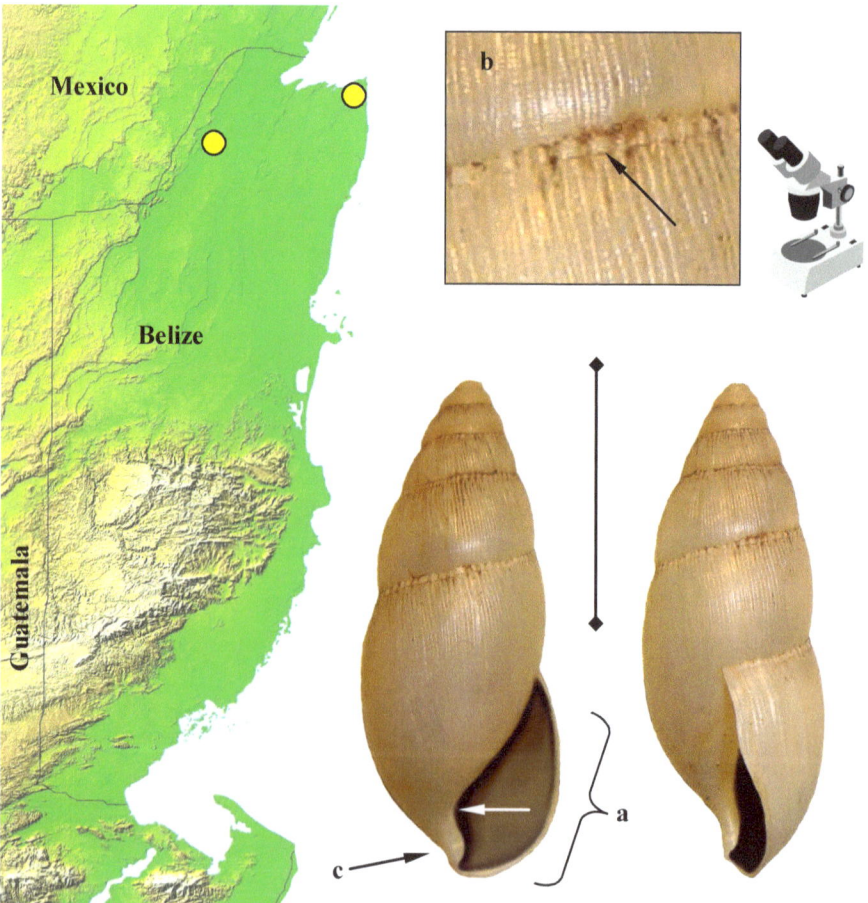

# Rosy Marauder                           SPIRAXIDAE

*Euglandina fosteri* (new species)

**Height**: 35-38 mm, Diameter: 14-16 mm

**Description**: Oval-shape; lip simple; shell with 6 whorls; imperforate; surface dull-glossy; pale rosy tan color; transverse and spiral striae are well developed and can be seen with a hand lens of 10X; the embryonic whorl is smooth (a), but the later protoconch whorls becoming distinctly ribbed (b); at the suture line there are notable thickened calcium deposits.

**Similar Species**: Most like *Euglandina cylindracea* (page 142) but having a different body shape; *E. cylindracea* is more bullet-shaped while *E. fosteri* is oval-shape, has a longer aperture and has a less concave columellar margin (c); *E. ghiesbreghti* is larger and has a smooth protoconch (b).

**Habitat**: Karst foothills under leaf litter; base of limestone cliffs.

**Status**: Rare; **ENDEMIC to BELIZE** found Belize District, Altun Ha Maya Archaeological Site, Runaway Creek and Shipstern Nature Preserves.

**Specimen**: Belize, Belize District, Tiger Sandy Bay. Holotype UF 505431, Paratype UF 504432 from same location (not pictured).

**Type Locality**: Tiger Sandy Bay, Belize District, Belize (17°17'60"N, 88°31'0"W).

**Etymology**: Named in honor of award-winning filmmakers, the late Richard Foster & Carol Farneti Foster for their significant contributions to the world of conservation through film including never-before filmed interactions of *Sibon* snakes and land snails.

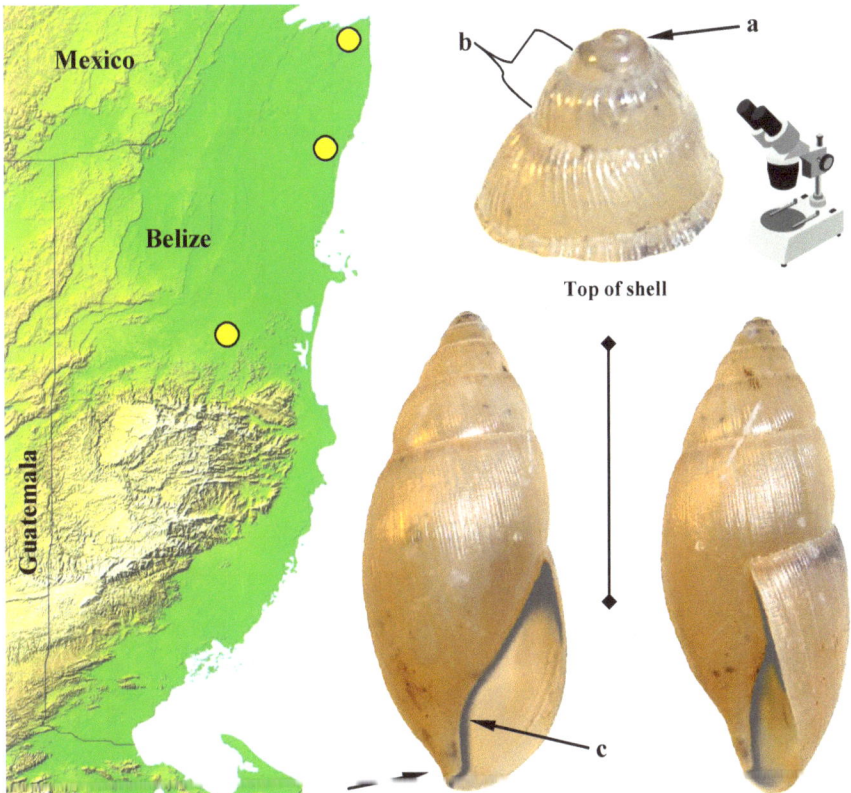

Top of shell

143

# Cuming's Marauder

## SPIRAXIDAE

*Euglandina cumingi* (Beck, 1827)

**Height**: 44 mm tall, 21 mm wide

**Description**: Shell oval-shape, obese (wide) in form; lip simple; shell with 6-7 whorls; imperforate; shell thin but solid with light vertical streaks on all whorls fading in old and weathered shells; surface dull-glossy; transverse striae well developed and form thickened calcium deposits at the sutures; spiral striae also well developed; last whorl broadly rounded.

**Similar Species**: *Euglandina ghiesbreghti* is much smaller, more narrow in form; *E. fosteri* is smaller, narrower in form with a rosy color, lacks vertical streaks and most notably, protoconch whorls are distinctly ribbed whereas *E. cumingi* protoconch whorls are smooth.

**Habitat**: Karst hills.

**Status**: Rare; found across the Petén region of Guatemala including Tikal, Maya Archaeological Site; in Belize, reported from Barton Ramie Estate, Cayo District only.

**Specimen**: Mexico, UF 241165.

**Type Locality**: Type locality unknown.

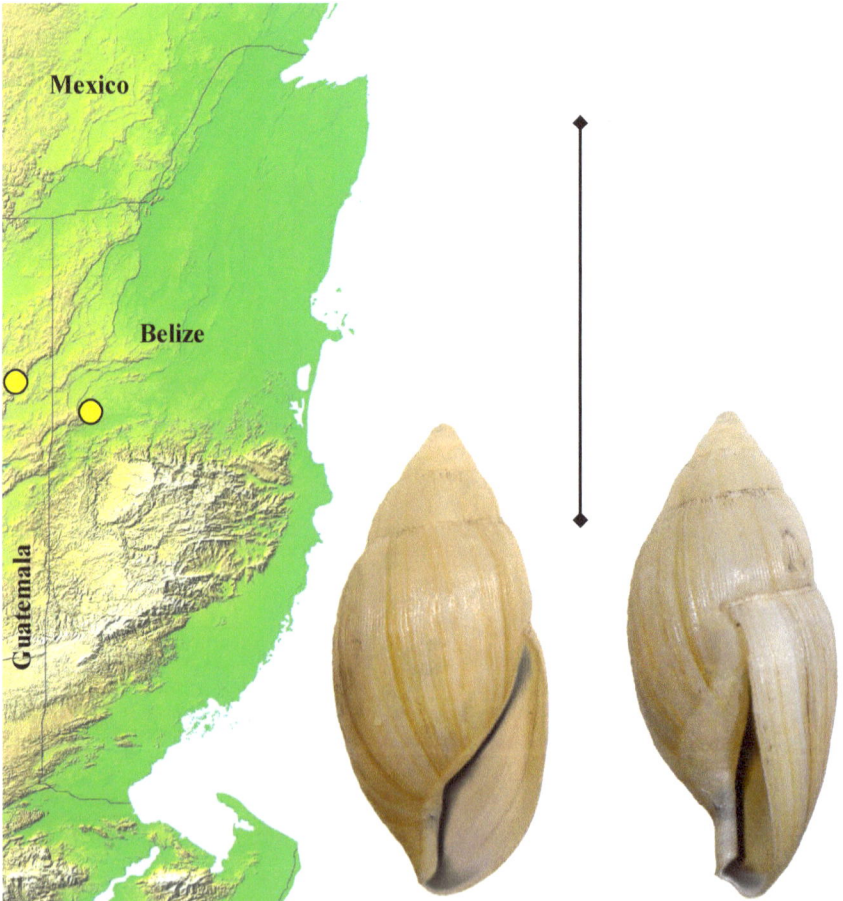

144

# Titan Marauder                    SPIRAXIDAE

*Euglandina titan* Thompson, 1987
**Height**: 100-115 mm, Diameter: 70 mm
**Description**: <u>Shell enormous</u> (size of an orange); oval-shape; lip simple; shell with 6-7 whorls; imperforate; shell thick (solid), rosy-buff; surface dull-glossy; suture lines without calcium deposits; last whorl broadly rounded as in *Drymaeus* and *Orthalicus* shells.
**Similar Species**: *Orthalicus* and other *Euglandina* species are significantly smaller. *Euglandina titan* is the largest living land snail reported within the scope of the book. No other land snail comes close to its enormous size!
**Habitat**: Found in upper elevation (800 meters) rainforests where it preys upon frogs and salamanders it traps in bromeliads (pers. comm. Fred Thompson, 2010).
**Status**: Not yet reported from Belize but close to its borders; currently known only from the type locality.
**Specimen**: Guatemala, Izabal Dept., Montanas del Mico, 7.4 mi. WSW, Puerto Santo Tomas, Holotype UF 35307.
**Type Locality**: Montanas del Mico, Guatemala.

Shown here life size

A Mayan Marauder overpowers a Sulfur Cone; there will be no escape!

# *Euglandina* Species Compared (shells to scale)

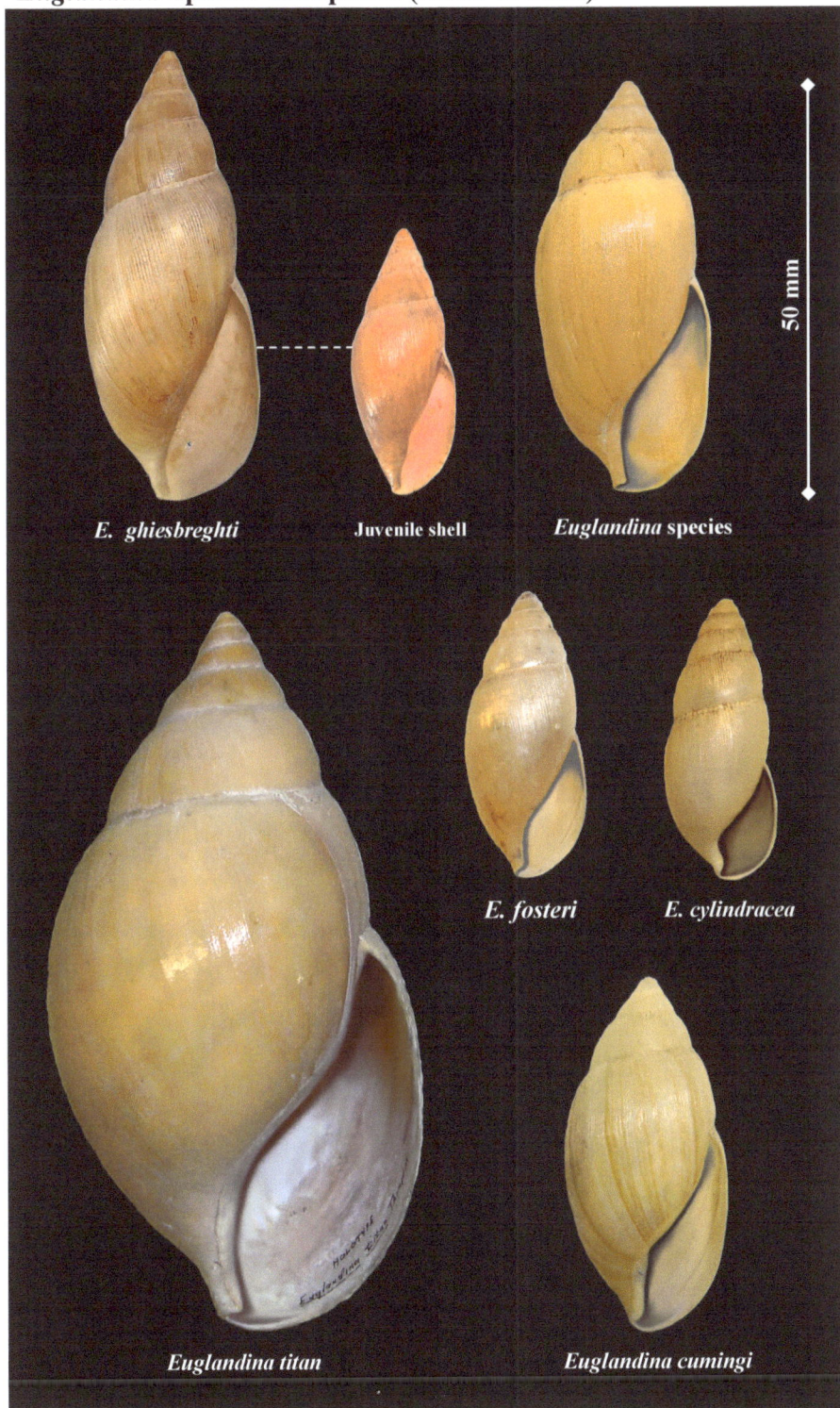

E. ghiesbreghti      Juvenile shell      *Euglandina* species

50 mm

E. fosteri      E. cylindracea

*Euglandina titan*      *Euglandina cumingi*

# Common Aggressor                    SPIRAXIDAE

*Varicoglandina monilifera* **(Pfeiffer, 1845)**

**Height**: 22 mm, Diameter: 10 mm

**Description**: <u>Oval-shape, wide</u>; lip simple; shell with 6 whorls; imperforate; shell surface glossy; <u>random dark vertical stripes with whitish borders</u>; without spiral striae; <u>columellar plait funnel-shaped, curling-up at the end (a)</u>; last whorl not broadly rounded; carnivorous.

**Similar Species**: Most similar to *Streptostyla* species but with dark, bold vertical stripes bordered by white and a columellar plait that is funnel-shaped (a); *Streptostyla* have an open columellar plait, page 149 (c); *Euglandina* species have longer, more narrow shells.

**Habitat**: Karst hills and talus slopes covered by tropical rainforests; under leaves and during wet weather climbing on trees.

**Status**: Common in the Maya Mountains.

**Specimen**: Belize, Toledo District, Bladen Nature Reserve (Dourson collection).

**Type Locality**: Coban, Guatemala.

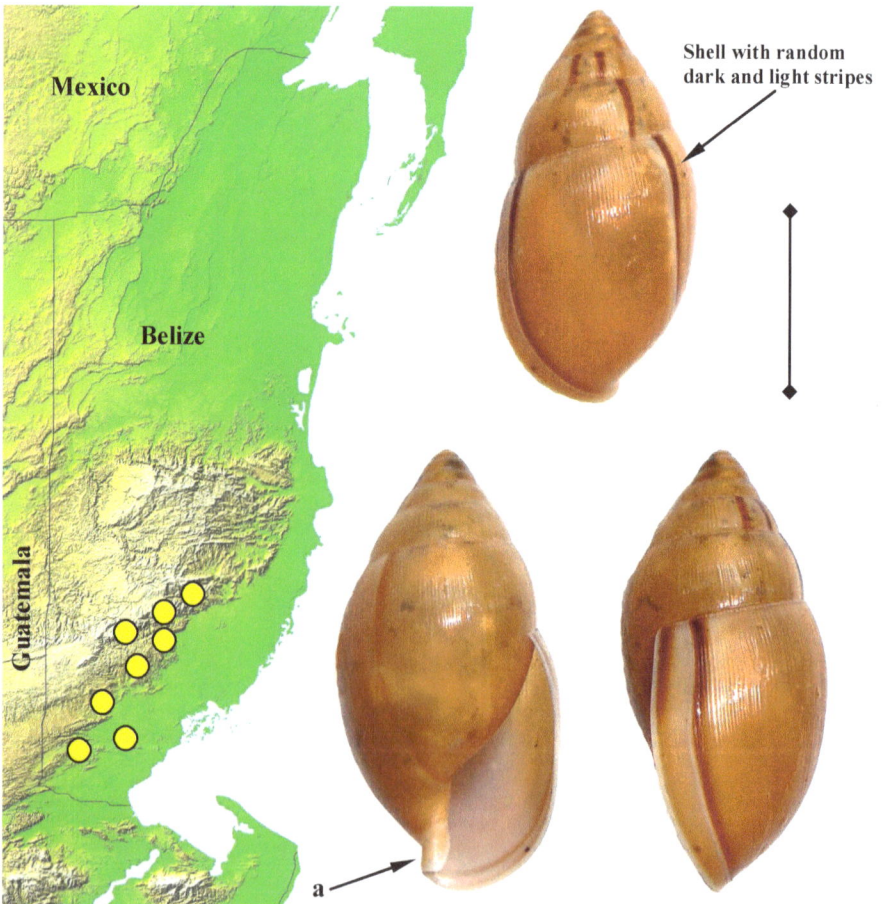

Mexico

Belize

Guatemala

Shell with random dark and light stripes

a

148

# Showy Oval          SPIRAXIDAE

*Streptostyla nigricans* **(Pfeiffer, 1845)**

**Height**: 25-40 mm, Diameter:10 mm

**Description**: Oval-shape, wide; lip simple; shell with 5-6 whorls; imperforate; shell surface glossy with light vertical stripes and light horizontal bands below the suture lines of the shell; without spiral striae; columellar plait open, not funnel-shaped (c) last whorl broadly rounded; carnivorous.

**Similar Species**: *Streptostyla lattrei* is generally smaller and more narrow in form; *S. delibuta* is a plain shell and without notable color streaks; *Orthalicus* species are larger and carry vivid color patterns that cover the entire shell.

**Habitat**: Found on hillsides of wet limestone under leaves and around rocks.

**Status**: Uncommon; found in southern Belize in Columbia River Forest Reserve and surrounding areas. Specimens collected from Belize are around 5-10 mm smaller than specimens from Guatemala.

**Specimen**: Figures (a) from Belize, Toledo District, karst hills surrounding the town of San Jose and figures (b) from Guatemala, Finca de la Providencia, UF 106004.

**Type Locality**: Verapaz, Guatemala.

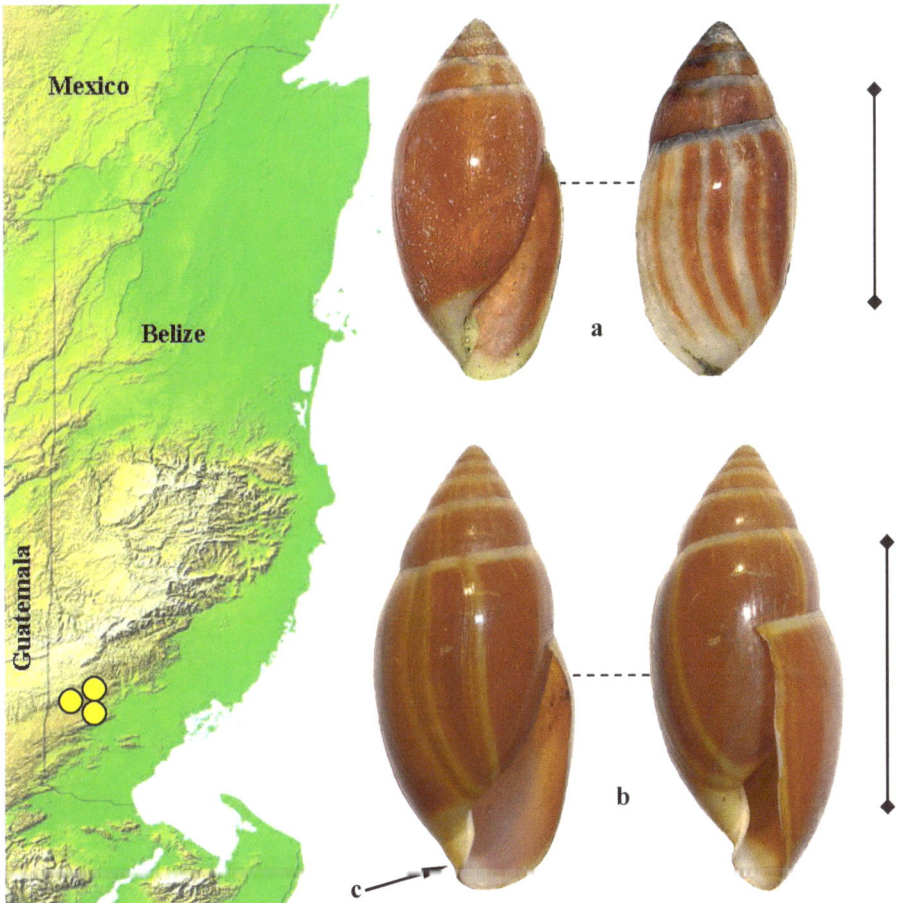

# Thomson's Oval SPIRAXIDAE

*Streptostyla cf. thomsoni* Ancey, 1888

**Height**: 21-30 mm, Diameter: 10-12 mm

**Description**: Oval-shape variable, wide; lip simple; shell with 7-8.5 whorls; imperforate; shell surface very glossy with an oily luster and light vertical streaks; live animal whitish with two lateral brownish bands and a brownish tail; without spiral striae; last whorl not broadly rounded as in *Drymaeus*.

**Similar Species**: Similar to *Streptostyla nigricans* only in size and form, lacking the more vivid stripes; *Drymaeus* species are more or less the same size but are cone-shape (see page 165 for shell comparisons.)

**Habitat**: Hilly limestone in the Maya Mountains; under leaves and the interstitial spaces of large flat rocks.

**Status**: Rare, reported from BNR and Rio Frio Cave; little information exists regarding *S. thomsoni* including adequate images; bottom photos and opposite page represent the only photographs of the species.

**Specimen**: Belize, Toledo District, Bladen Nature Reserve (Dourson collection).

**Type Locality**: Utila Island, Honduras.

150

Thomson's Oval, *Streptostyla cf. thomsoni*, Bladen Nature Reserve, Belize

# Plain Oval                                    SPIRAXIDAE

### *Streptostyla delibuta* (Morelet, 1851)

**Height**: 22-26 mm, Diameter: 9-10 mm

**Description**: <u>Oval-shape, narrow</u>; lip simple; shell with 6-7 whorls; imperforate; shell surface glossy with or without <u>(usually without) light vertical streaks</u>; shell corneous-green; <u>under a strong lens and good lighting, faint spiral striae may be visible, a rare feature in *Streptostyla*</u>; carnivorous.

**Similar Species**: Most like *Streptostyla cf. thomsoni* but its shell and aperture are more narrow in form; it is less elongate than *S. lattrei*; *S. nigricans* is larger and has a broader build.

**Habitat**: The species can be found in karst hills under leaf litter where it hunts and eats other land snails.

**Status**: Common; throughout the Maya Mountains except in the highest elevations.

**Specimen**: Belize, Toledo District, Bladen Nature Reserve (Dourson collection).

**Type Locality**: Verapaz, Guatemala.

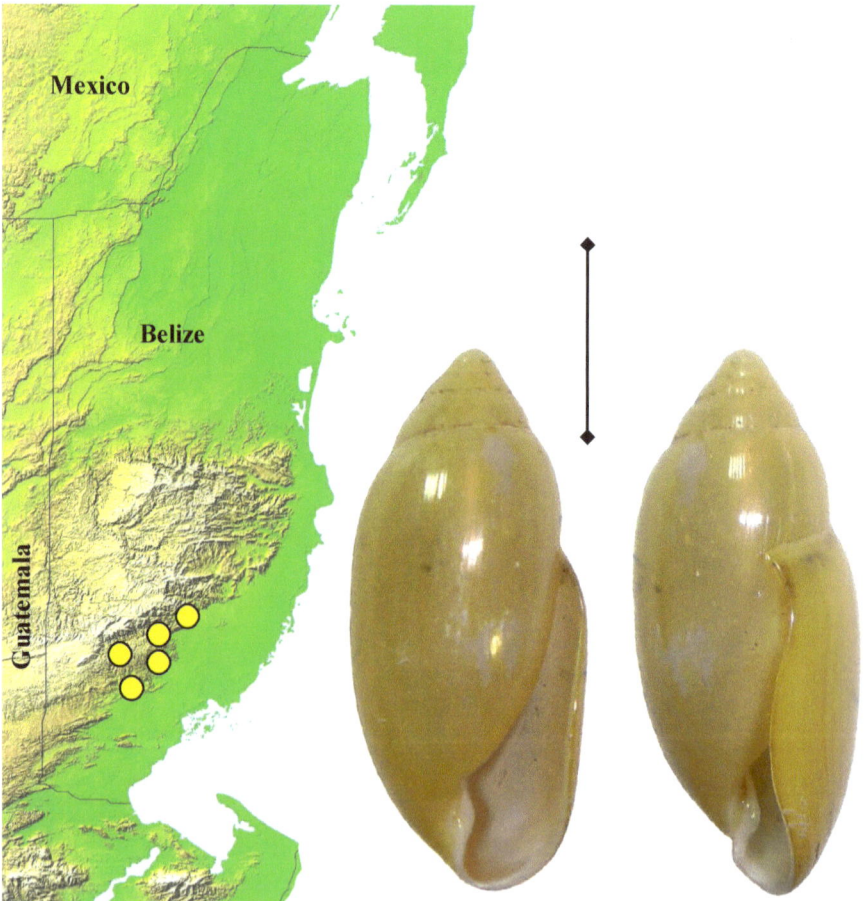

# Striped Oval        SPIRAXIDAE

*Streptostyla lattrei* **(Pfeiffer, 1845)**

**Height**: 22-25 mm, Diameter: 10 mm

**Description**: <u>Oval-shape, narrow</u>; lip simple; shell with 5-6 whorls; imperforate; shell surface especially glossy or glass-like with multiple, narrow light vertical stripes; without spiral striae; <u>the last whorl has only a very slight curve on either end with a nearly straight center (a)</u>; carnivorous.

**Similar Species**: *Streptostyla nigricans* is larger and wider in form; *S. delibuta* is a plain shell usually lacking notable color streaks.

**Habitat**: The species can be found up on trees or on the ground of karst hills where its hunts and eats other land snails.

**Status**: Common; currently the species appears restricted to the east side of the Maya Mountains in southern Belize.

**Specimen**: Belize, Toledo District, Bladen Nature Reserve (Dourson collection).

**Type Locality**: Not reported.

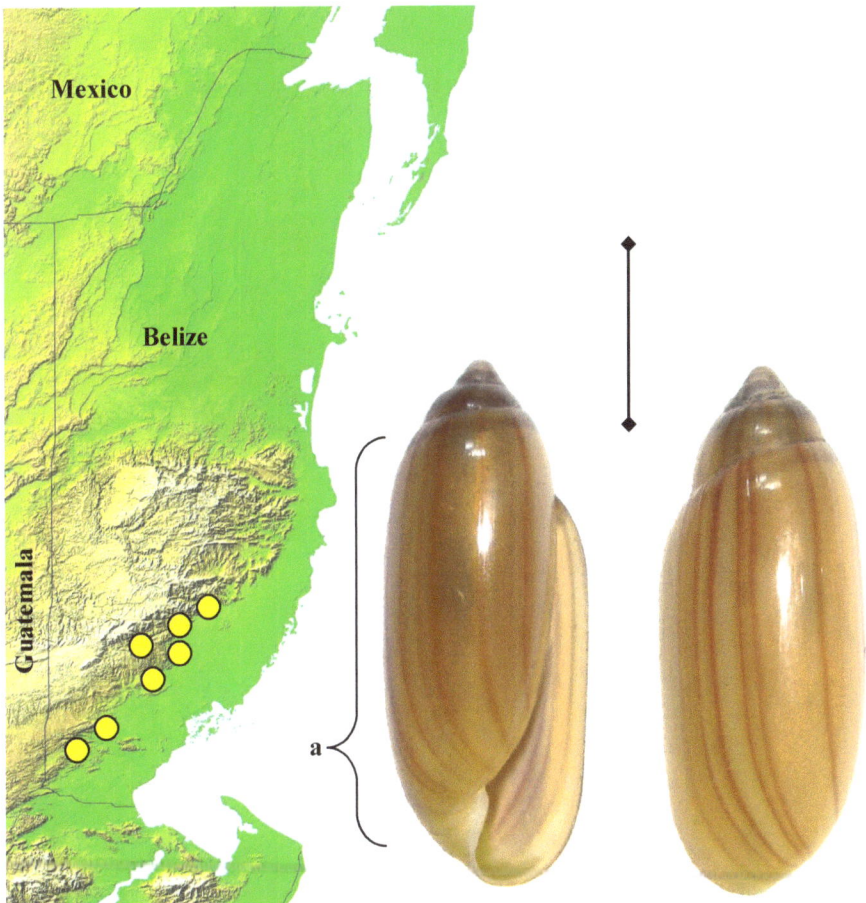

# Glossy Oval                                                          SPIRAXIDAE

## *Streptostyla cf. labida* (Morelet, 1851)

**Height**: 22-27 mm, Diameter: 12 mm

**Description**: <u>Oval-shape, narrow</u>; lip simple; shell with 7-8 whorls; imperforate; shell surface very glossy; without spiral striae; shell corneous-yellow, <u>last whorl with faint pale greenish streaks (a)</u>; live animal whitish with a brownish anterior; tail white; last whorl broadly roundish.

**Similar Species**: Other *Streptostyla* species are less pointed and more perfectly oval in form; most similar to *S. nigricans* in build but more narrow in form with only very faint streaks; it resembles *S. turgidula guatemalensis* in miniature (a species of Guatemala).

**Habitat**: Karst hills and talus slopes covered by tropical rainforests.

**Status**: Rare; this species appears to be restricted to southern Belize, currently known from only three areas, two near San Jose and one location near San Felipe.

**Specimen**: All specimens from Belize, Toledo District, 2 miles north of San Felipe (Dourson collection).

**Type Locality**: Alta Verpaz, Guatemala.

Glossy Oval, *Streptostyla cf. labida* (undetermined), Bladen Nature Reserve.

# Karst Oval                                    SPIRAXIDAE

## *Streptostyla ventricosula* (Morelet, 1849)

**Height**: 12-14 mm, Diameter: 5-6 mm

**Description**: <u>Oval-shape, wide</u>; lip simple; shell with 5-6 whorls; imperforate; shell surface glossy, pale yellow without any notable color bands or streaks and without spiral striae; last whorl broadly rounded; carnivorous.

**Similar Species**: Most similar to *Streptostyla meridana* in terms of shell size and form but is pale-amber not pale yellow and has a notably wider build; *S. dysoni* is smaller, narrower and has a more pointed apex; *S. ventricosa* (not illustrated) has a slimmer build and different color; other *Streptostyla* species are much larger in overall size; (see page 165 for shell comparisons.)

**Habitat**: The species can be found in karst hills under leaf litter, talus slopes and near cave entrances.

**Status**: Uncommon; reported from the Yucatan Region of Mexico; in Belize currently known from the areas around Rio Frio Cave and Shipstern Nature Preserve.

**Specimen**: Belize, Cayo District, mouth of Rio Frio Cave (Dourson collection).

**Type Locality**: Merida, Mexico.

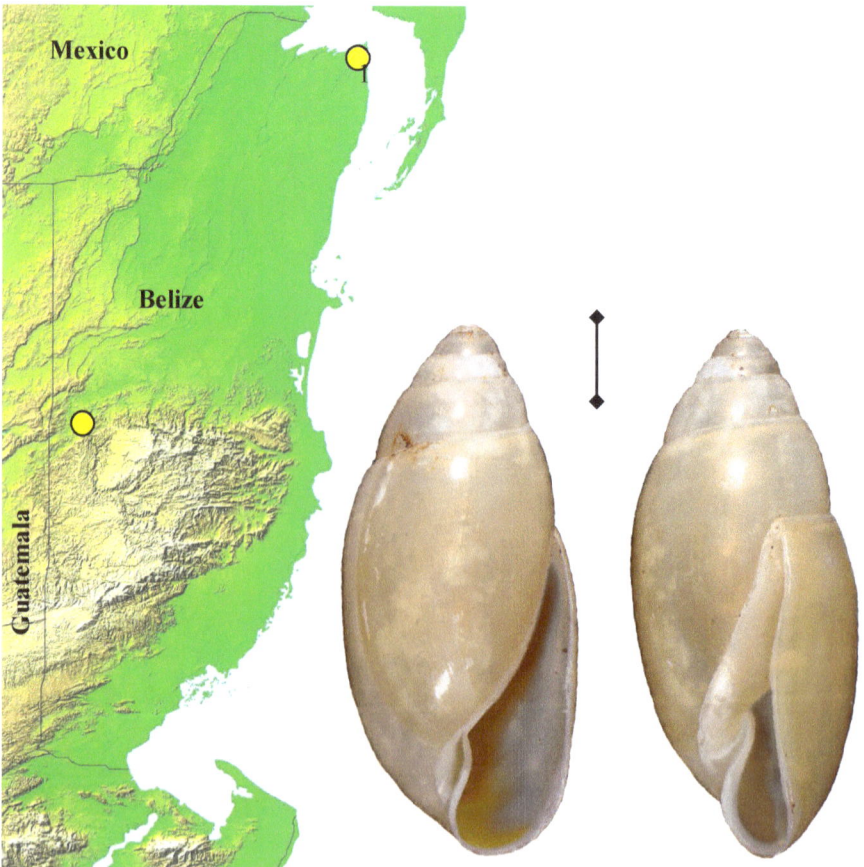

# Lesser Oval

*Streptostyla meridana* (Morelet, 1849)

**Height**: 10-12 mm, Diameter: 5-6 mm

**Description**: <u>Oval-shape, narrow</u>; lip simple; shell with 6 whorls; imperforate; shell surface glossy, <u>pale yellow</u>; without any notable color features and without spiral striae; last whorl broadly rounded; carnivorous.

**Similar Species**: Most like *Streptostyla dysoni* but is larger, has a yellow tint and is slightly wider in build with a more pointed apex; other *Streptostyla* are much larger in overall size; (see page 165 for shell comparisons.)

**Habitat**: The species can be found in karst hills under leaf litter and may be especially frequent in rocky regions.

**Status**: Common; found in the north at Shipstern Nature Preserve, Peccary Hills, and the southern Maya Mountains.

**Specimen**: Figure (a) Syntype from NHMUK 1893.2.4.6-8, figures (b) & (c) from Belize, Toledo District, Bladen Nature Reserve (Dourson collection).

**Type Locality**: Merida, Mexico.

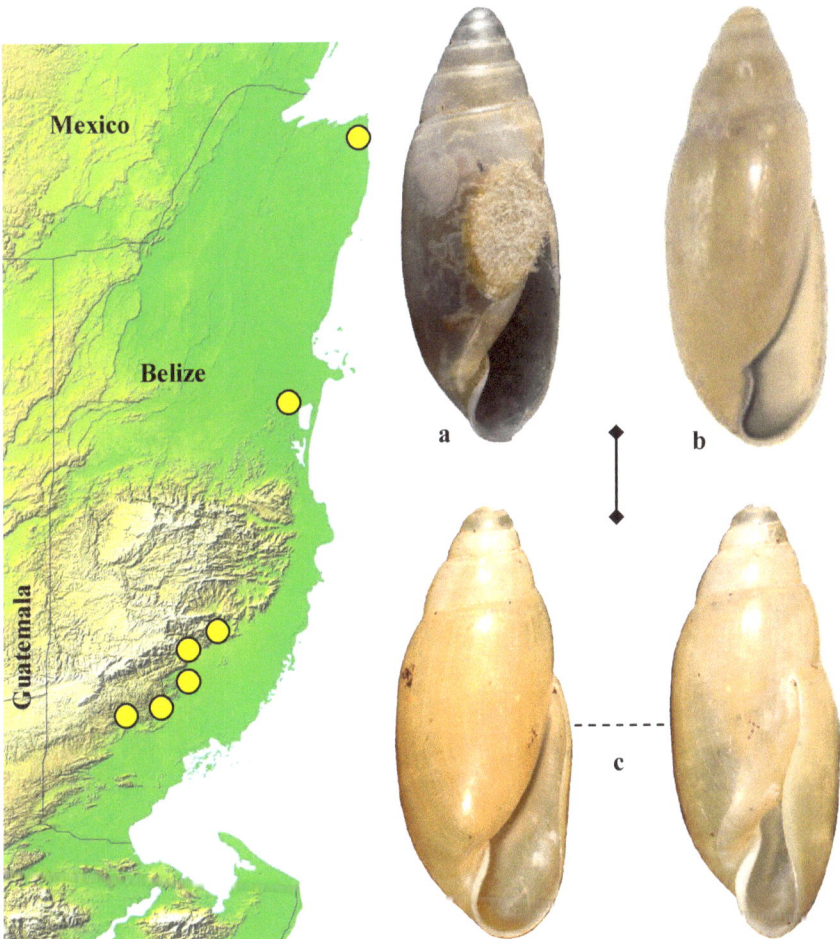

a

b

c

# Limestone Oval

## SPIRAXIDAE

### *Streptostyla* species (undetermined)

**Height**: 13-15 mm, Diameter: 5 mm

**Description**: <u>Oval-shape, narrow</u>; lip simple; shell with 6 whorls; imperforate; shell surface glossy, bronze; without any notable color features and without spiral striae; last whorl broadly rounded; the build of this shell somewhat suggests a small species of *Euglandina*.

**Similar Species**: Most similar to *Streptostyla meridana* in terms of shell form but is slightly larger, bronze (not pale yellow) and more narrowly built; *S. ventricosula* is around the same size but has a notably wider build; *S. dysoni* is smaller and a different color; other *Streptostyla* are much larger in overall size; (see page 165 for shell comparisons.)

**Habitat**: The species can be found in karst hills under leaf litter.

**Status**: Uncommon; found in the southern portion of the Maya Mountains and karst outcrops of southern Belize.

**Specimen**: Belize, Toledo District, Bladen Nature Reserve (Dourson collection).

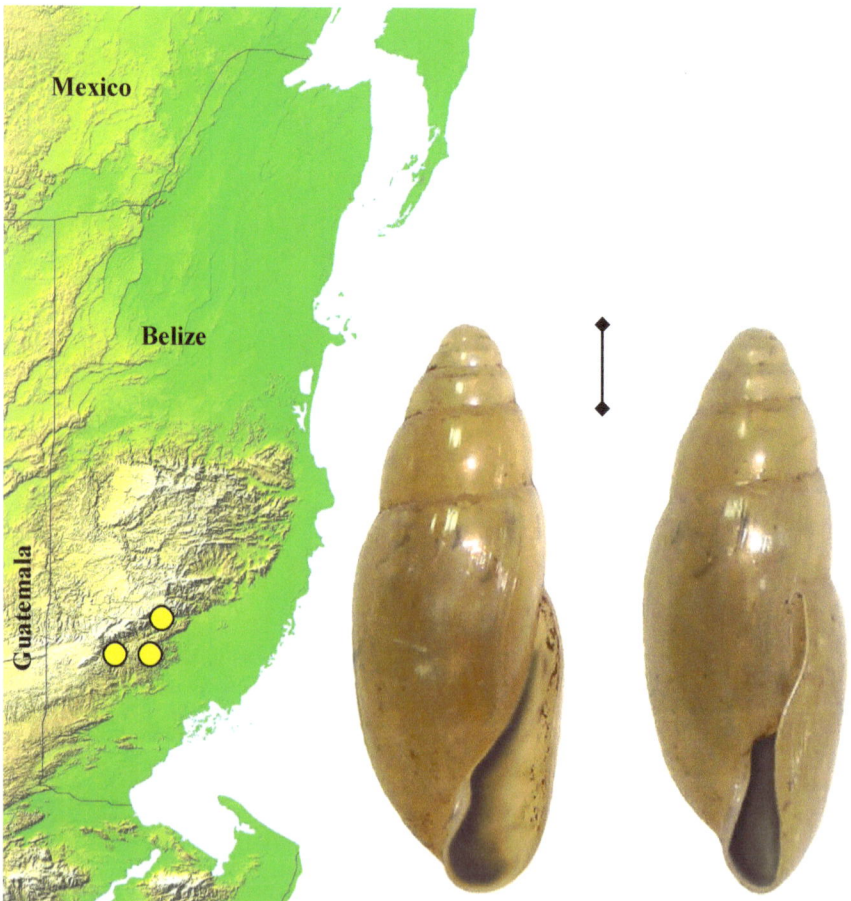

# Pygmy Oval                                    **SPIRAXIDAE**

## *Streptostyla dysoni* (Pfeiffer, 1846)

**Height**: 9 mm, Diameter: 6 mm

**Description**: <u>Oval-shape, narrow</u>; lip simple; shell with 6 whorls; imperforate; shell surface glossy and translucent, <u>pearly-white</u>; without any notable color features and without spiral striae; last whorl broadly rounded; carnivorous.

**Similar Species**: Most similar to *Streptostyla meridana* in terms of shell form but is smaller, pearly-white (not pale yellow) and is more narrow in build with a more pointed apex; *S. ventricosula* is larger and has a notably wider build; *Streptostyla meridana cobanensis* is larger and a bronze color not pearly-white as in *S. dysoni*; other *Streptostyla* are much larger in overall size; (see page 165 for shell comparisons.)

**Habitat**: Found in karst hills under leaf litter.

**Status**: Uncommon; scattered distribution throughout Belize.

**Specimen**: Belize, Toledo District, Ramos Creek, Bladen Nature Reserve (Dourson collection).

**Type Locality**: Honduras.

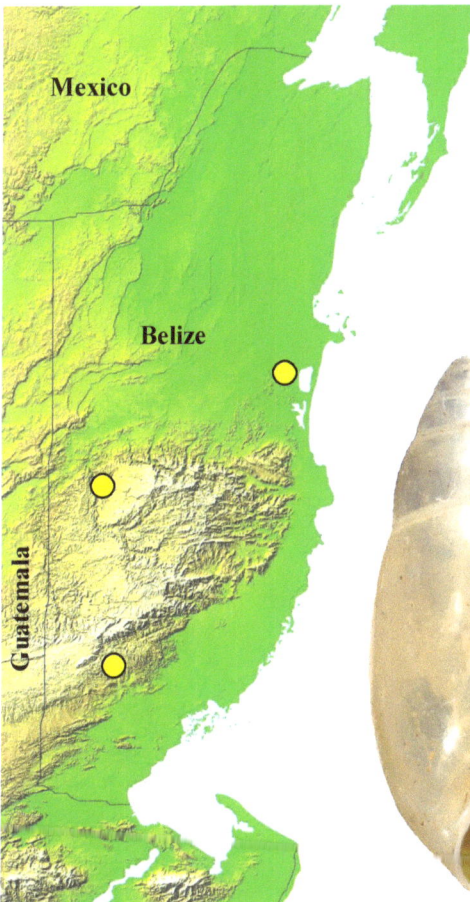

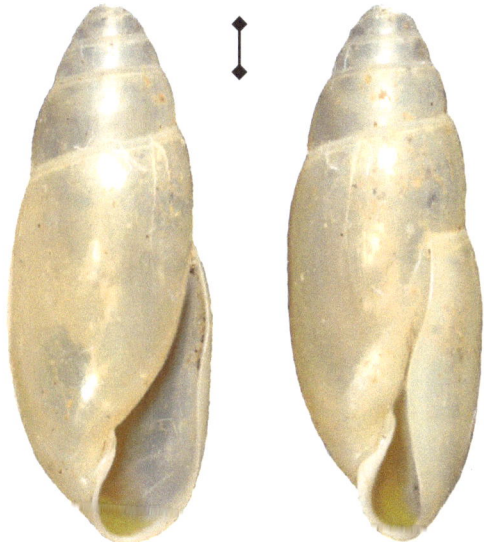

# Black-banded Oval

## SPIRAXIDAE

*Streptostyla ligulata* **(Morelet, 1849)**

**Height**: 15 mm, Diameter: 7 mm

**Description**: <u>Cone-shape, narrow</u>; lip simple; shell with 6-7 whorls; imperforate; shell surface glossy, translucent and thin; shell straw-yellow; <u>one encircling dark-black band that is most prominent on the body whorl (a)</u>; this band distinctive on live and fresh dead shells but may be a faded characteristic on old and weathered ones; without spiral striae; last whorl broadly rounded; carnivorous.

**Similar Species**: No other small *Streptostyla* shells display a bold black band; (see page 165 for shell comparisons.)

**Habitat**: The species can be found in upper elevation moist, karst hills under leaf litter and around talus.

**Status**: Rare; in Belize, known only from one site in southern Belize around San Jose.

**Specimen**: Belize, Toledo District, 5 miles west of San Jose (Dourson collection).

**Type Locality**: Petén, Guatemala.

# Separating *Streptostyla* from *Salasiella*

**Streptostyla**                    **Salasiella**

Similar looking species like *Streptostyla* and *Salasiella* are most easily distinguished by the columellar plait shape: ***Streptostyla*** columellar plait is twisted outward (a) while the plait of ***Salasiella*** is gently curved inward (b).

 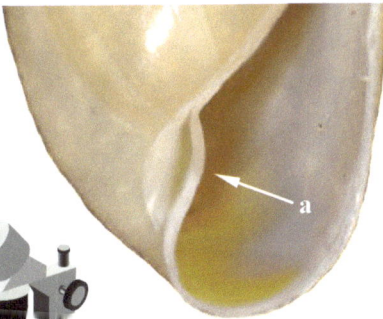 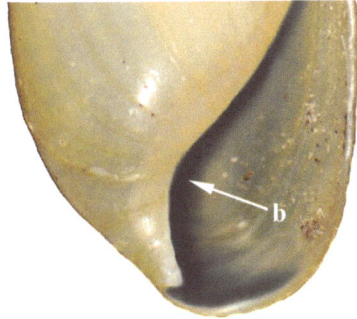

**Streptostyla**                    **Salasiella**

# Common Salasiella                     SPIRAXIDAE

*Salasiella cf. pulchella* (Pfeiffer, 1856)

**Height**: 11 mm, **Diameter**: 4.5 mm

**Description**: <u>Oval-shape, narrow</u>; lip simple; shell with 6 whorls; imperforate; shell surface glossy, pale yellow, thin and translucent; without any notable color features or with faint vertical streaks; transverse striae poorly developed and widely spaced; without spiral striae; <u>columellar plait simple, not twisted (a), see page 161</u>; last whorl broadly rounded; carnivorous.

**Similar Species**: Stands closest to the *Streptostyla dysoni* in terms of shell form but is larger and most importantly, <u>has a columellar plait that is not twisted as in *Streptostyla* species</u>, (see page 161.)

**Habitat**: The species can be found in karst hills under leaf litter and in talus slopes of the Bladen River watershed.

**Status**: A common leaf litter species in southern Belize.

**Specimen**: All specimens from Belize, Toledo District, Bladen Nature Reserve (Dourson collection).

**Type Locality**: Chiapas, Mexico.

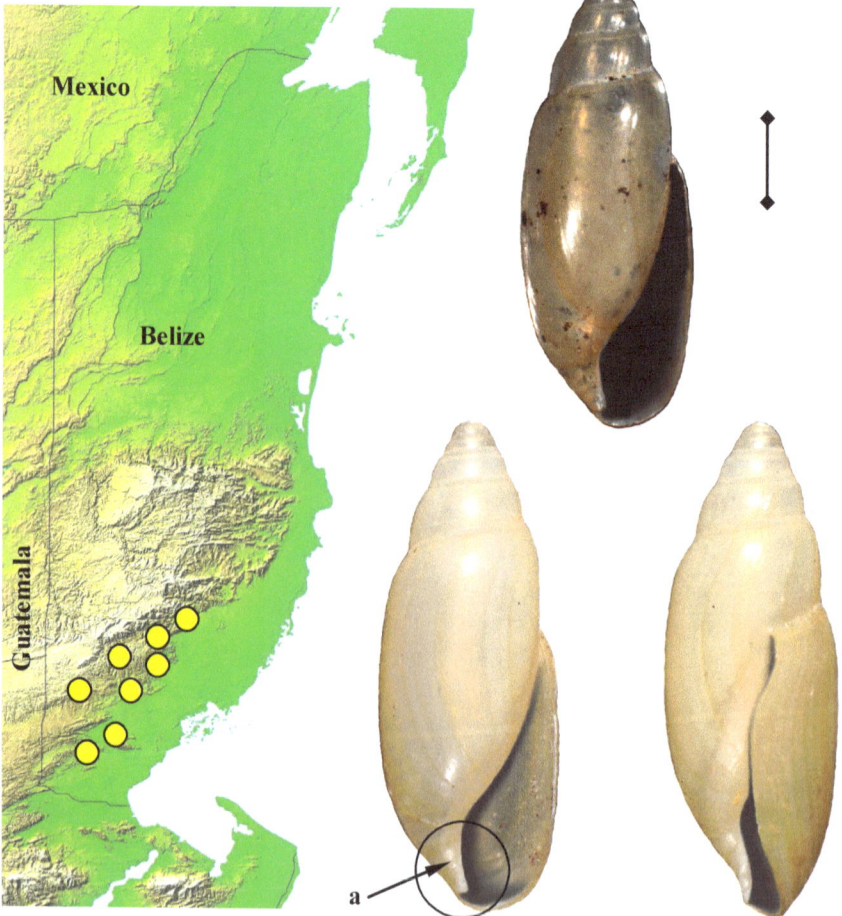

# Tiny Salasiella                          **SPIRAXIDAE**

## *Salasiella modesta* (Pfeiffer, 1862)

**Height**: 4 mm, Diameter: 1.5 mm

**Description**: <u>Oval-shape, narrow</u>; lip simple; shell with 5 whorls; imperforate; shell surface glossy, thin and translucent without any notable color features or with faint vertical streaks; transverse striae poorly developed and widely spaced; without spiral striae; <u>columellar plait simple, not twisted (a), see page 161</u>; last whorl broadly rounded; a carnivorous species.

**Similar Species**: Most similar to *Salasiella cf. pulchella* but only half the size, having a proportionally wider aperture and a thinner, more translucent shell.

**Habitat**: The species can be found in karst hills under leaf litter and in talus slopes of the Bladen River watershed.

**Status**: Common where it occurs in the Bladen Nature Reserve of the southern Maya Mountains.

**Specimen**: All specimens from Belize, Toledo District, Forest Hill, Bladen Nature Reserve (Dourson collection).

**Type Locality**: Verapaz, Guatemala.

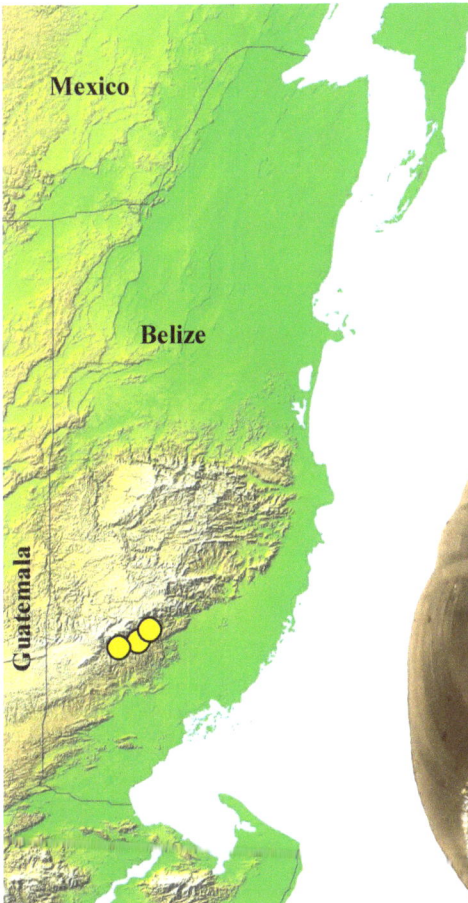

The internal shell

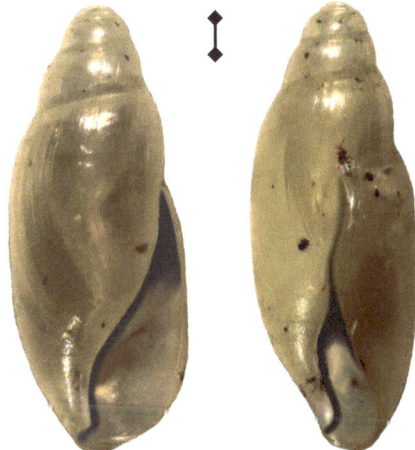

163

# Guatemalan Salasiella                    SPIRAXIDAE
*Salasiella guatemalensis* Pilsbry, 1920

**Height**: 9-10 mm, Diameter: 4 mm

**Description**: <u>Oval-shape, narrow</u>; lip simple; shell with 5.5 whorls; imperforate; shell surface glossy, thin and translucent, <u>pale yellow</u>; without any notable color features or with faint vertical streaks; transverse striae poorly developed and widely spaced; without spiral striae; <u>columellar plait simple,</u> last whorl rounded with a slight shoulder (a); carnivorous.

**Similar Species**: Stands closest to *Salasiella cf. pulchella* in size and form but has a slightly wider build and less pointed apex.

**Habitat**: The species can be found in karst hills under leaf litter and among talus slopes.

**Status**: Rare. In Belize, known only from around San Jose, Toledo District.

**Specimen**: Figure (a) from Gulan, Guatemala, Syntype ANSP 114838, (image courtesy of Academy of Natural Sciences Philadelphia) and figure (b) Belize, Toledo District; road to San Jose, just past Crique Jute (Dourson collection).

**Type Locality**: Gulan, Guatemala.

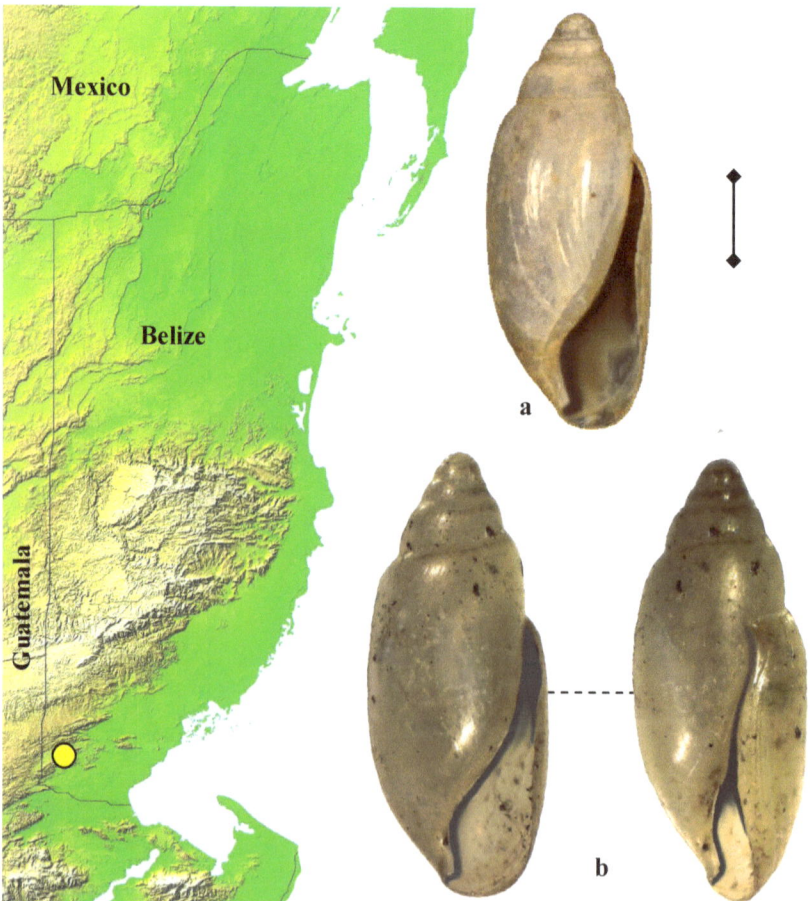

164

# Similar SPIRAXIDAE Shells Compared (shells proportionate)

*Varicoglandina monilifera*      *Streptostyla cf. thomsoni*      *Streptostyla delibuta*

20 mm

*Streptostyla nigricans*    *Streptostyla cf. labida*    *Streptostyla lattrei*    *Streptostyla ligulata*

*Streptostyla ventricosula*    *Streptostyla* species (undetermined)    *Streptostyla meridana*    *Salasiella cf. pulchella*    *Streptostyla dysoni*    *Streptostyla modesta*

10 mm

165

# Foothills Mayaxis                                    SPIRAXIDAE

## *Mayaxis martensiana* (Pilsbry, 1920)

**Height**: 7-11 mm, Diameter: 5-7 mm

**Description**: Cylinder-shape; lip simple; shell with 9 whorls; imperforate; shell surface glossy, thin and translucent without any notable color features; spiral striae absent; whorls have wide, nearly flat riblets that create fine notches at the suture; the riblets are as wide as or wider than their interspaces (Thompson 1995); the protoconch contains two narrow whorls, the first of which is smooth and the second bears heavy axial ribs; last whorl blockish; carnivorous.

**Similar Species**: *Pseudosubulina* species are similar in shell form but the riblets are farther apart, have a more tapered apex and a more compact or square-like aperture.

**Habitat**: The species can be found in karst hills under leaf litter.

**Status**: Rare to uncommon in southern Belize.

**Specimen**: Belize, Toledo District, Bladen Nature Reserve (Dourson collection).

**Type Locality**: Livingston, Guatemala.

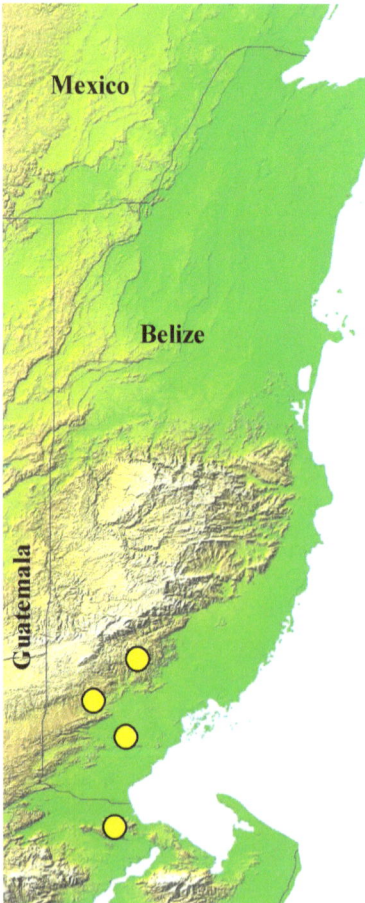

# Blue Creek Shaft                                SPIRAXIDAE

*Pseudosubulina juancho* (new species)

**Height**: 10 mm, Diameter: 5-6 mm

**Description**: Cylinder shape; lip simple; shell with 13 whorls; imperforate; shell surface glossy, thin and translucent, without any notable color features; spiral striae absent, the sharp-pointed riblets are not as wide as their interspaces, and are broadly spaced becoming farther apart on the final whorl; the protoconch contains two narrow whorls (both smooth); last whorl rounded; a carnivorous species that hunts and eats other land snails.

**Similar Species**: *Pseudosubulina juancho* differs from *P. berendti* by its more elongated shell, its more widely spaced riblets (especially on the body or final whorl) and by a smaller aperture.

**Habitat**: Limestone talus covered in rainforests at around 300 meters.

**Status**: Rare, ENDEMIC to BELIZE currently known from only a few locations in southern Belize.

**Specimen**: Belize, Toledo District, Blue Creek Cave on limestone hillside above cave entrance Holotype UF 505433, Paratypes UF 505434 near San Jose, Belize, (not pictured).

**Type Locality**: Hillside around Blue Creek Cave, Toledo District, Belize (16° 12'26N, 89°2'57").

**Etymology**: In honor of Juan Cho, for his contribution to the environment by promoting and utilizing sustainable organic agricultural practices in the Toledo District of Belize to produce organic chocolate.

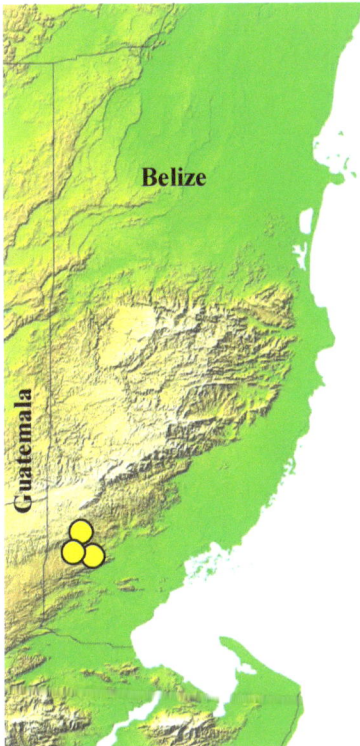

# Comparing the Genus *Volutaxis* to *Rectaxis*

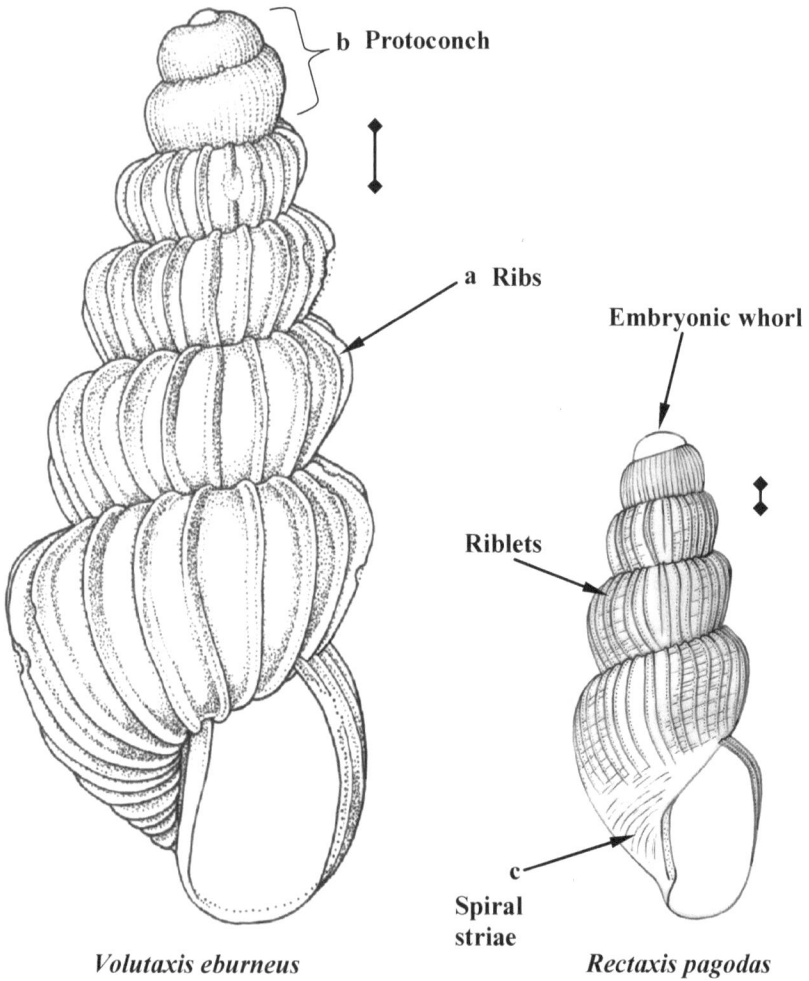

b Protoconch

a Ribs

Embryonic whorl

Riblets

c
Spiral
striae

*Volutaxis eburneus*

*Rectaxis pagodas*

Illustrated above are *Volutaxis eburneus* and *Rectaxis pagodas* (Thompson 2010) from Costa Rica and Panama, Central America. When comparing **Volutaxis** to **Rectaxis,** its important to look at size and micro-ornamentation of the shells. *Volutaxis* species are 6-9 mm tall, have notably raised ribs (figure a) and no spiral striae where as *Rectaxis* species are usually less than 3 mm tall, have lower, finer ribs or riblets and are with delicate spiral striae (figure c). A strong scope will be required to see these fine features. Note the smoother first and second protoconch whorls on *Volutaxis* (figure b), illustrated above. Both are imperforate. Illustrations courtesy of Florida Museum of Natural History, USA.

# Maya Mountain Barb                    SPIRAXIDAE

*Volutaxis similaris* **(Strebel, 1882)**

**Height**: 6-7 mm, Diameter: 2-3 mm

**Description**: Cylinder-shape; lip simple; shell with 7-8 whorls; imperforate; shell surface very glossy and translucent in live and fresh shells, bleaching with age; without color features; spiral striae absent; the body whorl contains moderately spaced, angled ribs that become tighter on later whorls; the protoconch contains two narrow whorls, the first smooth and the second with low riblets; last whorl rounded; carnivorous.

**Similar Species**: *Volutaxis* species differ from *Mayaxis* by having widely spaced ribs and a more tapered shell; differs from *Pseudosubulina* by being shorter and containing fewer whorls; differs from *Rectaxis* by being much larger without any traces of spiral striae.

**Habitat**: Limestone talus and under leaf litter at most elevations.

**Status**: Common; found in scattered locations throughout Belize.

**Specimen**: Belize, Toledo District, Bladen Nature Reserve (Dourson collection).

**Type Locality**: Veracruz, Pacho, Mexico.

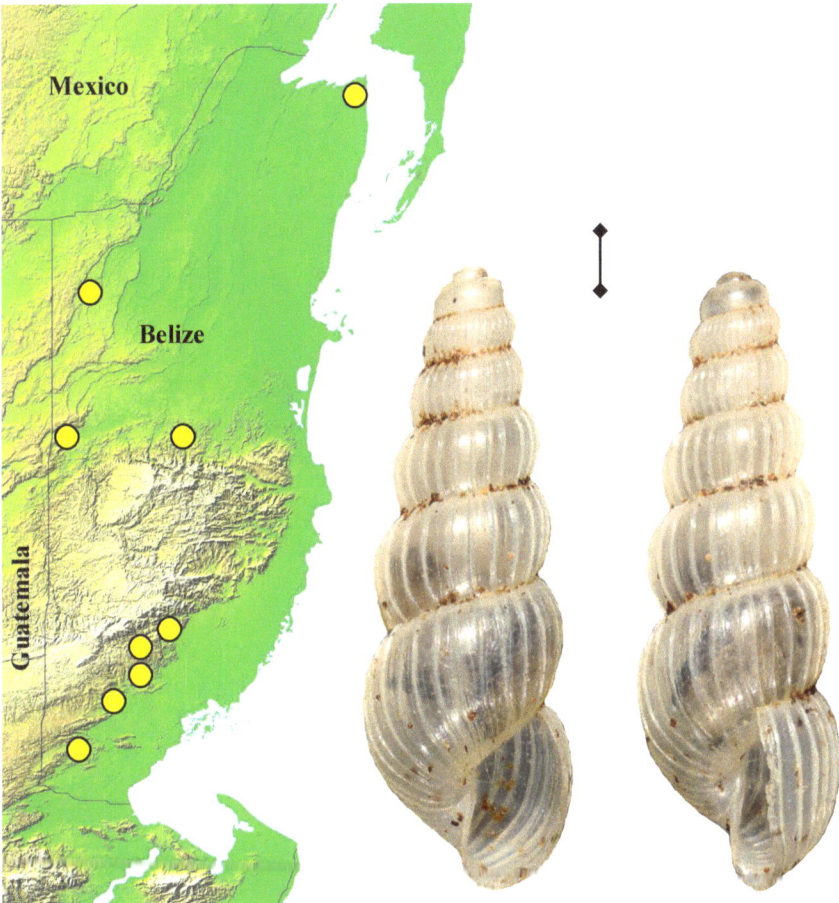

# Common Barb                     SPIRAXIDAE

## *Volutaxis sulciferus* (Morelet, 1851)

**Height**: 8 mm, Diameter: 3-4 mm

**Description**: Cylinder-shape; lip simple; shell with 8 whorls; imperforate; shell surface very glossy and translucent corneous in live and fresh shells, bleaching with age; without color features; spiral striae absent, body whorl contains moderately spaced, angled ribs that become tighter on later whorls; the protoconch contains two narrow whorls, the first smooth and the second bearing low riblets; last whorl slightly squared (a); carnivorous.

**Similar Species**: Similar to *Volutaxis livingstonensis* but only slightly smaller with more crowded riblets and a more curving, reflected aperture (b) but this feature is variable.

**Habitat**: Found under leaf litter at lower elevations.

**Status**: This species is reported from Rio Frio Cave by Haas and Solem (1960) and Macal River Gorge; specimens found by the authors from Shipstern Nature Preserve remain in question.

**Specimen**: Belize, Cayo District, .5 miles downriver from Black Rock Lodge in limestone sinkhole (Dourson Collection).

**Type Locality**: Palenque, Mexico.

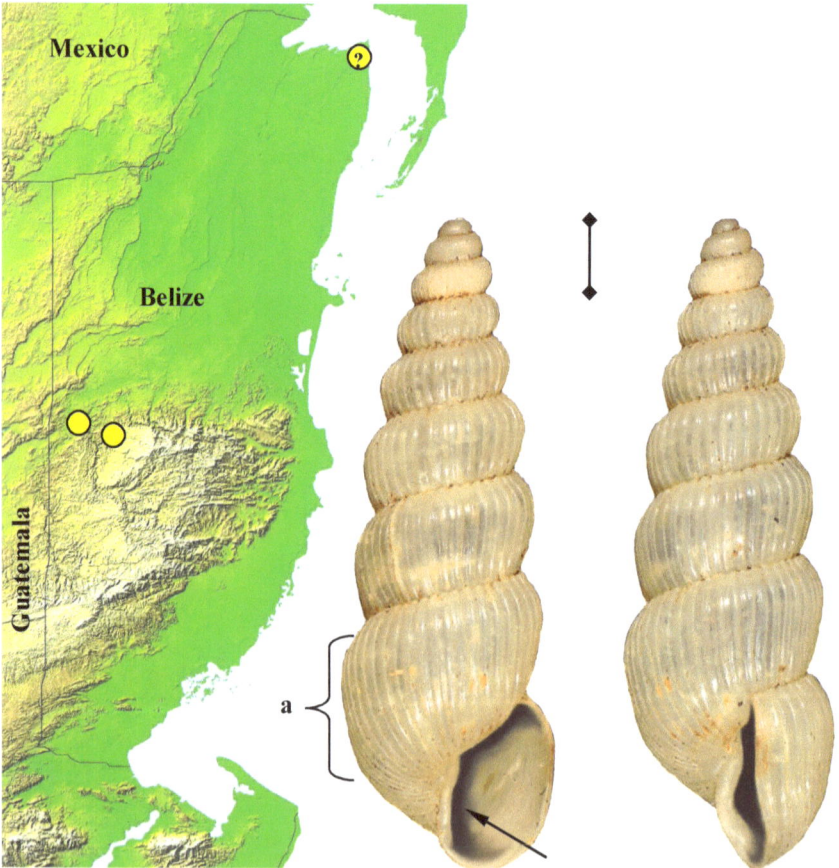

SPIRAXIDAE

*Volutaxis sulciferus sulciferus* (Morelet, 1851) Holotype NHMUK 1893.2.4.1153.

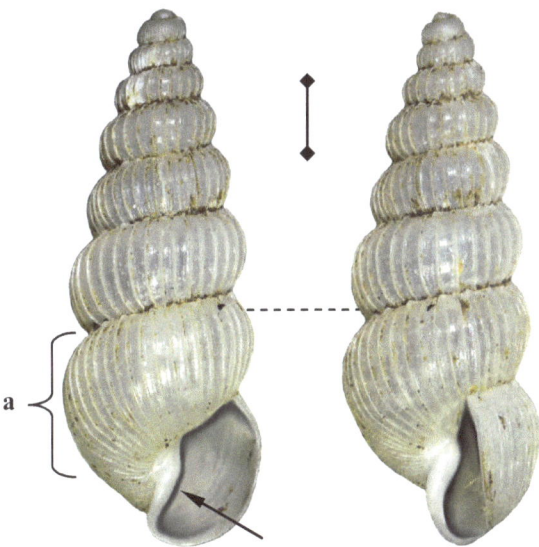

Above specimen was labeled as *Volutaxis sulciferus sulciferus,* UF 77801 from Mexico, Oaxaca, 9.2 km northeast of Valle Nacional, 800 feet in elevation. Note the more roundish last whorl (a). It may actually represent a different species than *Volutaxis sulciferus* but this warrants further investigation

# Livingston Barb                                      SPIRAXIDAE

## *Volutaxis livingstonensis* (Pilsbry, 1920)

**Height**: 9-10 mm, Diameter: 2-3 mm

**Description**: <u>Cylinder-shape, wide</u>; lip simple; shell with 9-10 whorls; imperforate; shell surface glossy, very pale-yellow; apex and protoconch whorls are mostly smooth while remaining whorls have well developed, evenly-spaced riblets (Pilsbry 1920); no notable color features; <u>without spiral striae</u>; last whorl rounded; carnivorous.

**Similar Species**: *Volutaxis sulciferus* is similar in size and build but is more obese, having closer riblets and a more curving aperture; it resembles *V. longior* but has a broader shell and is slightly taller.

**Habitat**: Karst hills under leaf litter and in talus formations.

**Status**: Rare, in Belize reported only from the Bladen Nature Reserve.

**Specimen**: Figure (a) from Guatemala, mountains west of Livingston, Lectotype ANSP 107537, (images courtesy of Academy of Natural Sciences Philadelphia); figure (b) from Belize, Toledo District, Bladen Nature Reserve (Dourson collection).

**Type Locality**: Livingston, Guatemala.

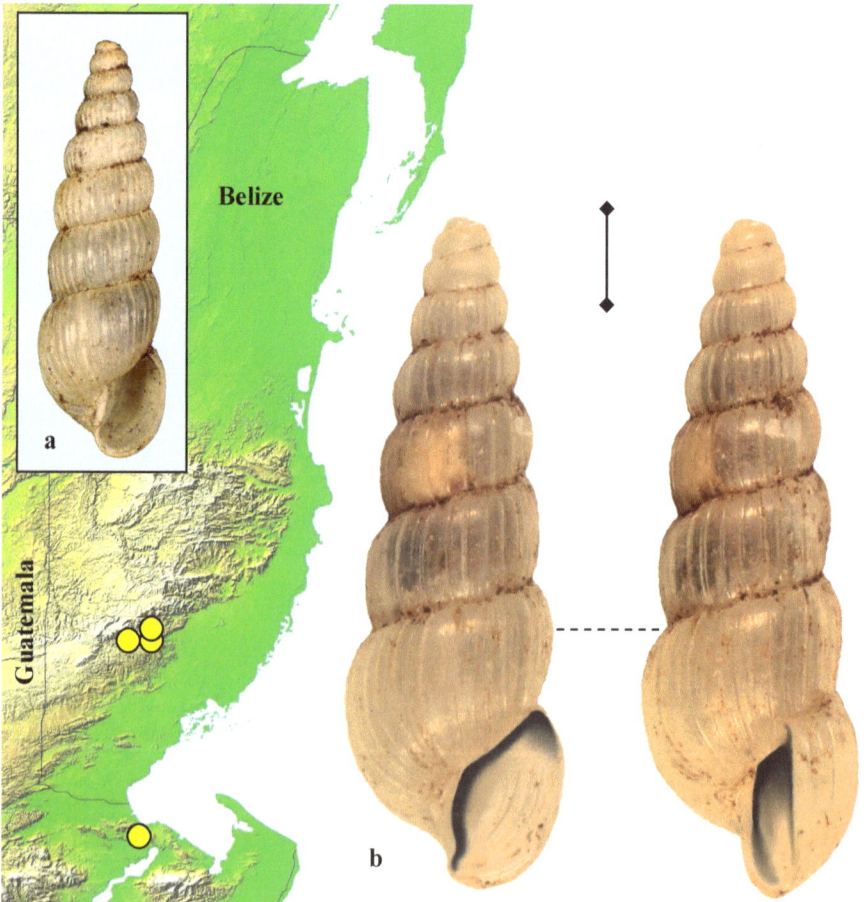

# Long Barb                                       SPIRAXIDAE

*Volutaxis longior* (Pilsbry, 1920)

**Height**: 8 mm, Diameter: 2 mm

**Description**: <u>Cylinder-shape, narrow</u>; lip simple; shell with 9-10 whorls; imperforate; shell color is corn silk; apex and protoconch whorls are mostly smooth while remaining whorls have well developed, evenly-spaced riblets (Pilsbry 1920); <u>without spiral striae</u>; last whorl rounded; carnivorous.

**Similar Species**: *Volutaxis longior* is most similar to *Volutaxis livingstonensis* but is slightly smaller and has a more narrow build; *Rectaxis* are smaller.

**Habitat**: The species can be found in karst hills under leaf litter and on talus slopes covered in broad leaf rain forest.

**Status**: Uncommon; known only from southern Belize.

**Specimen**: Figure (a) from Guatemala, mountains west of Livingston, ANSP 107536, (images courtesy of Academy of Natural Sciences Philadelphia) and figure (b) from Belize, Toledo District, Bladen Nature Reserve (Dourson collection).

**Type Locality**: Livingston, Guatemala.

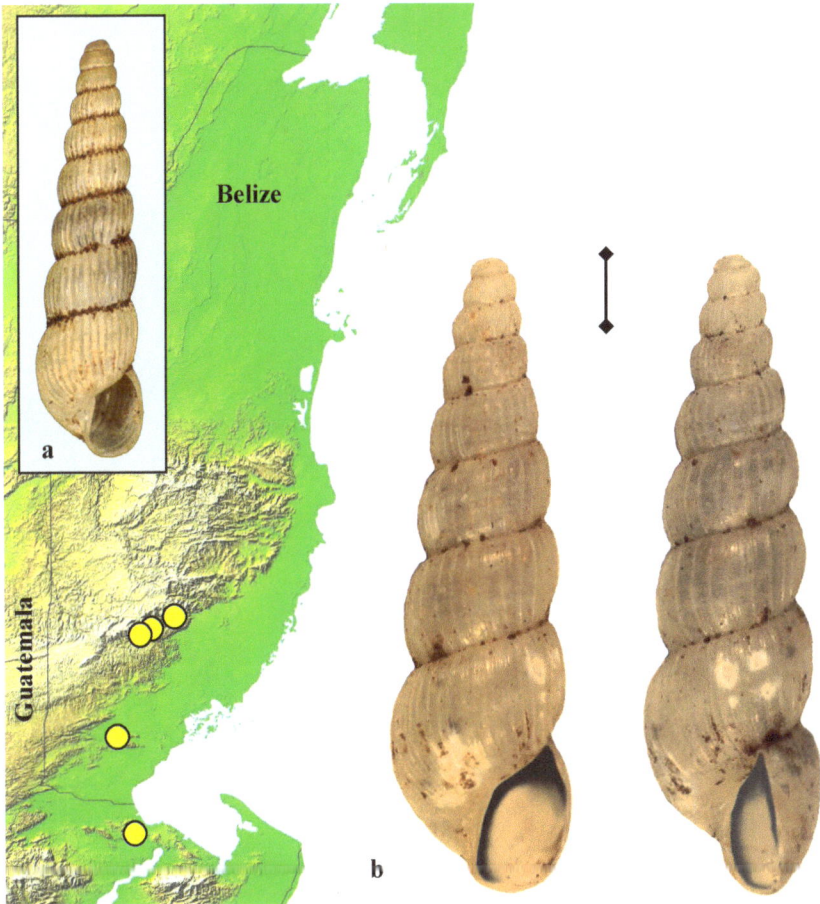

1) ***Rectaxis funibus***, Belize, Toledo District, Bladen Nature Reserve, 200 meters. Specimens are 2.5-3 mm, have 5-6 whorls and are pellucid in color.

2) ***Rectaxis aff. funibus***, Belize, Toledo District, Bladen Nature Reserve, Oak Ridge in montane forest, 1100 meters, GPS (16°31'1"N, 88° 55'43W). Specimens have an atypical color (pearly white), 5-6 whorls and are 3-3.5 mm (larger by around 1 mm). May represent an ecological form or new species.

3) ***Rectaxis aff. funibus***, Belize, Orange Walk District, rocky outcrop around Gallon Jug area, GPS (17°33'33.5196N, 89° 2' 21.3648W). Specimens have a more narrow build and have 6-7 whorls (one more whorl). May represent an ecological form or new species.

4) ***Rectaxis alvaradoi***, Belize, Cayo District,4.9 mi SE San Ignacio on Cristo Rey Rd, Duffy Bank area, 110 meters, UF 146256.

Further collections and studies are needed on figures 2 and 3 to make a final determination; they are not listed in the species accounts (only here).

# Common Splinter                     SPIRAXIDAE

*Rectaxis funibus* (**Goodrich & van der Schaile, 1937**)

**Height**:2.5-3 mm, Diameter: .75 mm

**Description**: Cylinder-shape; lip simple with a slight reflection; shell with 5-6 whorls; imperforate; shell surface glossy and translucent in live and fresh shells, bleaching with age; shell color pellucid; embryonic whorl smooth; remaining whorls with moderately spaced, rounded riblets, crossed by even finer spiral striae (a), a strong scope required to see this fine micro-feature; last whorl rounded.

**Similar Species**: *Rectaxis* species differ from *Volutaxis* (a species often found with *Rectaxis*) by being fully 6 mm smaller and most importantly, having faint spiral striae crossing the riblet formations of the shell.

**Habitat**: Found in limestone talus and under leaf litter at most elevations.

**Status**: Common, one of the common small leaf litter gastropods in Belize; also reported from Petén region of Guatemala.

**Specimen**: Belize, Toledo District, Bladen Nature Reserve (Dourson collection).

**Type Locality**: El Paso de Los Cadallos, Guatemala.

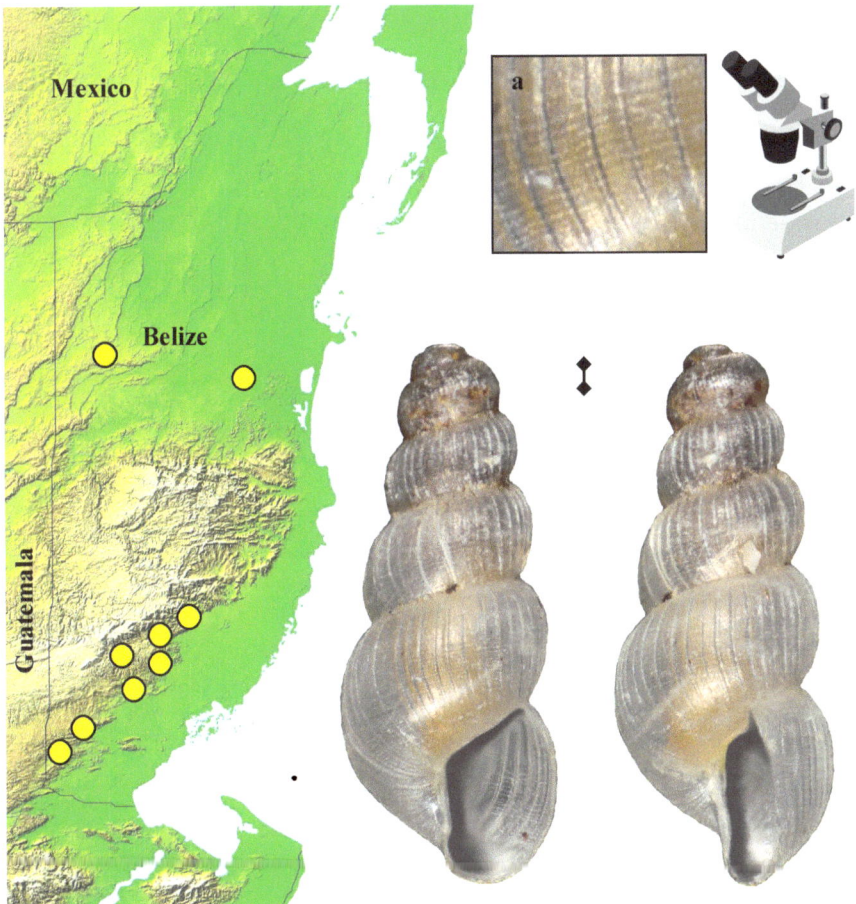

175

# Petén Splinter
## SPIRAXIDAE
### *Rectaxis alvaradoi* (Goodrich & van der Schalie, 1937)

**Height**:3.5-4 mm, Diameter: 1.5 mm

**Description**: Cylinder-shape; lip simple with a slight reflection; shell with 6-7 whorls; imperforate; shell surface glossy white and translucent; embryonic whorl smooth; remaining whorls with moderately spaced, rounded riblets, crossed by even finer spiral striae (a); last whorl rounded.

**Similar Species**: *Rectaxis* species differ from *Volutaxis* (a species often found with *Rectaxis*) by being around 6 mm smaller and most importantly, having faint spiral striae crossing the transverse striae formations of the shell; *Rectaxis funibus* is fully 2 mm smaller.

**Habitat**: Found among limestone talus on hillsides, in ravines and under leaf litter at most elevations.

**Status**: Uncommon to common in northern half of Belize; the type locality is Limestone Knoll, 5 miles north El Paso de Caballo, Petén Guatemala.

**Specimen**: Belize, Cayo District,4.9 mi SE San Ignacio on Cristo Rey Rd, Duffy Bank area, 110 m., UF 146256.

**Type Locality**: El Paso de Los Cadallos, Guatemala.

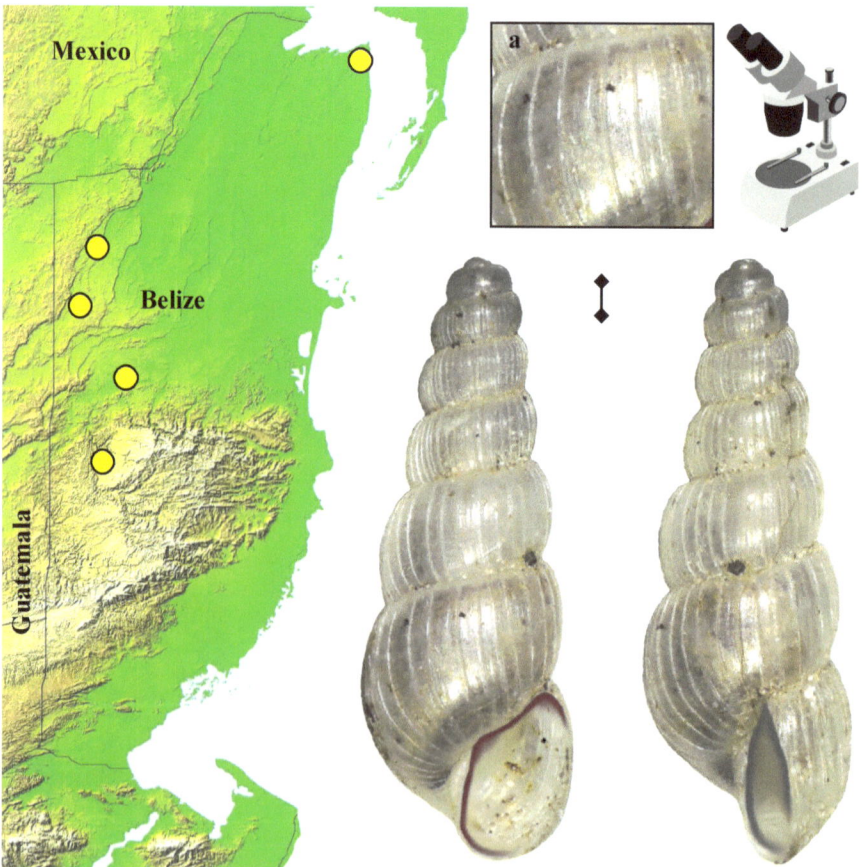

# Montane Splinter                    SPIRAXIDAE

*Rectaxis breweri* (new species)

**Height**: 3 mm, Diameter: 1.8 mm

**Description**: Cylinder-shape; lip simple with a slight reflection; shell with 6 whorls; imperforate; shell surface glossy, amber color; <u>embryonic whorl smooth; remaining whorls with moderately spaced, rounded riblets, crossed by even finer spiral striae (a), the spiral best seen on the base</u>; last whorl rounded.

**Similar Species**: *Rectaxis alvaradoi* (page 176) is larger but more narrow in form, white not amber and is a species of lower elevation forests; *Rectaxis funibus* (page 174) is around the same height but much narrower in form.

**Habitat**: Believed to be restricted to montane oak forests above 1000 meters in the Maya Mountains of Belize.

**Status**: Rare, ENDEMIC to BELIZE precious little is known of this apparently range restricted land snail; known only from the type locality.

**Specimen**: Belize, Toledo District, Oak Ridge, Bladen Nature Reserve, elevation 1100 m, Holotype UF 505456.

**Type Locality**: Oak Ridge, Bladen Nature Reserve, Belize (16°31'1"N, 88°55'43"W).

**Etymology**: Named in honor of Steven Brewer, an extraordinary and passionate botanist who has spent countless hours exploring and cataloguing Belize's outstanding plant life.

177

# Similar Shell Shapes Compared (proportionate)

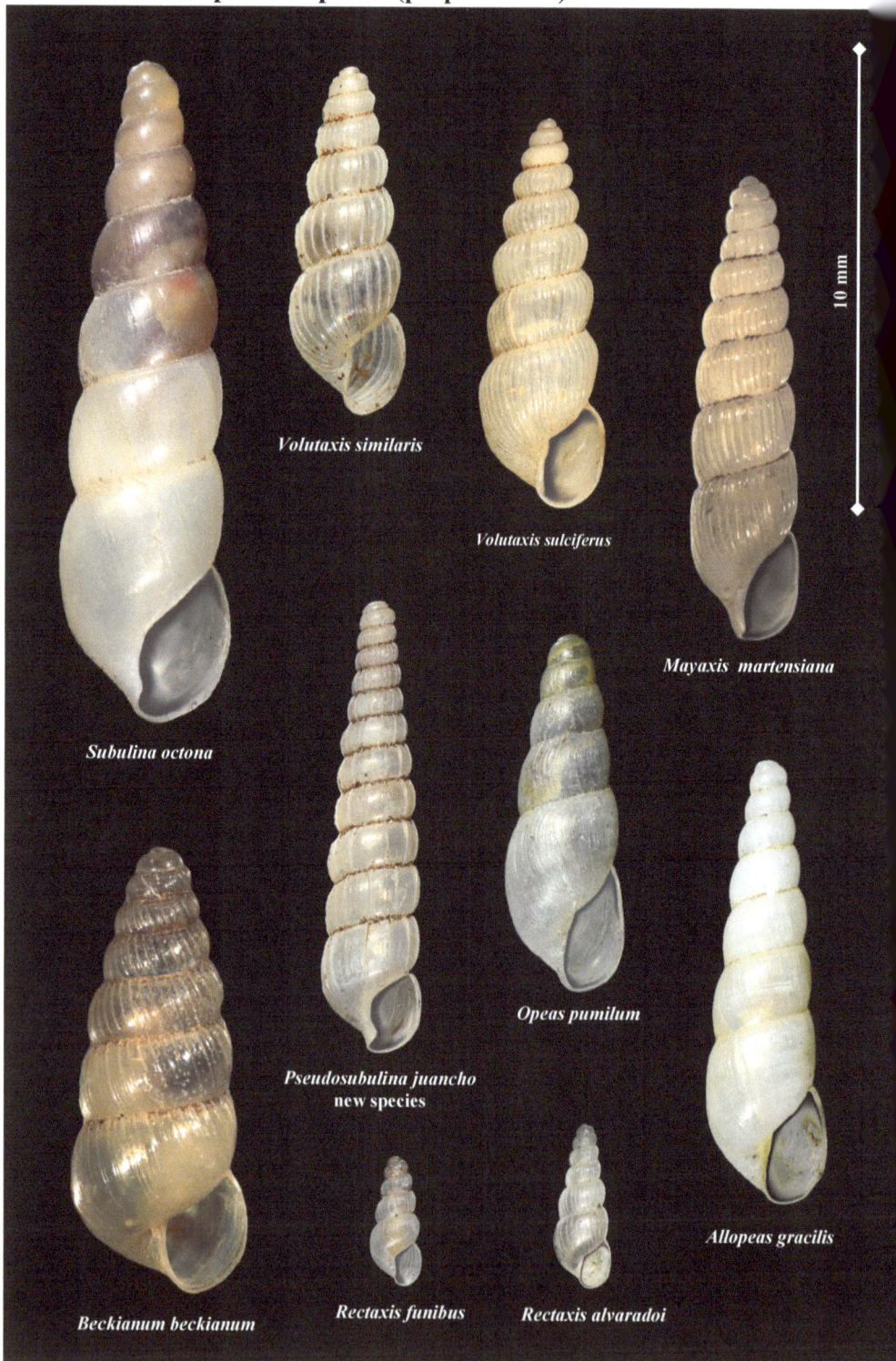

Volutaxis similaris

Volutaxis sulciferus

10 mm

Mayaxis martensiana

Subulina octona

Pseudosubulina juancho
new species

Opeas pumilum

Allopeas gracilis

Beckianum beckianum

Rectaxis funibus

Rectaxis alvaradoi

178

# Family SUBULINIDAE Fischer & Crosse, 1877

**DISTRIBUTION:** Tropical America east through Africa and southeast Asia.

**TAXONOMY:** Numerous genera. Most Middle American SUBULINIDAE were previously grouped in the genus *Leptinaria*. A species critical to this classification was *Allopeas gracilis* which was placed variously within *Opeas* and then *Lamellaxis*, which in turn was considered a subgenus of *Leptinaria*. H. B. Baker (1945) pointed out the anatomical differences between *Opeas*, *Beckianum*, *Allopeas*, *Leptopeas*, *Lamellaxis* and *Leptinaria*. The anatomies of only a few species among these genera were known. Authors continued to unite most of the Middle American species within *Leptinaria* because of the lack of discrete shell differences between it and *Lamellaxis*, *Leptopeas* and *Allopeas*. This classification recognizes *Opeas*, *Allopeas*, *Leptopeas*, *Lamellaxis*, and *Beckianum* as separate genera because of their known anatomies (excerpted from Thompson 2011). Seven genera containing ten recognized species are currently reported from Belize. Four species (known only from caves) are discussed in Chapter 7 (page 191).

**Important Note:** *Beckianum, Allopeas, Leptopeas & Opeas* species are what we call the "**trashy four**" land snails of tropical regions around the world due to their abundance in disturbed habitats. Although these snails live in natural, undisturbed habitats (usually in small numbers), they thrive in more disturbed habitats where they often attain their greatest abundance. These land snails have been widely introduced across the globe.

## Genera Included:
(in order of appearance in text)

*Allopeas*
*Opeas*
*Leptopeas*
*Beckianum*
*Lamellaxis*
*Leptinaria*
*Subulina*

# Traveling Tramp                                    SUBULINIDAE

*Allopeas gracilis* **(Hutton, 1934)**

**Height**: 10 mm, Diameter: 3 mm

**Description**: Cylinder-shape, strongly tapering to the apex; lip simple; shell with 7-9 whorls; perforate; shell surface glossy and translucent in live animals and fresh dead shells; densely sculptured with irregular but weakly developed transverse striae, strongest near the suture; the retraction of the outer lip at the suture is shallow (a); outer lip thin; last whorl rounded.

**Similar Species**: Perhaps most like *Opeas pumilum* but the retraction of the outer lip at the suture is shallow (a) whereas in *O. pumilum,* it is deep (figure a) opposite page.

**Habitat**: A species of karst foothills usually found in large numbers under leaf litter and around rock structure but also common in degraded habitats such as city and suburbs.

**Status**: Common where it occurs; found at four sites in Belize.

**Specimen**: Mexico, El Abra, UF 00110734.

**Type Locality**: Mirzapur, India.

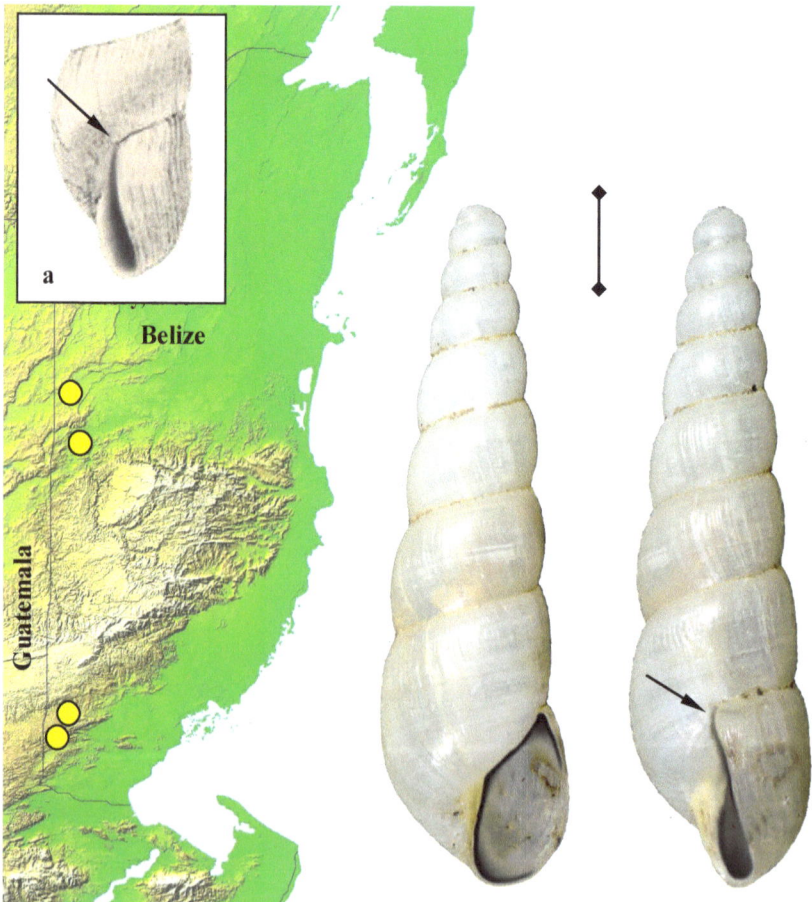

# Common Opeas                           **SUBULINIDAE**

## *Opeas pumilum* (Pfeiffer, 1840)

**Height**: 5-7.2 mm, Diameter: 2 mm

**Description**: Cylinder-shape; lip simple; shell with 6-7 whorls; umbilicus minutely perforate; shell surface glossy and translucent; color pale corneous to almost white, live animal greenish yellow; densely and sharply sculptured with irregular and rather strongly developed transverse striae, the retraction of the outer lip at the suture is deep (a); outer lip thin; last whorl rounded.

**Similar Species**: *Allopeas gracilis* is taller and the retraction of the outer lip at the suture is shallow not deep as in *O. pumilum*.

**Habitat**: A species of karst foothills usually found in high numbers under leaf litter and around rock structure but also a species of waste places.

**Status**: Common in the southern half of Belize.

**Specimen**: Figure (b) from Costa Rica, Limon Prov. Buenos Aires, UF 198418; figure (c) from Belize, Toledo District, 5 miles west of San Jose (Dourson collection) and figure (d) Belize, Cayo District, Benque, UF 146254.

**Type Locality**: Bristol, England.

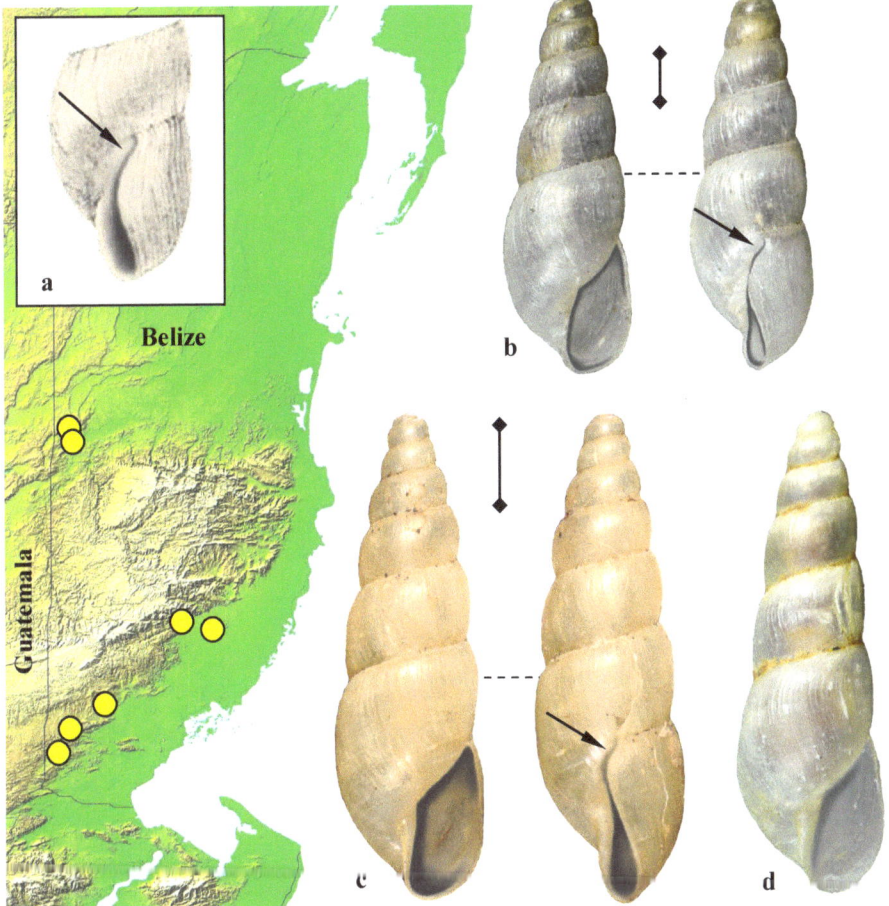

181

# Pigmy Leptopeas                      SUBULINIDAE

*Leptopeas micra* **(Orbigny, 1835)**

**Height**: 6 mm, Diameter: 2.5 mm

**Description**: Cylinder-shape; lip simple; shell with 8 whorls; nearly umbilicate; shell surface glossy and translucent in live animals and fresh shells; color
yellowish-white; transverse striae rather weakly developed; without spiral
striae; last whorl rounded.

**Similar Species**: Most like *Opeas pumilum* but being smaller and stouter in
form with a less pointed apex; *Lamellaxis* species have smoother shell surfaces.

**Habitat**: Found in a variety of habitats including rocky soils and under leaf
litter of limestone foothills covered in jungle but also a species of disturbed
habitats.

**Status**: Uncommon, in Belize reported from Cayo District, Cristo Rey Road,
Duffy Bank 4.9 miles southeast of San Ignacio.

**Specimen**: Niue Island, UF 426787.

**Type Locality**: Santa Cruz de la Sierra, Bolivia.

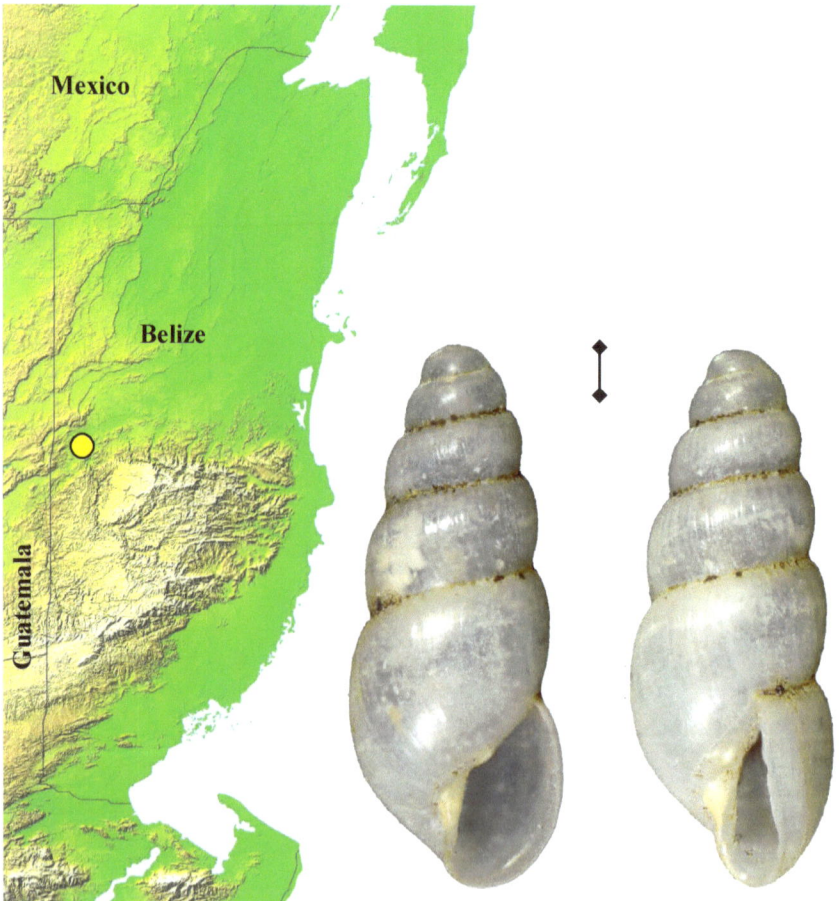

# Yucatan Leptopeas                    SUBULINIDAE

## *Leptopeas yucatanense* (Pilsbry, 1906)

**Height**: 6-7 mm, Diameter: 3 mm

**Description**: Cylinder-shape; lip simple; shell with 7-8 whorls; very narrowly rimate; shell surface glossy and translucent with the axis showing faintly through; color yellowish-corneous; transverse striae low and rounded and at the sutures forming raised callus deposits; without spiral striae; last whorl roundish.

**Similar Species**: *Lamellaxis* are smoother without the somewhat unique calcium deposits at the suture line but otherwise similar in build; *Pseudosubulina* species have better developed transverse striae.

**Habitat**: A species of rocky limestone hilltops covered in rainforest and in and around limestone talus slopes.

**Status**: Uncommon; found in scattered locations throughout Belize but likely more common than current records indicate.

**Specimen**: Belize, Orange Walk District, around the town of Gallon Jug (Dourson collection).

**Type Locality**: Ticul, Mexico.

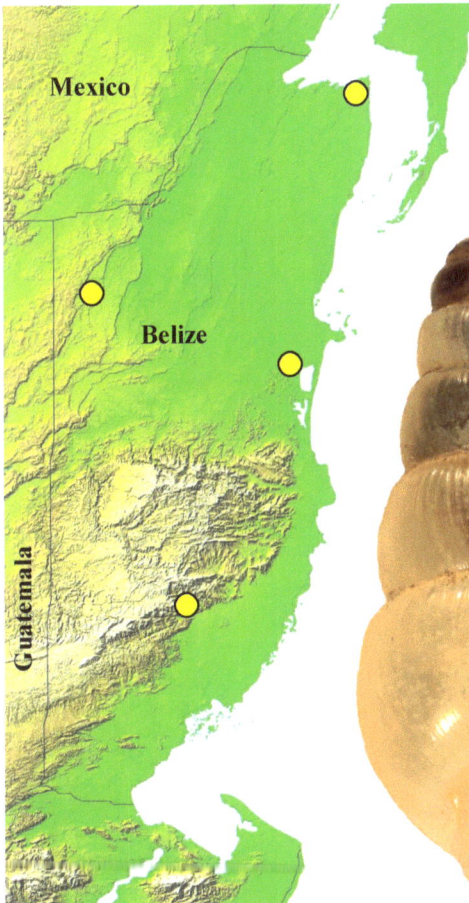

# Common Leptopeas                    SUBULINIDAE

*Leptopeas cf. guatemalense* (Strebel, 1882)

**Height**:10 mm, Diameter: 3-4 mm

**Description**: Cylinder-shape; lip simple; shell with 8-9 whorls; imperforate; shell thin, glossy and translucent in live animals and fresh shells; color of shell is ash gray; transverse striae are poorly developed, strongest on the last whorl; without spiral striae; last whorl rounded.

**Similar Species**: *Leptopeas yucatanense* is shorter with better developed transverse striae and a different color; *Pseudosubulina* and *Mayaxis* species have well developed transverse striae or riblets.

**Habitat**: A species of limestone foothills usually found in low numbers under leaf litter and overhanging limestone rock features covered in tropical rain forest.

**Status**: Common; in some locations can be quite abundant.

**Specimen**: Belize, Toledo District, Bladen Nature Reserve (Dourson collection).

**Type Locality**: Alta Verpaz, Coban Guatemala.

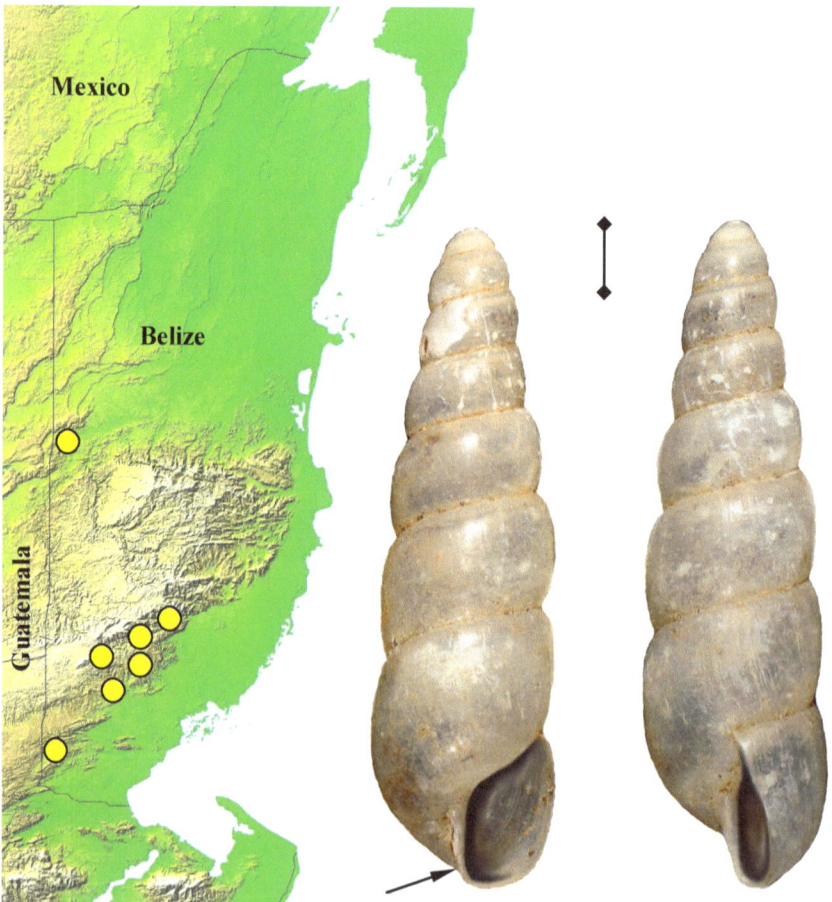

184

# Tropical Tramp                   SUBULINIDAE

*Beckianum beckianum* (Pfeiffer, 1846)

**Height**:8-12 mm, Diameter: 4 mm

**Description**: Cylinder-shape, lip simple; shell with 8-9 whorls (taller form with 2 or 3 additional whorls); perforate; shell surface glossy and translucent in live animals and fresh shells; <u>transverse striae are well developed</u>; without spiral striae; last whorl rounded.

**Similar Species**: Most similar to *Lamellaxis fordianus* but taller and most importantly, having a proportionately smaller aperture; *Leptopeas* species have similar body builds but less defined transverse striae.

**Habitat**: Found in a variety of habitats including rocky soils and under leaf litter of limestone foothills covered in jungle; also a species of disturbed habitats, hence the common name, Tropical Tramp.

**Status**: Common, the species likely occurs throughout Belize.

**Specimen**: Figure (a) from Belize, Toledo District, Bladen Nature Reserve and Figure (b) from Belize, Cayo District, Rio Frio Cave (Dourson collection).

**Type Locality**: Polvon, Nicaragua.

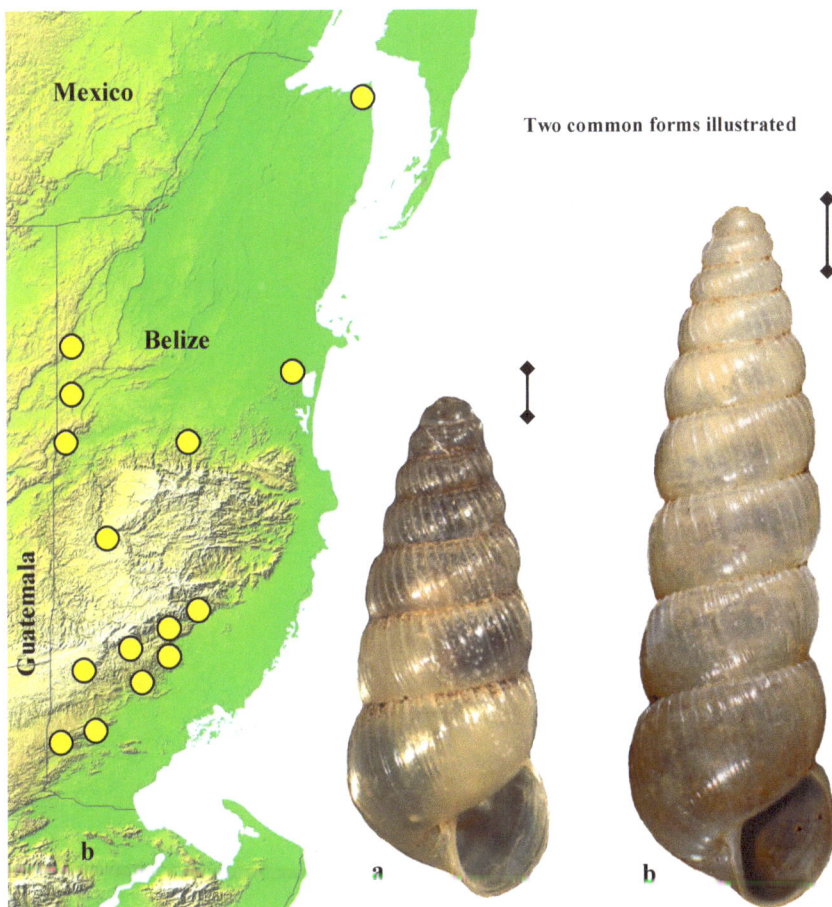

Two common forms illustrated

# Ancey's Lamellaxis                    SUBULINIDAE

## *Lamellaxis cf. fordianus* (Ancey, 1886)

**Height**: 6-7.3 mm, Diameter: 3.6

**Description**: Cone-shape; lip simple; shell with 5-6 whorls; imperforate; shell pellucid (colorless); <u>under a 10X hand lens well developed transverse striae or low thin riblets can be observed (a)</u>; without notable color features and without spiral striae; last whorl rounded.

**Similar Species**: *Leptinaria lamellata* is similar to *L. cf. fordianus* but has a larger and wider build, a smoother shell and has a deeply entering lamellae (figure a, page 187).

**Habitat**: Found in karst hills under leaf litter and the entrances of caves.

**Status**: Relatively common in Belize; reported from Rio Frio Cave and the Bladen Nature Reserve; generally distributed throughout lower elevations of the Maya Mountains.

**Specimen**: Belize, Toledo District, Bladen Nature Reserve (Dourson collection).

**Type Locality**: Atila Island, Honduras.

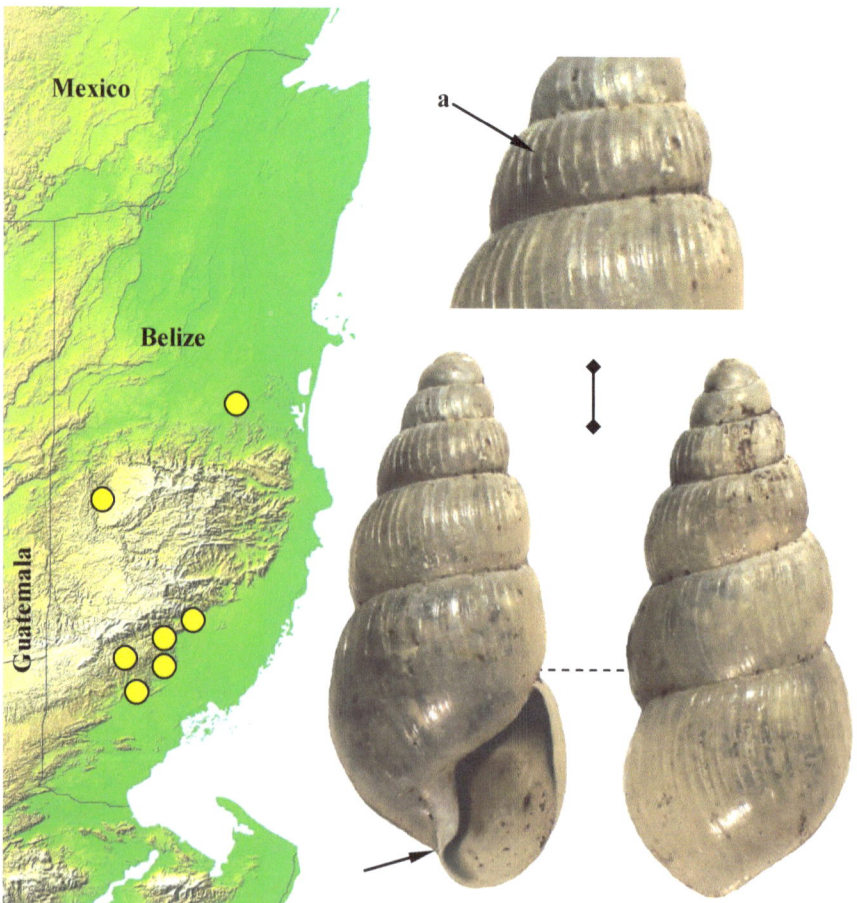

# Obese Funnel                           SUBULINIDAE

## *Leptinaria lamellata* (Potiez & Michaud, 1838)

**Height**:10-12 mm, Diameter: 4-5 mm

**Description**: Cone-shape; lip simple; shell with 5-6 whorls; imperforate; shell thin, glossy and translucent; in shells up to 6 whorls there is a deeply entering lamellae (a) that may be absent in some adult shells; transverse striae are poorly developed and nearly smooth on the last whorl; columellar plait is notably wide and twisted (b); last whorl rounded.

**Similar Species**: *Lamellaxis* species are usually narrow in form; *Leptinaria livingstonensis* has well developed transverse striae and is more narrow in build.

**Habitat**: A species of limestone foothills, under leaf litter, around rock structure and talus slopes.

**Status**: Common, found throughout lower elevations of the Maya Mountains.

**Specimen**: All figures from Belize, Toledo District, Bladen Nature Reserve (Dourson collection).

**Type Locality**: Not reported.

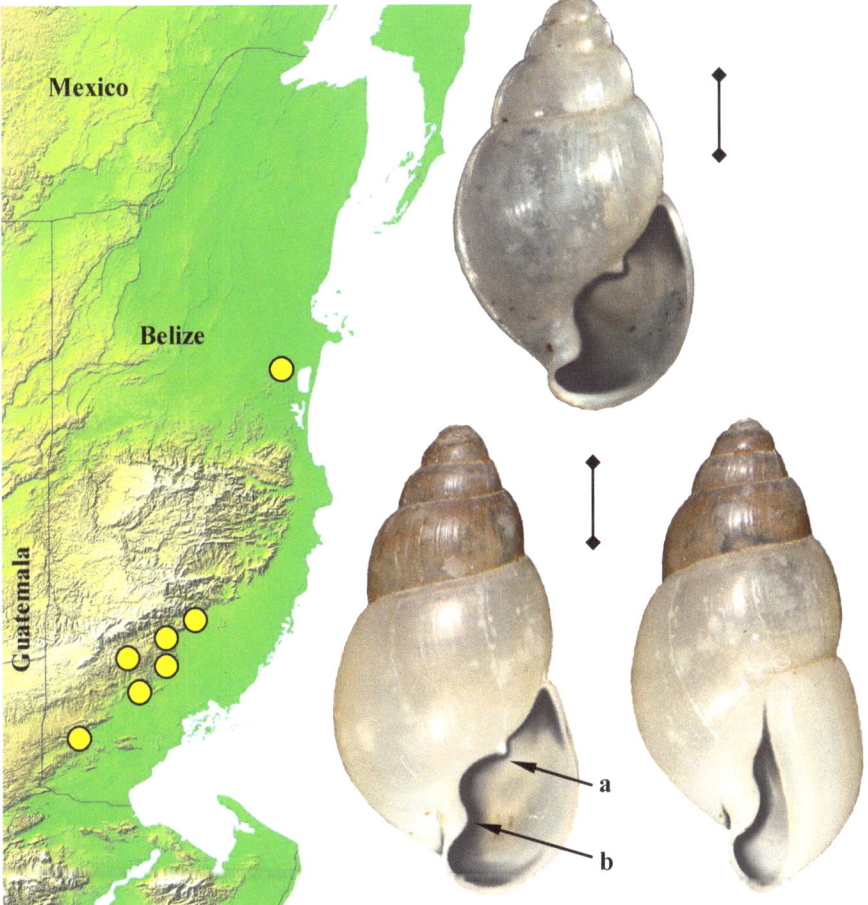

187

# Ribbed Funnel                                    SUBULINIDAE

## *Leptinaria livingstonensis* Hinkley, 1920

**Height**: 9.5-11 mm, Diameter: 4 mm

**Description**: Cone-shape; lip simple; shell with 6-7 whorls; imperforate; shell thin, glossy and translucent; with or without a deeply entering lamellae (a); transverse striae are moderately developed on the last whorl; columellar plait is notability wide and twisted; last whorl rounded.

**Similar Species**: Most similar to *Lamellaxis fordianus* in build but larger, having a deeply entering lamellae (*Lamellaxis* are without lamellae) and smoother surface; *Leptinaria lamellata* has a much wider build.

**Habitat**: A species of town dumps and other disturbed sites; in natural habitats found in limestone foothills under leaf litter.

**Status**: Reported from Livingston, Guatemala where it was found in rubbish piles along with *Subulina octona* (Hinkley 1920); not yet reported from Belize, but expected to occur.

**Specimen**: Guatemala, Izabal Dept. Livingston, UF 00110792.

**Type Locality**: Livingston, Guatemala.

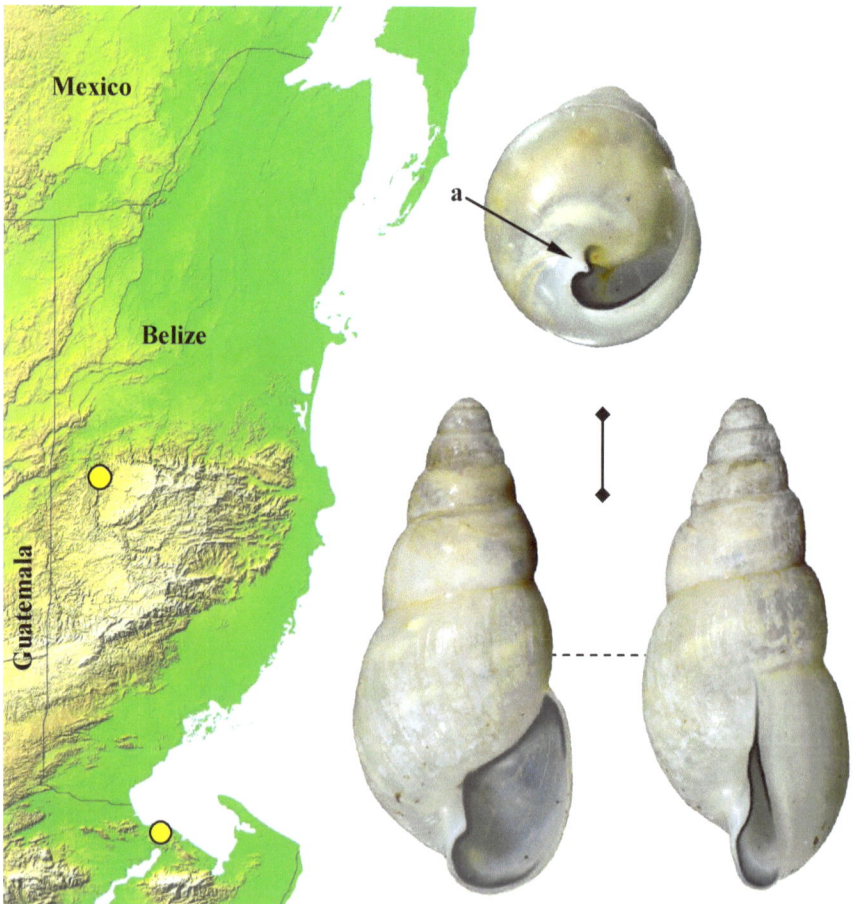

188

# Glossy Subulina                 SUBULINIDAE

*Subulina octona* **(Bruguiére, 1789)**

**Height**:18 mm, Diameter: 3-5 mm

**Description**: Cylinder-shape; lip simple; shell with 8-9 whorls; imperforate; shell surface glossy and translucent in live animals and fresh shells; <u>transverse striae are faint; no spiral striae present; columellar plait "S" shaped (a)</u>; last whorl rounded.

**Similar Species**: Most like *Allopeas* and *Leptopeas* species much taller by around 10 mm; other cylinder-shape shells of similar size such as *Mayaxis* and *Volutaxis* will have stronger shell ornamentation features present.

**Habitat**: Found in a variety of habitats including rocky soils and under leaf litter as well as around degraded habitats and settlements.

**Status**: Common, a species that should occur throughout Belize in appropriate habitat; found on nearly every continent.

**Specimen**: Belize, Toledo District, limestone hill around San Jose (Dourson collection).

**Type Locality**: Probably South America (Thompson 2011).

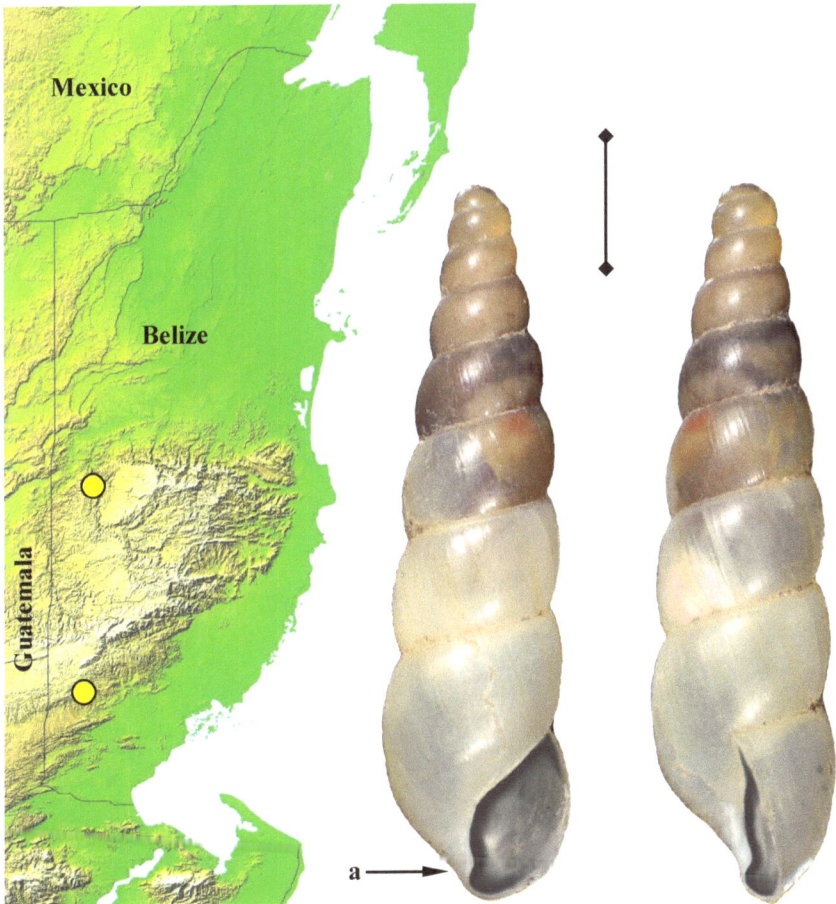

Actual size

**The internal shell of Tiny Salasiella,** *Salasiella modesta* (Pfeiffer, 1862)

# 7 Troglobytic Land Snails Found in Limestone Caves

Caves are unique habitats unlike anything found on the surface of the planet. They contain whole ecosystems in which species live, breed and die without ever seeing the light of day. Since these creatures spend their entire lifetime in total darkness, they are usually without pigment or functional eyes and are referred to as troglobitic organisms. These are some of the most highly specialized animals found on Earth. Most are extraordinarily rare, a result of isolation from surface environments. Troglobitic snails are highly sensitive to changes in climate and human pollution. These gastropods walk (actually crawl) a thin line between life and death, entirely dependent on the waste material (guano) of crickets and bats and other foods brought into the caves during flood events. We conservatively estimate that a dozen or more troglobitic land snails thrive in Belizean caves (all expected to be new to science). Four species have been documented and described thus far but countless caves in Belize await further investigation! The snails in this chapter represent the first troglobitic land snail species to be described from Central America cave systems.

## The Families and Genera Included in Chapter 7

### SUBULINIDAE

# Belize Cave Snail                    SUBULINIDAE

*Opeas marlini* **(new species)**

**Height**:7-8 mm, Diameter 2.5 mm

**Description**: Cylinder-shape; lip simple but in fully adult shells reflecting widely near the columella insertion; shell with 7-8 whorls; perforate; shell thin, glossy and translucent in live animals and fresh shells; live animal white; all whorls have deeply arching, crowded, transverse striae that develop thickened deposits of calcium at the suture line; the retraction of the outer lip at the suture is deep (a); without spiral striae. Although shells are common in some caves, a result of slow decay, live animals are rarely observed.

**Similar Species**: *Opeas pumilum* (page 181) has a less tapered build, a more narrow aperture and straighter, less curved transverse striae.

**Habitat**: This is a strict cave obligate land snail, found only in dry portions of the cave which do not receive flood water.

**Status**: The most wide ranging cave snail in Belize; ENDEMIC to BELIZE found in Swiss Cheese Cave (BNR), Blue Creek Cave, St. Hermann's Cave (narrow form), Xuan's Cave & Tiger Sandy Bay Caves .

**Specimen**: Belize, Toledo District, Swiss Cheese Cave, Bladen Nature Reserve. Holotype UF 505441. Paratypes UF 505442 (not pictured.)

**Type Locality**: Swiss Cheese Cave, Bladen Nature Reserve, Belize (16°33'33"N, 88°44'15"W).

**Etymology**: Named in honor of Jacob Marlin, founder and executive director of Belize Foundation for Research and Environmental Education. Jacob has dedicated most of his life to passionately protecting the Bladen Nature Reserve.

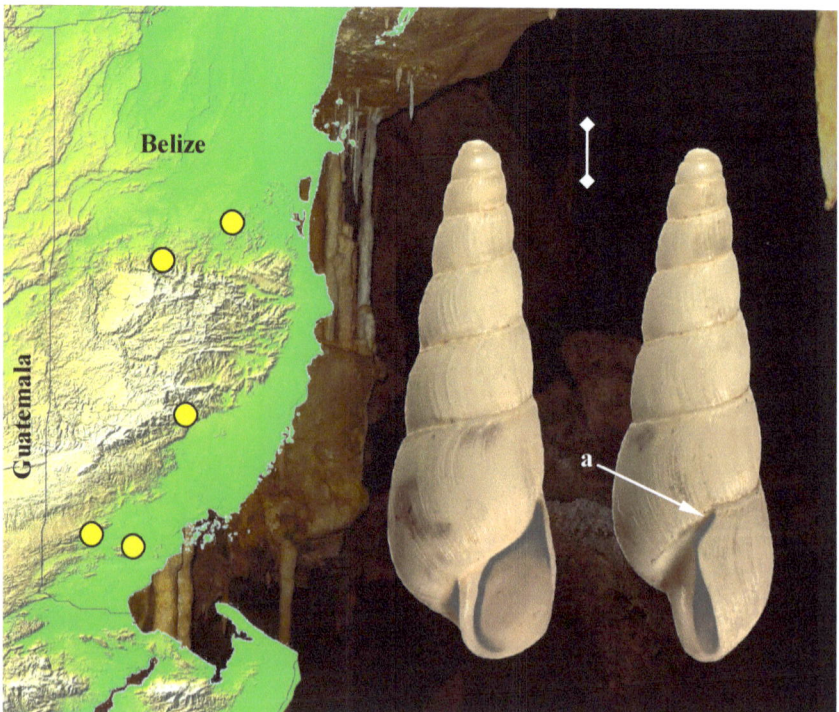

**Belize Cave Snail,** *Opeas marlini*: Troglobitic land snails spend their entire life in the total darkness of caves and therefore, are usually without pigment and functional eyes or no eyes at all. While shells are moderately common, a result of stable conditions inside the cave; live individuals are exceedingly rare, the below images of the only ones ever photographed! The species has been observed feeding on the carcasses of bats, but also likely feeds on the guano of bats and cave crickets.

Live animal white and eyes are much reduced

Frothy-mucus believed to ward off attacking cave arthropods

Notable thickened calcium deposits at the suture lines

Transverse striae are crowded and deeply curved

193

# Mayan Cave Snail

## SUBULINIDAE

*Leptopeas corwinii* (new species)

**Height**: 10-11 mm, Diameter: 3 mm

**Description**: Cylinder-shape; lip simple; shell with 8-9 whorls; perforate; shell thin, glossy and translucent; transverse striae are moderately developed, strongest near the sutures; without spiral striae; last whorl rounded.

**Similar Species**: *Leptopeas yucatanense* (page 183) is smaller, has fewer whorls, a more elongated aperture and a yellowish tinge (*L. corwinii* is whitish); *Leptopeas micra* (page 182) has a much shorter build.

**Habitat**: This is a strict cave obligate land snail, found only in dry portions of the cave which are not subject to flooding; like most troglobitic snails, generally found around and on flow stone formations and under bat roosting sites where it likely feeds on bat guano.

**Status**: Rare; ENDEMIC to BELIZE currently known from Blue Creek Cave, Saint Herman's Cave and Tiger Sandy Bay Caves; in Blue Creek Cave, shells are moderately common but live individuals have not been observed.

**Specimen**: Belize, Belize District, Tiger Sandy Bay Holotype UF 505439. Paratypes UF 505440 (not pictured).

**Type Locality**: Tiger Sandy Bay Cave systems, Belize District, Belize (17° 17'60"N, 88°31'0"W).

**Etymology**: Named in recognition of Jeff Corwin, a conservation biologist who conducted research at Blue Creek Cave for a Master's degree and continues to educate the public about the natural world through his outstanding wildlife films.

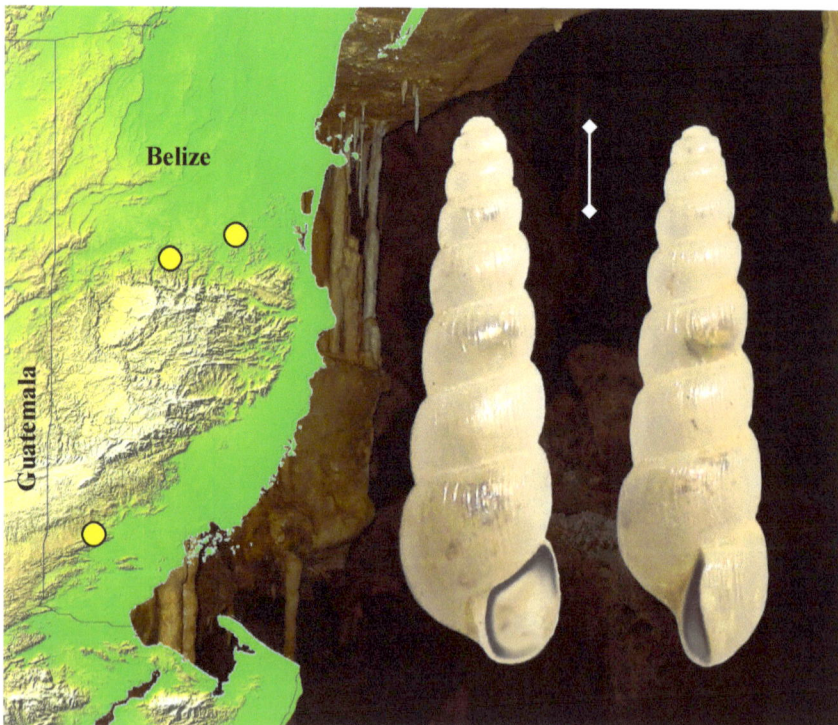

# Macal River Cave Snail                SUBULINIDAE

*Lamellaxis matola* **(new species)**

**Height**:14-16 mm, Diameter 5 mm

**Description**: Cylinder-shape; lip simple; shell with 8-9 whorls; perforate; shell thin, glossy and translucent; <u>transverse striae are well developed in the beginning and middle whorls, becoming weakest on the last whorl</u>; the columellar plait is notably twisted; last whorl rounded.

**Similar Species**: *Lamellaxis fordianus* (page 186) is around 6-7 mm smaller, containing 3-4 less whorls and is a species of the surface.

**Habitat**: This is a strict cave obligate gastropod, found in upper passageways of the cave which do not experience flooding; shells have been found on a few wet flowstone formations; live individuals have not been observed.

**Status**: Rare; **ENDEMIC to BELIZE** known only from a sinkhole in the Macal River Gorge.

**Specimen**: Belize, Cayo District, unnamed sink hole above road and the Macal River, a few miles downstream (north) of Black Rock Lodge, Holotype UF 505435. Known only from the type locality. Paratypes UF 505436, (not pictured).

**Type Locality**: A sink hole above the Macal River, Cayo District, Belize (17° 3'7"N, 89°4'3"W).

**Etymology**: Named in honor of Sharon Matola, founder and director of the Belize Zoo—the best little zoo in the world— for her dedication and perseverance in protecting Belize's abandoned and injured wildlife and commitment to educating Belizeans and visitors about the incredible wildlife of Belize.

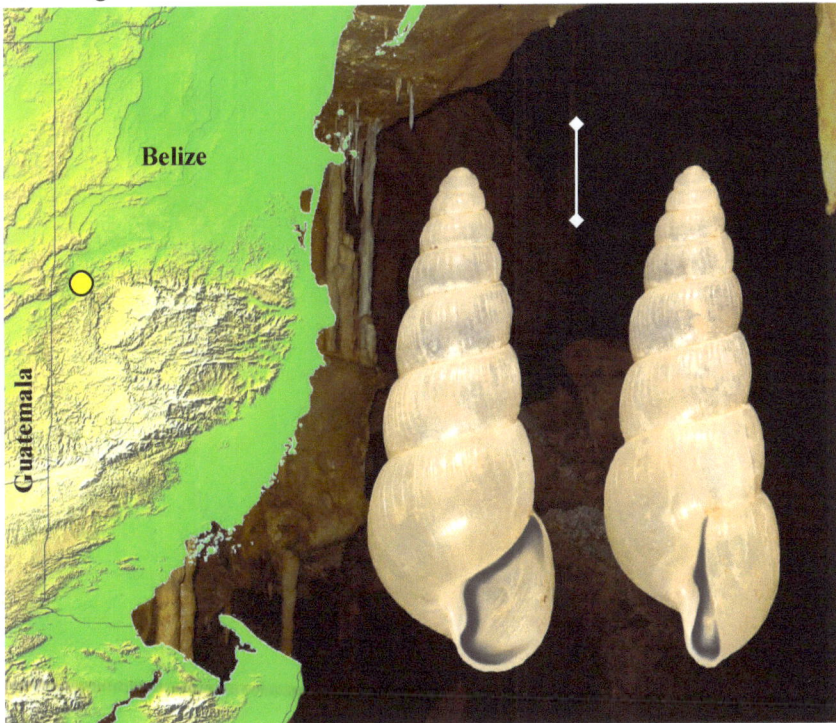

# Blue Creek Cave Snail                    SUBULINIDAE

*Leptinaria doddi* (**new species**)

**Height**:15-16 mm, Diameter 7 mm

**Description**: Cone-shape; lip simple; shell with 7-8 whorls; perforate; shell thin, glossy and translucent; <u>in shells up to 6 whorls there are 1 or 2 deeply entering lamellae (a)</u> which may be absent in adult shells; poorly developed transverse striae; the columellar plait is notably twisted; last whorl rounded; the species likely feeds on the guano of both bats and cave crickets.

**Similar Species**: *Leptinaria livingstonensis* (page 188) has a slightly wider build, smaller by around 4 mm and is a species of the surface; *Leptinaria lamellata* (page 187) is smaller with a notably wider build.

**Habitat**: A strict cave obligate found only in upper dry passageways of Blue Creek Cave on flowstones (page 197) that do not experience flooding.

**Status**: Rare; ENDEMIC to BELIZE; known only from Blue Creek Cave.

**Specimen**: Belize, Toledo District, upper entrance to Blue Creek Cave, specimen from a large room Holotype UF 505437, Paratypes UF 505438, (not pictured).

**Type Locality**: Blue Creek Cave, Toledo District, Belize (16°12'25"N, 89° 2'57"W).

**Etymology**: Named in honor of Frederick Dodd, founder of International Zoological Expeditions (IZE) who had the foresight to purchase and protect the wild jungles surrounding Blue Creek Cave, the only known location for the globally rare Blue Creek Cave Snail.

Habitat of the Blue Creek Cave Snail

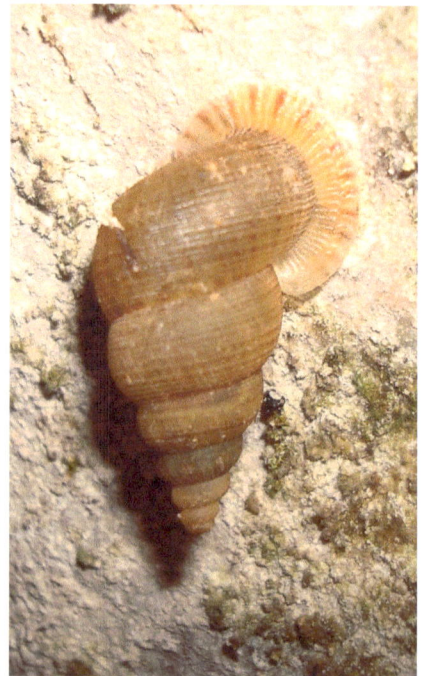

While land snails are active during rainy weather, most species go into aestivation (a kind of hibernation) when conditions begin to dry out—in Belize usually March through June. Selection of sites for aestivation is generally based on where the snail is located when the dry season arrives but any number of surfaces may be utilized including trees, under leaves or in the shade zones of caves. Snails may form an epiphragm (a cellophane-like covering) over the aperture or possess an operculum which helps preserve the snail's critical body moisture.

# 8 Shells Taller than Wide, Greater than 5 mm Tall, Succiniform

This section includes adult land snails taller than wide and greater than 5 mm tall. These land snails, although considered terrestrial, can live in wet conditions and are often referred to as amphibious, living close to water but not in it. One representative species from each family is illustrated to the right below.

## The Families and Genera Included in Chapter 8

**SUCCINEIDAE**
*Succinea*.........................page 202

Does my helix make me look fat?

**Guatemalan Amber Snail,** *Succinea guatemalensis*

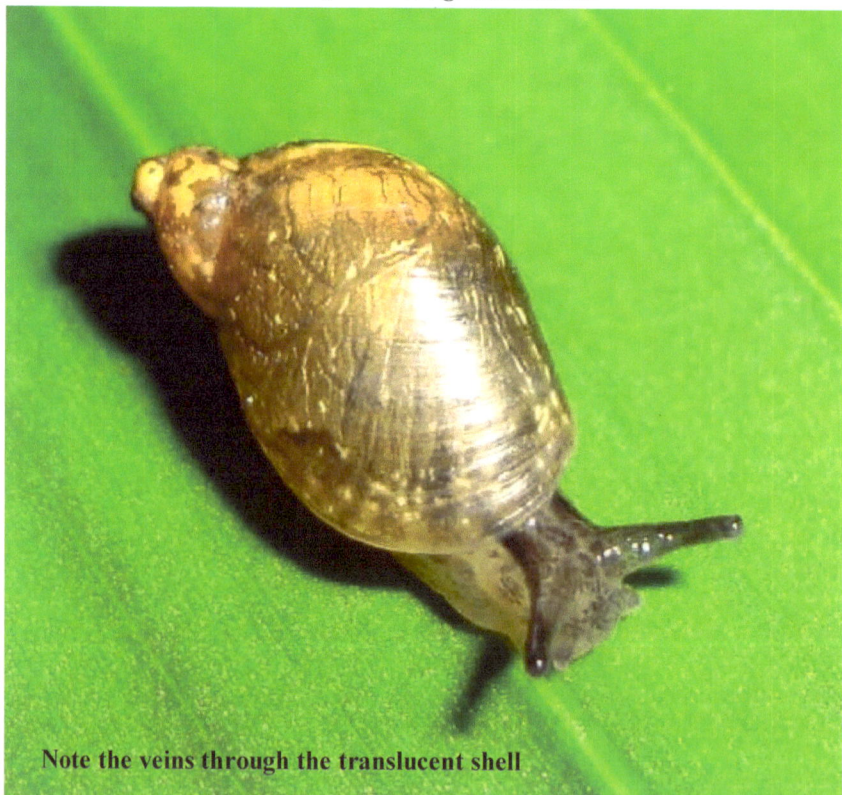

Note the veins through the translucent shell

**DISTRIBUTION:** Northern Europe, North America, Hawaii and Samoa.
**TAXONOMY:** Three subgenera are recognized (Thompson 2011). One genera containing two species are currently known from Belize. These land snails are often referred to as amphibious or semi-aquatic, usually found living near water edges but not actually in it. Shells are paper thin and do not weather well which reduces the chances of documenting their presence. There are likely more species in Belize awaiting discovery.

# Genera Included:
(in order of appearance in text)

## *Succinea*

# Guatemalan Amber Snail          SUCCINEIDAE
*Succinea guatemalensis* **Morelet, 1849**

**Height**:10 mm, Diameter: 7-8 mm

**Description**: Succiniform; lip simple; aperture large, about 3/4 the size of shell; shell with 3 whorls; perforate; shell nearly paper thin, glossy and translucent and very fragile; color of shell a light amber; live animal bluish-white with scattered black spots; transverse striae are moderately developed on all three whorls; last whorl broadly rounded.

**Similar Species**: *Succinea luteola luteola* is longer and more narrow in form, with a thicker, less fragile shell.

**Habitat**: A species of low wet places, under leaf litter; occasionally in upland sites; often referred to as amphibious, living around water but not in it.

**Status**: Uncommon in Belize, currently known from three sites in Belize.

**Specimen**: Figure (a) *Succinea guatemalensis*, Syntype NHMUK 1893.2.4.943 and Figure (b) from Belize, Toledo District, 5 miles west of San Jose next to main paved road (Dourson collection).

**Type Locality**: Guatemala, on banks of small streams, Dept. Guatemala

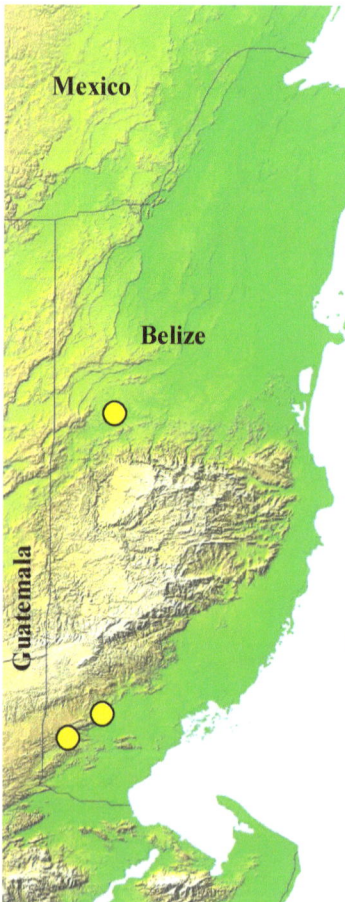

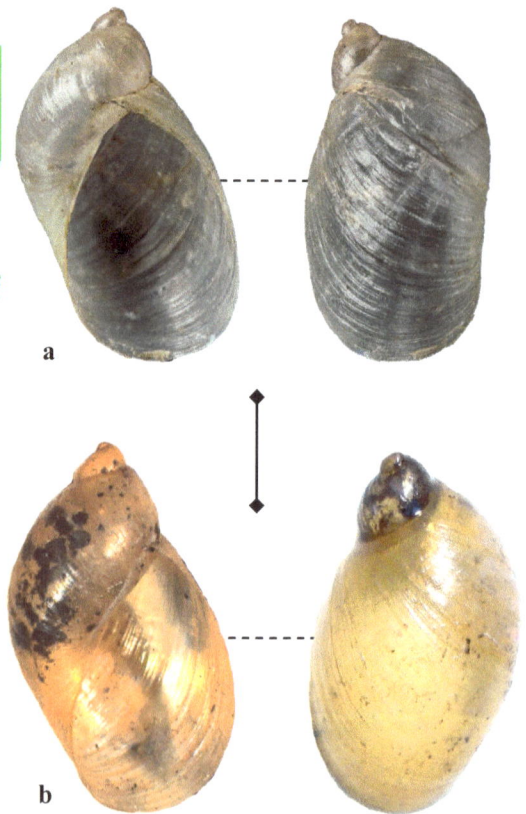

202

# Island Amber Snail SUCCINEIDAE

*Succinea luteola luteola* **Gould, 1849**

**Height**:10 mm, Diameter: 5.5 mm

**Description**: Succiniform; lip simple; <u>aperture large, about half the size of shell</u>; shell with 3-4 whorls; perforate; <u>shell thin, glossy and translucent and fragile</u>; color of shell a light amber; transverse striae are moderately developed on all whorls; last whorl broadly rounded.

**Similar Species**: Similar to *S. guatemalensis* but longer in form with a thicker shell; no other land snail genera in Belize have the body build of the genus, *Succinea*.

**Habitat**: A species of low wet places, under leaf litter; occasionally in upland sites; these terrestrial gastropods are often referred to as amphibious, living around water but not in it.

**Status**: Uncommon; reported from Blackadore Caye, but not yet reported on the mainland, the overall distribution in Belize remains unknown.

**Specimen**: Mexico, Coahuila, 16 km. W of Castanos 800 m, UF 251597.

**Type Locality**: Galveston, Texas, USA.

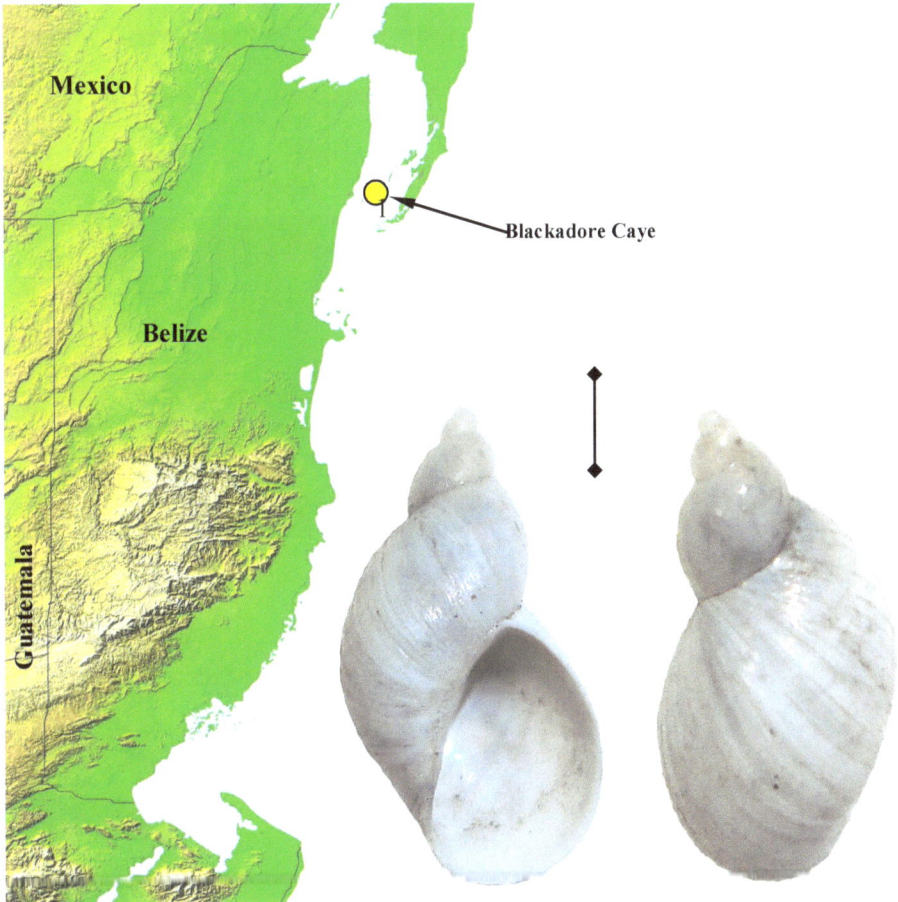

# Hanging by a Thread

Dangling from a limestone rock overhang by a thin mucus thread (a), an aestivating juvenile Pearly Tuba, *Halotudora kuesteri,* is safe from most predators except perhaps small, hovering bats. This protective strategy is employed by most juvenile species of ANNULARIIDAE found in Belize. Shells dangling on lines may also be empty shells in use by spiders or harvestmen (Breure 2011).

# 9 Shells Taller than Wide, Greater than 5 mm Tall, Tuba-Shape

This section includes adult land snails taller than wide and greater than 5 mm tall, having a distinctive tuba shape. All species have an operculum and are often found hanging from limestone rock ceilings by a mucus thread of their own making (see opposite page). This is thought to be a protective strategy, making it difficult for predators to reach the shell. The ANNULARIIDAE of the Caribbean have recently been examined and revised by G. Thomas Watters. His work is referenced in the following species accounts.

## The Families and Genera Included in Chapter 9

**ANNULARIIDAE**

I have to wake up at the crack of dusk to look this good!

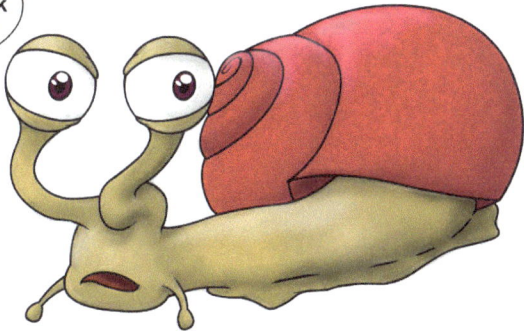

## Astonishing Shell Evolution in ANNULARIIDAE

A diverse and wide-ranging family of the Americas, ANNULARIIDAE comprise over 1500 described species, mostly occurring in the West Indies (Watters 2006). The Caribbean Isles have been especially favorable to the evolution of this family. Rich in limestone deposits, isolation from the mainland and fewer predators has put the West Indies on evolution-overdrive, manufacturing many spectacular species; Cuba containing many taxa. Much fewer are reported from Mexico, Central and South America, around 20 species, most of those living on the Caribbean slope. Nevertheless, Central America, including Belize, has some rather beautiful species that are illustrated on the following pages. One need only study the intricate shell ornamentation of *Blaesospira echinus* (below) to fully appreciate one of nature's greatest achievements in shell architecture. Like HELICINDAE, ANNULARIIDAE are some of the most strangely ornate terrestrial gastropods on the planet.

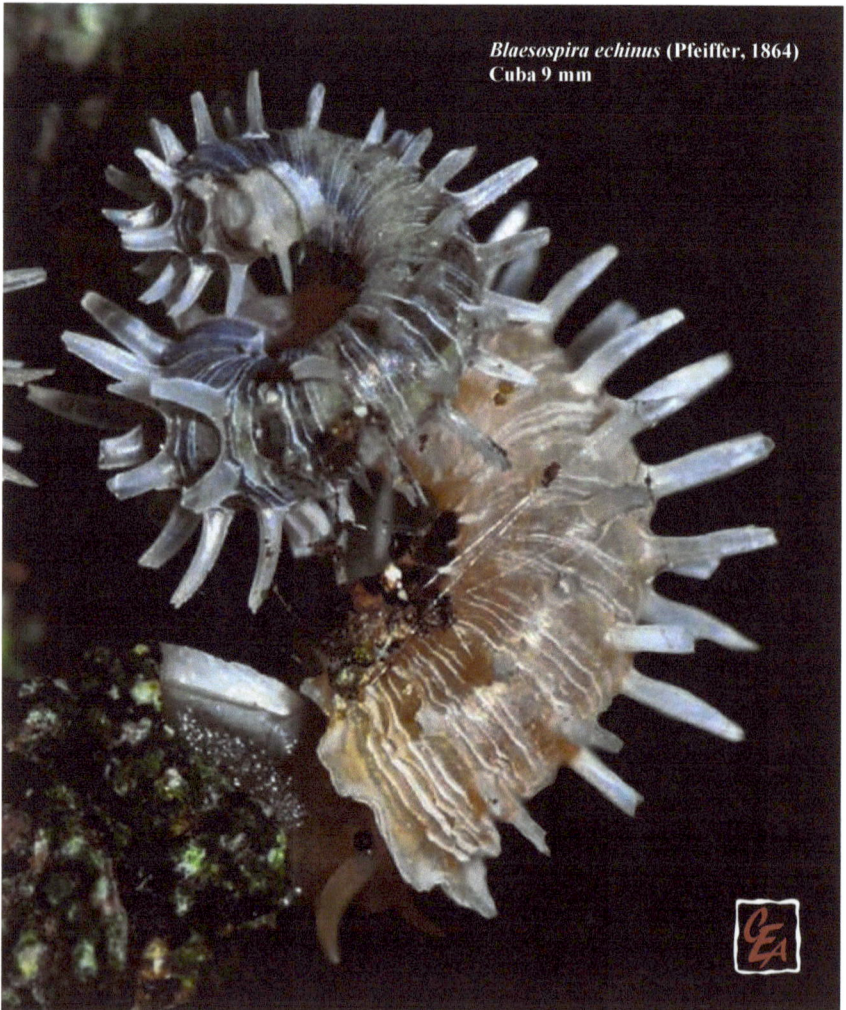

*Blaesospira echinus* (Pfeiffer, 1864)
Cuba 9 mm

# Family ANNULARIIDAE Henderson & Bartsch, 1920

**DISTRIBUTION:** West Indies, south Florida, Mexico south to Panama, Colombia and Venezuela.

**TAXONOMY:** Historically the classification of ANNULARIIDAE was based on structure of the operculum and upon breathing devices associated with the aperture. Henderson & Bartsch (1920) provided a comprehensive classification of the ANNULARIIDAE based on these features. Watters (2006-2014), reviewed the family and proposed a classification that differs from systems adapted by previous workers, de-emphasizing the importance previous authors placed upon opercula and apertural features. Watters has provided a wealth of information on the family, listing over 1500 species and subspecies. This is the first compilation of species-group taxa since Henderson & Bartsch and it is a monumental achievement by itself (Thompson 2011). Eighteen species are reported from Mexico and Central America (Watters 2014). The Yucatan Peninsula is the most speciose for annulariids. Farther south, species numbers decline dramatically with only two species reported from Honduras, one from each of Panama and Nicaragua, and none from El Salvador.

A total of six genera containing eight species are found in Belize. These are medium-sized gastropods that are generally tuba-shaped, exhibit intricate shell surface sculpture and are found living close to or on calcium rich soils. They can be especially abundant around limestone outcroppings where live animals are often seen clinging to rock surfaces.

# The Genera Included in this Section:
(in order of appearance in text)

*Halotudora*
*Paradoxipoma*
*Parachondria*
*Tudorisca*
*Choanopomops*
*Diplopoma*

# Belize Tuba                    ANNULARIIDAE

## *Halotudora gruneri* (Pfeiffer, 1846)

**Height**:20-30 mm, Diameter: 15 mm

**Description**: Tuba-shape; lip widely reflected; with an operculum; shell with 5 whorls, 7-8 whorls with entire teleoconch and protoconch intact (a), but most shells are without; perforate; shell thin but rather solid, glossy and translucent; sometimes with a conspicuous brown band visible just inside the aperture and behind the lip (figures b and c next page), in some lots, specimens have a curious splash of brown on the back side of the last whorl (d); also with broken encircling bands on most of the whorls, becoming widest and boldest on the lip; transverse striae are poorly developed and widely spaced on all whorls; delicate spiral striae present; last whorl rounded.

**Similar Species**: Most similar to *Halotudora kuesteri* but larger with a smoother shell surface; this is the largest ANNULARIIDAE in Belize.

**Habitat**: At the base of and on limestone cliffs and talus.

**Status**: Common, found in numerous sites across the country.

**Specimen**: Figure below from Belize, Ramos Creek, Bladen Nature Reserve, figure (e) from Belize, Toledo District, Bladen Nature Reserve (Dourson collection) and figure (f) from Belize, Toledo District, a limestone hill, 3.2 km. east of Blue Creek Village, UF 207876.

**Type Locality**: Listed only as Belize.

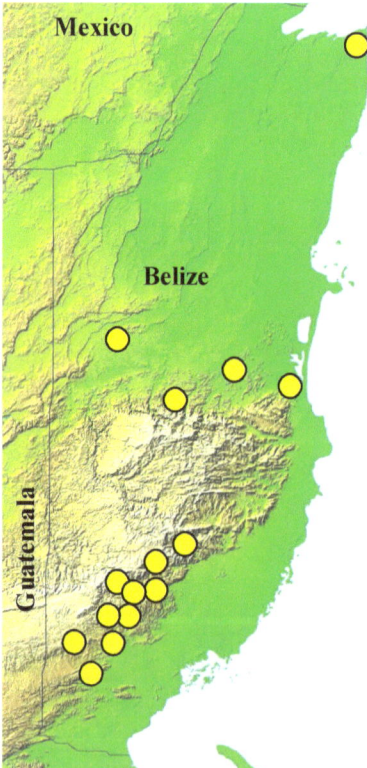

**Figure (e)** *Halotudora gruneri (*Belize)

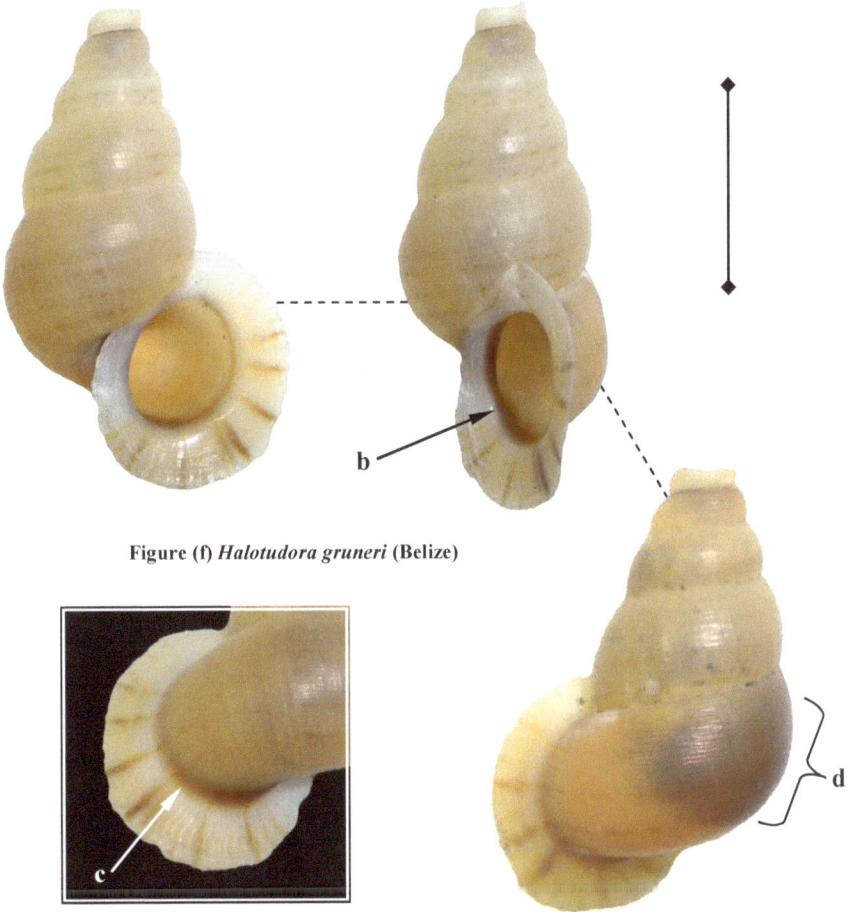

b

**Figure (f)** *Halotudora gruneri* (Belize)

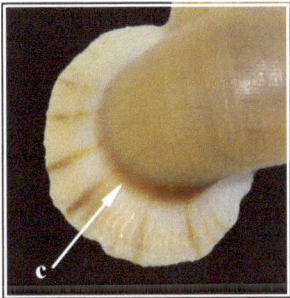

c

d

# Yucatan Tuba

# ANNULARIIDAE

*Halotudora gaigei* **Bequaert and Clench, 1931**

**Height**: 15-18 mm, Diameter: 11 mm

**Description**: Tuba-shape; lip widely reflected; shell with 5 whorls; perforate; with an operculum; shell thin, but rather solid, with faint broken color features, strongest on the last whorl; color of shell variable but most specimens found in Belize are a light buff; transverse striae forming low riblets widest on the first whorls, becoming more crowded on the last whorl, these riblets crossed with strongly developed spiral striae, all forming a net-like appearance (a) image from Bequaert & Clench (1933); last whorl rounded.

**Similar Species**: *Halotudora kuesteri* is larger, has weakly developed transverse striae and a widely reflected lip; *C. largillierti* has conspicuous calcium nodules at the sutures.

**Habitat**: Found in dry tropical forests that cover calcareous soils.

**Status**: Uncommon; currently reported from Orange Walk District, 5.4 km. W of Neuendorf and Gallon Jug.

**Specimen**: Belize, Orange Walk District, 5.4 km. W of Neuendorf, UF 207182.

**Type Locality**: Chichen Itza, Mexico.

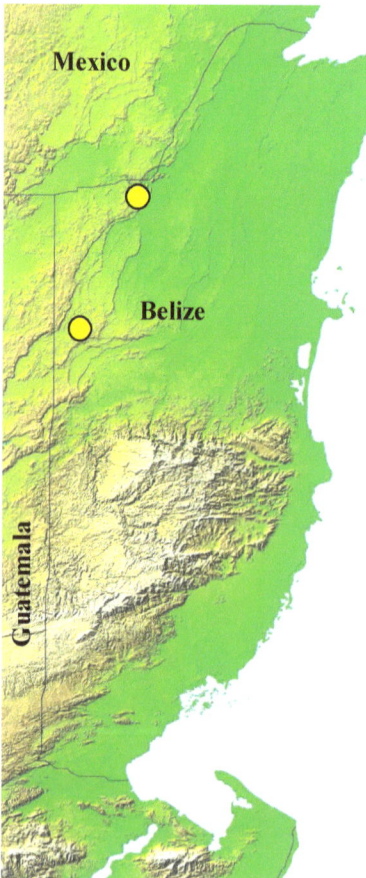

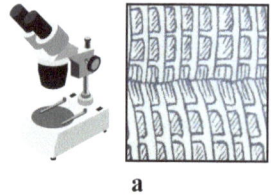

a

*Choanopomops largillierti* (Pfeiffer, 1846)

*Diplopoma rigidulum* (Morelet, 1851)

*Halotudora kuesteri* (Pfeiffer, 1852)

# Pearly Tuba                                    ANNULARIIDAE

## *Halotudora kuesteri* (Pfeiffer, 1852)

**Height**:10-15 mm, Diameter: 12 mm

**Description**: Tuba-shape; lip widely reflected; shell with 5 whorls; perforate; with an operculum; shell thin, but solid, with faint broken color features, strongest on the lip; color of shell variable but most specimens found in Belize are a light buff; shell surface covered in rows of broken, roundish or pyramid–shaped projections, opposite page; last whorl rounded.

**Similar Species**: *Halotudora gruneri* is larger, has weakly developed transverse striae and a widely reflected lip; *C. largillierti* has conspicuous calcium nodules at the sutures.

**Habitat**: Empty shells are found at the base of limestone cliffs while live snails are usually on the face or undersides of those same cliffs.

**Status**: Common; a common species where found.

**Specimen**: On opposite page, figures (a-c) from Belize, Toledo District, Bladen Nature Reserve (Dourson collection) shows typical shell texture and figures (d and e) showing an atypical shell texture from Belize, Toledo District, 2 miles east of San Benito Poite (Dourson collection).

**Type Locality**: Listed only as Belize.

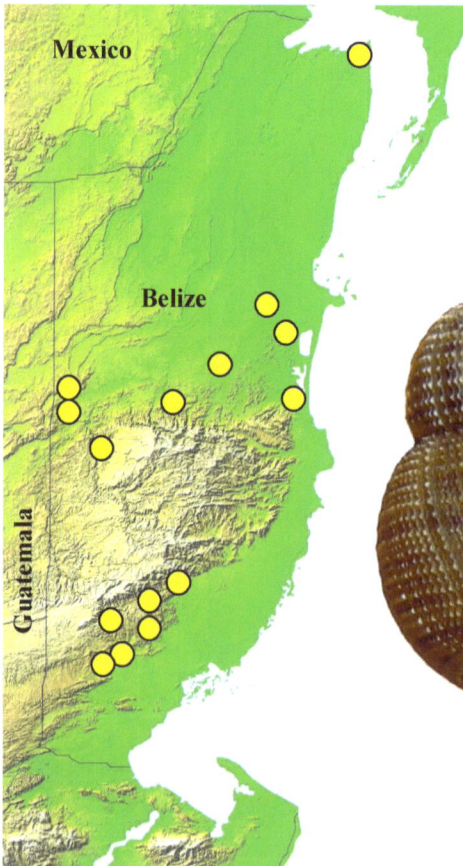

Figure (a) *C. kuesteri*    Figure (b) *C. kuesteri*

Two color morphs

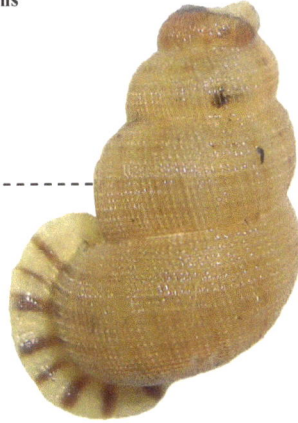

Figure (c) *C. kuesteri* (atypical form in Belize)

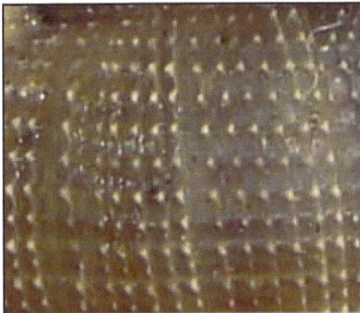

Figure (e)

213

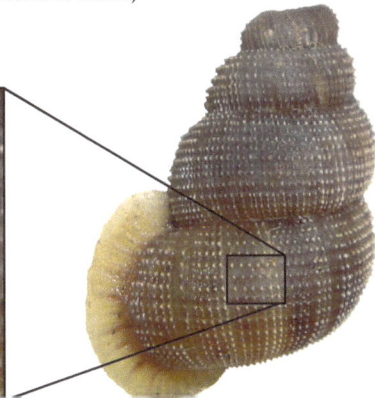

Figure (d) *C. kuesteri* (atypical form)

# Lipless Tuba                    ANNULARIIDAE

## *Paradoxipoma enigmaticum* Watters, 2014

**Height**:8-9.6 mm, Diameter: 4-5 mm

**Description**: Tuba-shape; lip simple or only slightly reflected in adults (a); shell with 5-6 whorls; openly perforate; with an operculum; shell thin, glossy; shell with faint broken color features on most of the whorls, transverse striae are well developed and closely spaced on all whorls; without spiral striae; last whorl bluntly angular (b).

**Similar Species**: *Halotudora kuesteri* is larger, has weakly developed transverse striae and a widely reflected lip, other ANNULARIIDAE in Belize have widely reflected lips; the angular whorls and lack of spiral sculpture on the base characterize this species.

**Habitat**: Known only from limestone outcrops in the vicinity of Gracie Rock.

**Status**: Rare, ENDEMIC to BELIZE; the species is known only from the Gracie Rock area. *Paradoxipoma enigmaticum* was formally described based on shell morphology.

**Specimen**: Belize, Belize District, Gracie Rock Paratypes UF 479322. (images by G. Thomas Watters).

**Type Locality**: Gracie Rock, Belize.

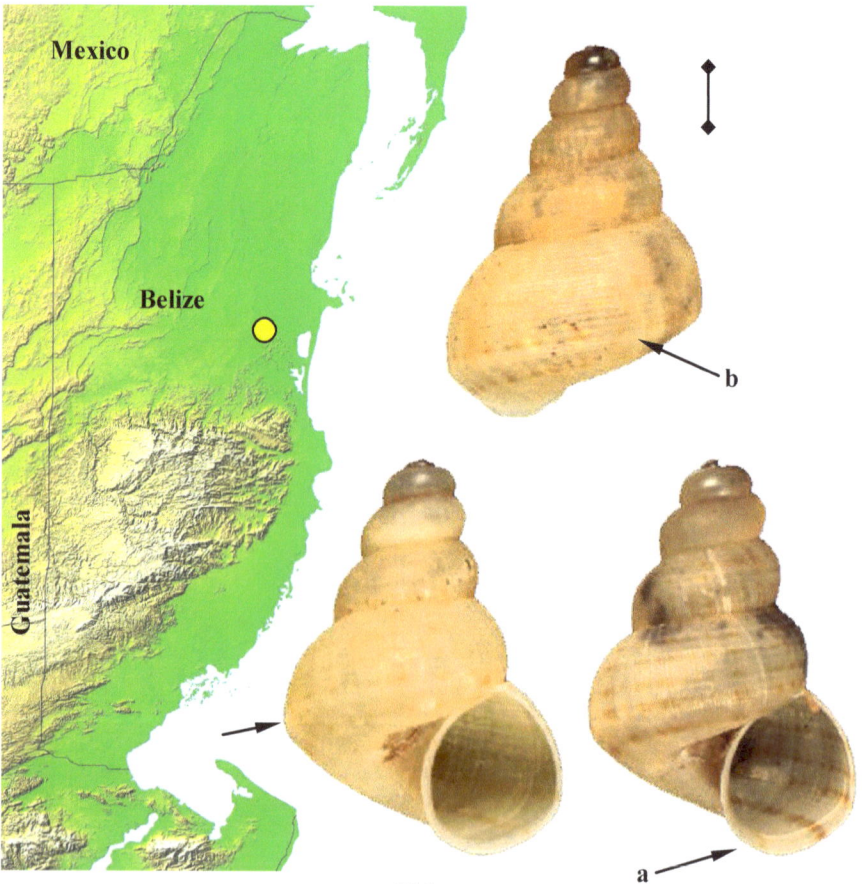

# Amber Tuba                                   ANNULARIIDAE

*Parachondria rubicundus* (Morelet, 1849)

**Height**:10-13 mm, Diameter: 5 mm

**Description**: Tuba-shape; lip reflected; shell with 5-6 whorls; perforate; with an operculum; shell thin, but rather solid, glossy; shell with faint broken color features on most of the whorls, <u>color of shell a light red or orange</u>, transverse striae are well developed and closely spaced on all whorls; without notable spiral striae; last whorl rounded.

**Similar Species**: *Halotudora kuesteri* is larger, has strongly developed but broken, transverse striae and a more widely reflected lip, *H. gruneri* is larger and has more notable spiral striae; *D. rigidulum* is smaller and has a lip that grows free from the last whorl; *C. largillierti* has conspicuous calcium nodules at the sutures.

**Habitat**: Empty shells are found at the base of limestone cliffs while live snails are usually on the face or undersides of same cliffs.

**Status**: Common, the species likely occurs throughout most of Belize.

**Specimen**: Belize, Toledo District, Bladen Nature Reserve (Dourson collection).

**Type Locality**: Petén, Vera Paz, Guatemala.

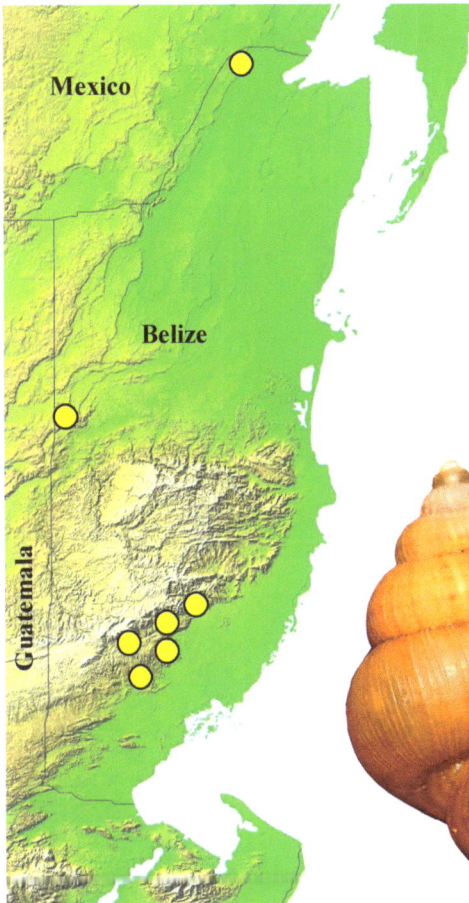

215

# Short Tuba                                    ANNULARIIDAE

*Tudorisca andrewsae* (Ancey, 1886)

**Height**:12 mm, Diameter: 9 mm

**Description**: Tuba-shape, <u>shell stature short</u>; lip reflected; shell with 4.5 whorls; like all ANNULARIIDAE in Belize, the shell is perforate (a); with an operculum; shell thin but solid, with faint broken color features on all whorls; well developed transverse and crossing spiral striae; last whorl roundish; female and male shells exhibit different shell morphology, female shells being notably taller; also note the drill hole (b) made by some unknown predatory invertebrate.

**Similar Species**: *Halotudora kuesteri* is larger, has weakly developed transverse striae and a widely reflected lip, *H. gaigei* is taller and more pointed in form, *C. largillierti* has conspicuous calcium knob-like nodules at the sutures.

**Habitat**: Found in dry tropical dry forests covering calcareous soils.

**Status**: Found throughout the Yucatan and Honduras; in Belize, the species has been found from the Sartenaja region only.

**Specimen**: Mexico, Quintana Roo, 6.4 km. E Xpujil,UF 19130.

**Type Locality**: Utila Island, Honduras.

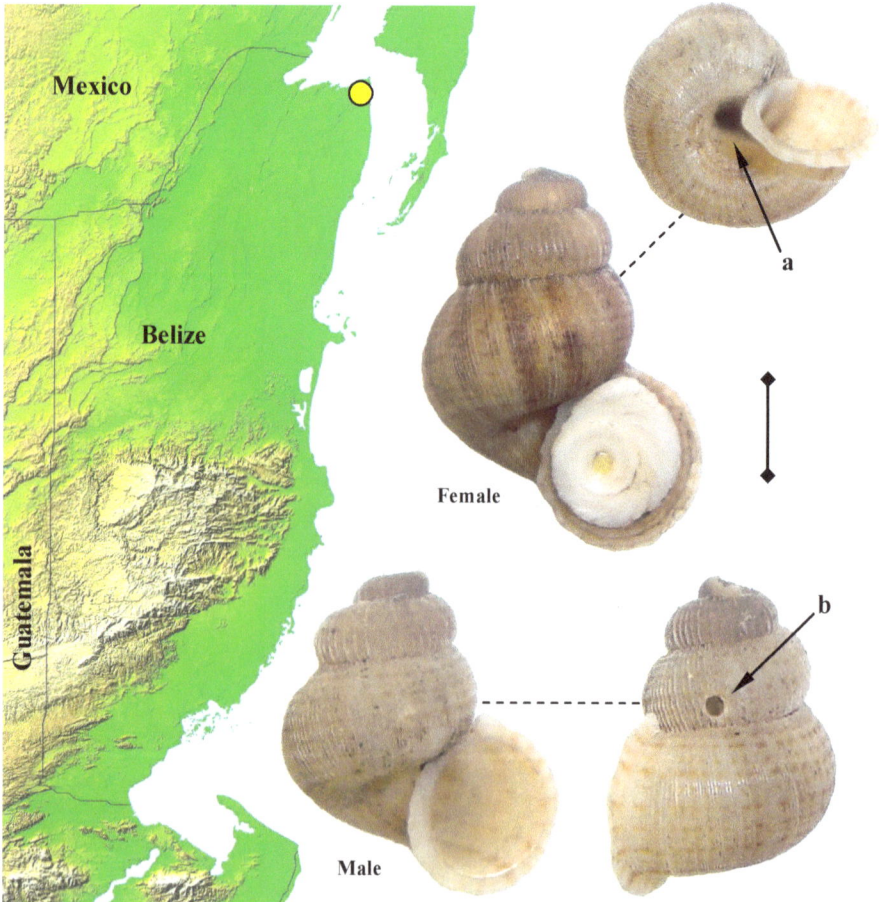

Female

Male

216

# Knobby Tuba                    ANNULARIIDAE

*Choanopomops largillierti* **(Pfeiffer, 1846)**

**Height**:15 mm, Diameter: 6-7 mm

**Description**: Tuba-shape; lip reflected; shell with 5 whorls; perforate; with an operculum; shell thin, but rather solid, with faint broken color features, strongest on the lip; color of shell variable but most specimens found in Belize are a khaki; there are well developed transverse striae on all whorls, which at the sutures, form heavy knob-like calcium deposits (a); spiral striae weaker but present on all whorls; last whorl rounded.

**Similar Species**: *Halotudora kuesteri* is around the same size but has a more widely reflected lip and lacks the well developed knob-like deposits at the sutures, *H. gaigei* is taller and more pointed in form.

**Habitat**: Found in rather dry tropical forests that cover calcareous soils in Northern Belize.

**Status**: Uncommon; appears to be restricted to northern Belize.

**Specimen**: Belize, Orange Walk District, Shipstern Nature Reserve. (Dourson collection).

**Type Locality**: Yucatan, Mexico.

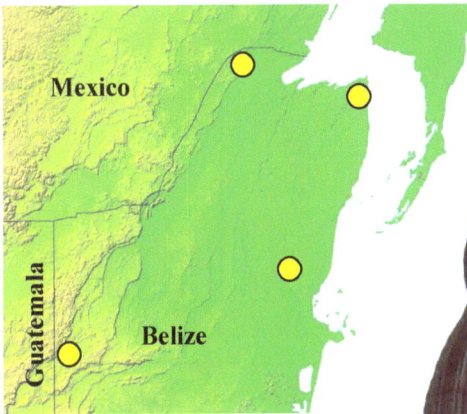

Shell enlarged to show detail of knob-like calcium deposits at the sutures

217

# Detached Tuba                    ANNULARIIDAE

## *Diplopoma rigidulum* (Morelet, 1851)

**Height**:10-12 mm, Diameter: 5 mm

**Description**: Tuba-shape; lip reflected; shell with 4.5 whorls; perforate; with an operculum; shell thin, with faint broken color features on all whorls; shell fulvous; well developed transverse striae or ribs that are sharp, pointed (a), expanding in size at the sutures; without spiral striae; last whorl roundish, hanging free from the shell body (b).

**Similar Species**: *Halotudora kuesteri* is larger, has broken transverse striae and a widely reflected lip, *H. gaigei* is larger and wider in form, the last whorl remaining attached to the body of the shell; *C. largillierti* has conspicuous calcium knob-like nodules at the suture (not blade–like as in *D. rigidulum*).

**Habitat**: Empty shells are found at the base of limestone cliffs while live snails are usually on the face or undersides of those same cliffs.

**Status**: Common and widespread in all but northern Belize.

**Specimen**: Belize, Toledo District, Bladen Nature Reserve (Dourson collection).

**Type Locality**: San Luis, Guatemala.

218

# ANNULARIIDAE Shells of Belize (proportionate)

*Halotudora gruneri* (Belize)

*Halotudora kuesteri,* two color morphs (Belize)

10 mm

*Choanopomops largillierti,* two color morphs (Belize)

*Tudorisca andrewsae,*
Shipstern Nature Preserve (Belize)

*Paradoxipoma enigmaticum*
(Belize)

*Parachondria rubicundus*
(Belize)

*Halotudora gulgel* (Belize)

*Diplopoma rigidulum*
(Belize)

The Detached Tuba, *Diplopoma rigidulum* (Morelet, 1851)

# 10 Shells Taller than Wide, Greater than 5 mm Tall, Barrel-Shape

This section includes adult land snails taller than wide and greater than 5 mm tall, having a distinctive barrel-shape. This grouping of land snails are without an operculum. While some genera have rather smooth shell surfaces, others have beautiful ribbed sculptures. Two representative species from the family are illustrated to the right below.

## The Families and Genera Included in Chapter 10

UROCOPTIDAE

*I hope there's no salt in here!*

### *Eucalodium* Locomotion

Although the thick shell of the Belize Drum, *Eucalodium belizensis,* provides a safe haven for the animal within from most predators such as bird's, the shell can also be a burden. Other land snails have a nearly proportionate shell to body weight ratio; their shells being lighter or equal to their body weight allowing them to crawl more fluidly. The Belize Drum, however, is outfitted with a shell that is substantially heavier than its own body weight and as a result, moves more awkwardly along; dragging its shell behind in short stop-and-go stages. First, it moves its body forward to an anchor spot followed by pulling its load to that same spot (similar to an inch worm). Below, the shell remains stationary in (figures 1, 2, and 3) and as the snail retracts into its shell (figures 4 and 5), the shell is pulled forward; this action repeated until the snail reaches its destination.

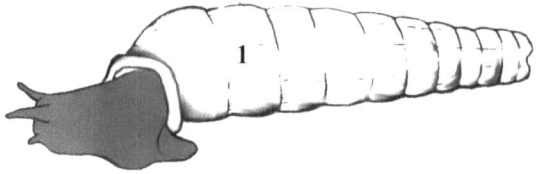

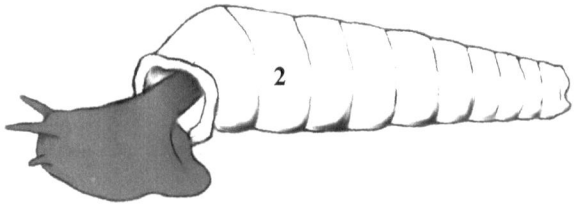

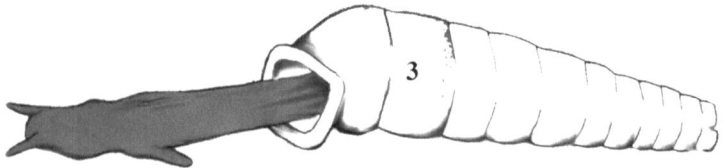

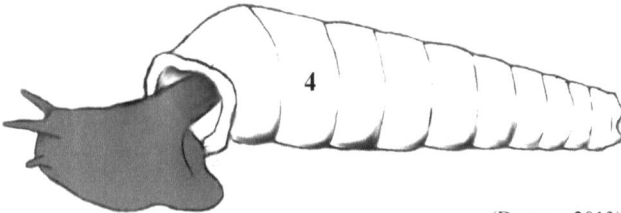

(Dourson 2013)

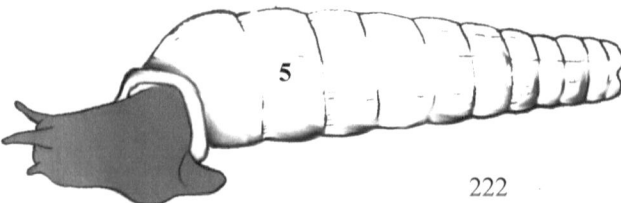

# Family UROCOPTIDAE Pilsbry, 1898

**DISTRIBUTION:** West Indies, Mexico, Guatemala, Honduras and Belize.

**TAXONOMY:** Land snails found in the family UROCOPTIDAE display a staggering conchological and taxonomic diversity. In general, the group is based on the presence of one or more of the following shell characters: 1) a high-spired shell with many narrow whorls 2) the presence of columellar sculpture 3) the breaking off or decollation (illustrated below) of the early whorls 4) a circular or squarish aperture with an expanded or reflected peristome or lip (Uit de Weerd 2008). Shells in this family are generally barrel-shaped with shell sculpture ranging from nearly smooth to strongly ribbed, most exhibiting intricate raised striae and often found living in close proximity to calcium or calcium rich soils. UROCOPTIDAE can be especially abundant around limestone outcroppings, sometimes found climbing rock features in wet weather. More than 500 species have been documented in Cuba alone, about one-third of the total number of land snail species known from the island. Central America including Mexico is also extremely rich in UROCOPTIDAE, the number of species yet to be determined as many new specie continue to be added to an ever-growing list. Snails of the *Eucalodium* genera, however, are exceedingly rare, known only from the type locality.

Five genera that includes a total of twelve described species and one undescribed species have been documented in Belize and are included here.

## The Genera Included in this Section:
(in order of appearance in text)

*Brachypodella*
*Epirobia*
*Coelocentrum*
*Eucalodium*
*Microceramus*

Decollation of shell

# Key to the *Brachypodella* of Belize

## 1) Riblets rounded and low or entirely absent on most whorls, only raised and sharp-pointed on final whorls

A) Whorls flat or very slightly convex; sutures shallow; shell smooth or nearly so except for the last whorl near the lip. ***B. levisa***

> This is the smoothest of the *Brachypodella* species in Belize

B) Whorls notably convex; 12 to 14 mm long with 16-21 whorls in truncate or 23 whorls in entire shells. ***B. subtilis***

> Differs from *B. morini*, *B. speluncae* and *B. dubia* in the much weaker riblet sculpture

## 2) Riblets sharp-pointed and raised on nearly all whorls except protoconch

A) Whorls very convex with crowded 25 to 30 or more riblets on the penult whorl; 11-15 mm long, with 15-18 whorls in truncate, 15-16 mm long with 22-24 whorls in entire shells. ***B. morini***

> Most similar to *B. speluncae* but with more crowded riblets, less reflected lip and more convex whorls; differs from *B. subtilis* in its strong sculpture

B) Whorls moderately convex with widely-spaced, 14-16 riblets on penult whorl; entire shell 15-20 mm long, with 12-14 whorls in truncate, 14-20 mm with 22 whorls in entire shells ***B. speluncae***

> Has stronger riblets that are more widely spaced than *B. subtilis* or *B. morini*, and less closely-spaced riblets than *B. dubia*

C) Whorls somewhat squared or flattened with moderately crowded, sharp-pointed riblets, 20 or more riblets on the penult whorl; 10-16 mm; 15-16 mm. long with 19-23 whorls in entire shells. ***B. dubia***

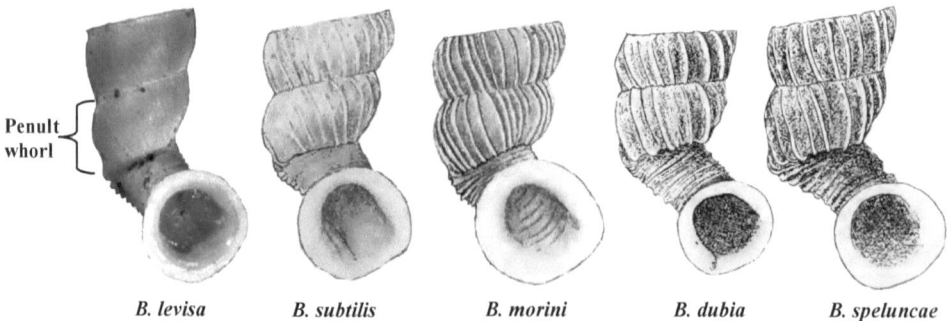

Penult whorl

*B. levisa*      *B. subtilis*      *B. morini*      *B. dubia*      *B. speluncae*

*B. levisa* image from (Dourson); B. *subtilis, B. morini* illustrations from Pilsbry 1904. *B. dubia* and *B. speluncae* illustrations from Pilsbry 1891.

## Brachypodella Species Compared

Whorls flat — *B. levisa,* Belize

*B. subtilis,* Belize

Whorls convex — *B. morini,* Guatemala

*B. speluncae,* Belize

5 mm

*B. dubia,* Mexico

225

**Brachypodella speluncae, Syntype**
**NHMUK 1893.2.4.338-340**

**Top view**

The Mayan Quill, *Brachypodella speluncae*, is reported to be used by the Maya to treat whooping cough. "Tu tut" (which broadly means snail in Kekchi Maya) are roasted live on a comal (a flat griddle) and mixed with tiny black stones gathered from clay deposits. The mixture is then ground to a powder, mixed with warm water and consumed to cure the uncontrolled whooping cough (pers. comm. Santiago Coc, 2009, San Pedro Columbia, Belize).

# Mayan Quill                    UROCOPTIDAE

*Brachypodella speluncae* Morelet, 1852

**Height**:15-17 mm, Diameter: 3 mm

**Description**: Cylinder-shape; lip reflected; 23 whorls; imperforate; shell thin and fragile; internal column is solid (a); without color features, shell color ivory; except for the protoconch, all whorls with well spaced, sharp-pointing, high riblets; without any notable spiral striae; last whorl squared and leaving the main shell body and extending outward (b).

**Similar Species**: *Brachypodella levisa* is of the same size and body form but without the distinguished ribs of *B. speluncae* excluding the last liberated whorl which is ribbed; *B. subtilis* is shorter and has a higher number of ribs in each whorl; it is not uncommon to find *B. speluncae, B. levisa* and *B. dubia* sympatrically.

**Habitat**: A species of shaded limestone rock faces and large boulders.

**Status**: A relatively common species in Belize.

**Specimen**: Belize, Toledo District, Bladen Nature Reserve (Dourson collection).

**Type Locality**: Cave Jovitsinal, Guatemala.

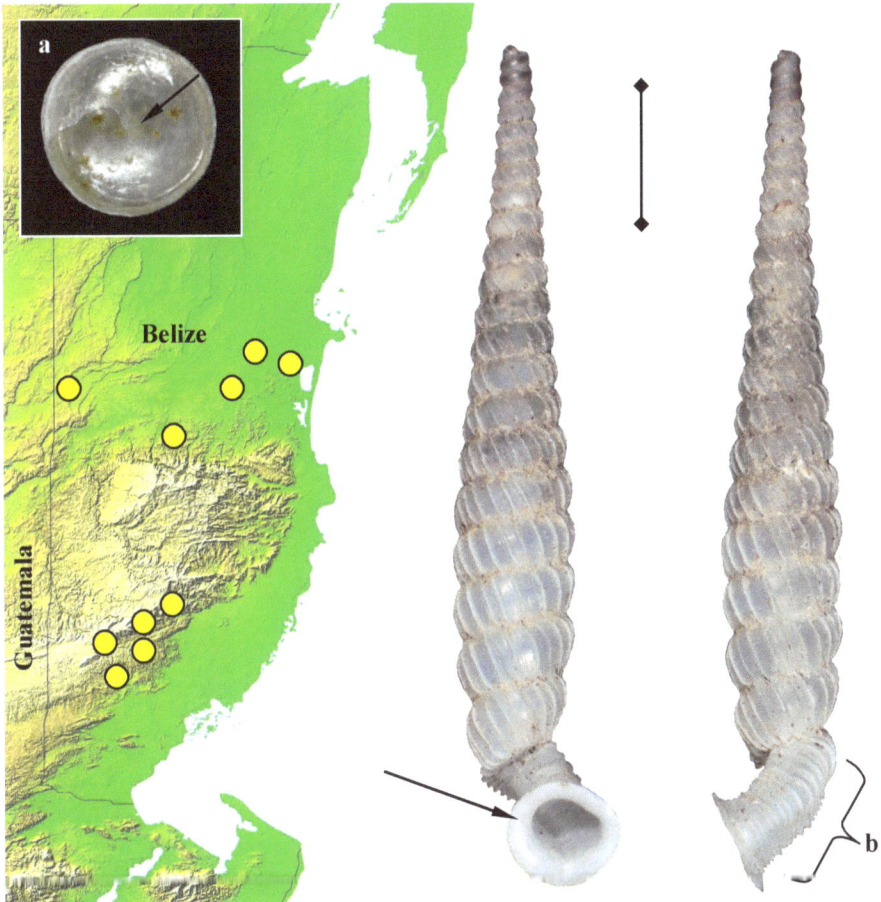

# Smooth Quill                              UROCOPTIDAE

## *Brachypodella levisa* (new species)

**Height**:15-22 mm, Diameter: 2-3 mm

**Description**: Cylinder-shape; lip reflected; 23 whorls; imperforate; shell thin and fragile; sutures shallow (not deep as in *B. speluncae*); shell color ivory; all whorls nearly smooth except for the last independent whorl which has riblets; without any notable spiral striae; the last ribbed whorl squared and leaving the main shell body and extending outward.

**Similar Species**: Most similar to *Brachypodella speluncae* (page 227) but has a smooth shell; *B. speluncae* is heavily ribbed; and the sutures in *B. levisa* are not as deep as in *B. speluncae*. The two species are occasionally found together with no intermediates or intergrades observed.

**Habitat**: A species of shaded limestone rock faces and large boulders covered in tropical rainforests.

**Status**: Rare, ENDEMIC to BELIZE known only from a few locations in southern Belize in the Maya Mountains.

**Specimen**: Belize, Toledo District, BNR (Dourson collection). Holotype UF 505443, (not pictured). Paratypes UF 505444 from same location, (not pictured).

**Type Locality**: Bladen Nature Reserve, Toledo District, Belize(16.552130° N, -88.719695°W).

**Etymology**: Levisa is Latin for smooth.

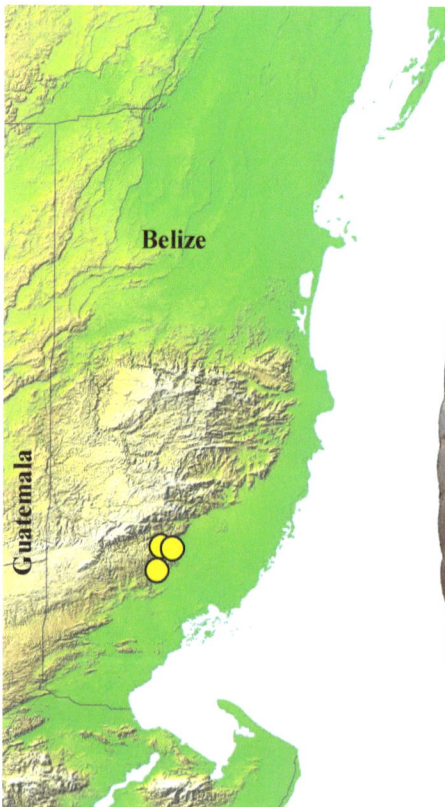

228

# Pigmy Quill                                    UROCOPTIDAE

## *Brachypodella subtilis* (Morelet, 1849)

**Height**:12-15 mm, Diameter: 2.2 mm

**Description**: Cylinder-shape; lip reflected; complete shell with 18-20 whorls; without top a few less whorls (a); imperforate; as with all *Brachypodella* species, internal column is solid (page 235); shell thin and fragile; suture deep; shell cream color; except for the protoconch, all whorls with numerous low, roundish riblets; without any notable spiral striae; the last whorl rounded and leaving the main shell body and extending outward.

**Similar Species**: *B. speluncae* is taller having a lower number of more prominent ribs per whorl.

**Habitat**: A species of shaded limestone rock faces and large boulders covered in tropical rain forests; live individuals are typically found adhering to the rock faces and small boulders while shells are found at the base of same.

**Status**: Uncommon to common where it occurs; found in scattered locations in Belize.

**Specimen**: Figure (b) *Brachypodella subtilis*, Syntype NHMUK 1893.2.4.335-337 and figure (c) from Belize, Toledo District, Punta Gorda, Big Hill UF 00135226.

**Type Locality**: Petén, Guatemala.

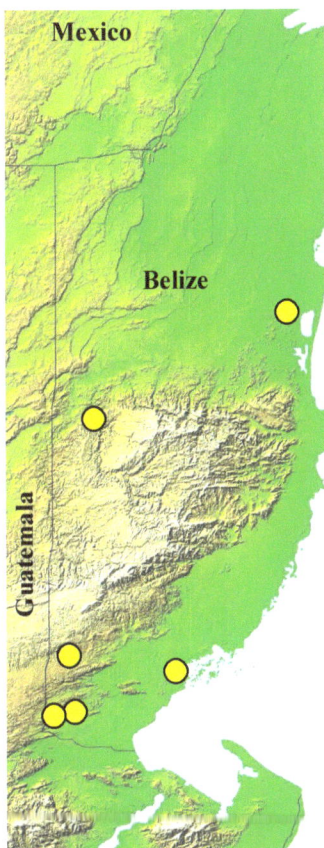

229

# Dubious Quill                    UROCOPTIDAE

## *Brachypodella dubia* (Pilsbry, 1891)

**Height**: 10-16 mm with full spire, 10-12 mm without, Diameter: 2 mm

**Description**: Cylinder-shape; lip reflected; 19-23 whorls in mature shells with complete spires, without spire 11-13 whorls below apical plug; whorls are squared not rounded; imperforate; except for the protoconch, all whorls with high and sharp-pointed riblets; the shell color a light beige (Thompson 1967); as in all *Brachypodella* species the internal column is solid (see page 235), this feature can be seen by breaking the shell open and viewing with a 10X hand lens.

**Similar Species**: Differs from *B. speluncae* by its low number of whorls below the apical plug, having a lower number of ribs per whorl, by its strong low axial ribs, its elongate conical shape and by its narrow peristome or lip (Thompson 1967).

**Habitat**:; Limestone boulders and cliff faces.

**Status**: Common where outcropping limestone occurs.

**Specimen**: Mexico, Campeche, 5.1 mi W Tikinmul, UF 19321.

**Type Locality**: Labna, Maya archeological site, Mexico.

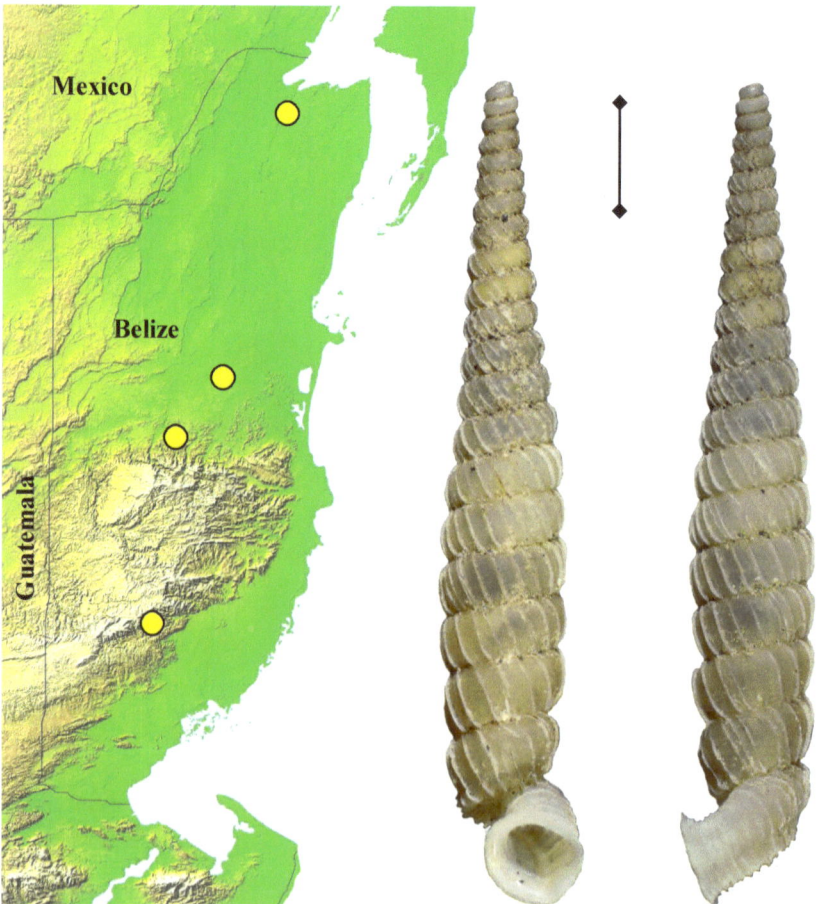

# Mountain Quill                    UROCOPTIDAE

## *Brachypodella morini* (Morelet, 1849)

**Height**: 12-15 mm tall

**Description**: Cylinder-shape, the upper half tapering to a narrow truncation or an entire apex, thin, corneous, with whitish riblets; surface lusterless, sculptured with oblique, hardly arcuate, thread-like riblets, parted by spaces of three or four times their width, and usually 25-30 in number on the penult whorl; whorls very convex, the last free, projecting forward, swollen at the periphery, strongly carinate below, concave above the keel; aperture sub vertical, rounded -ovate, the outer margin being a little pulled out (Pilsbry 1904).

**Similar Species**: Differs from *B. speluncae* by its smaller size and by its more crowded and lower riblets.

**Habitat**: A species of shaded limestone rock faces and large boulders covered in tropical rain forests.

**Status**: Not yet reported from Belize but known from the neighboring country of Guatemala.

**Specimen**: Guatemala, mountains of Cavech River, UF 159009.

**Type Locality**: Vera Paz, Guatemala

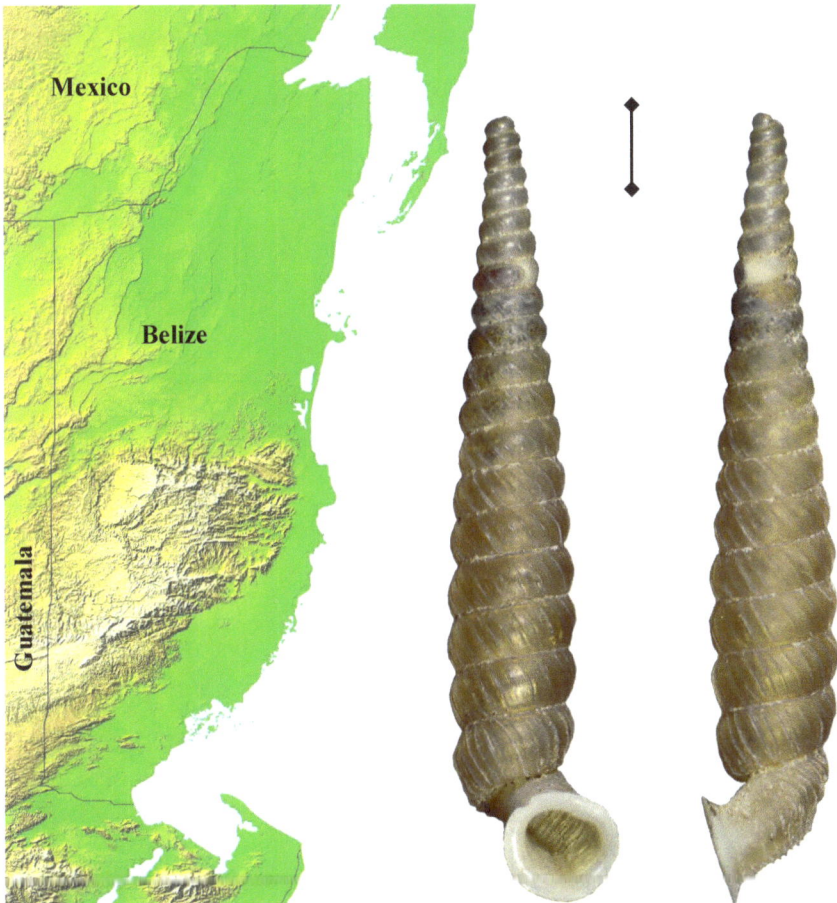

# Coban Quill                                    EPIROBIIDAE

## *Epirobia polygyrella* (Von Martens, 1863)

**Height**: 15 mm, Diameter: 3mm

**Description**: Cylinder-shape; lip reflected; 19-25 whorls in mature shells with complete spires. 11-13 whorls below apical plug without; unlike the center column of *Brachypodella* which are solid, the center column of *Epirobia* species are hollow (a).

**Similar Species**: Differs from *Brachypodella* primarily by its open center column (a), whereas in *Brachypodella* it remains solid (b).

**Habitat**: The species occurs in karst hills on shaded limestone ledges under broadleaf tropical rainforest.

**Status**: Although *Epirobia* species have not been reported from Belize, the genus is expected to occur in suitable habitat in southern portions of the country (pers. comm. Fred Thompson 2012). Specimens and images are provided here in the event *Epirobia* is discovered in Belize.

**Specimen**: Guatemala, Alta Verapaz Prov. 15 km. by road N of Coban, 1050 m, UF 189905.

**Type Locality**: Coban, Guatemala.

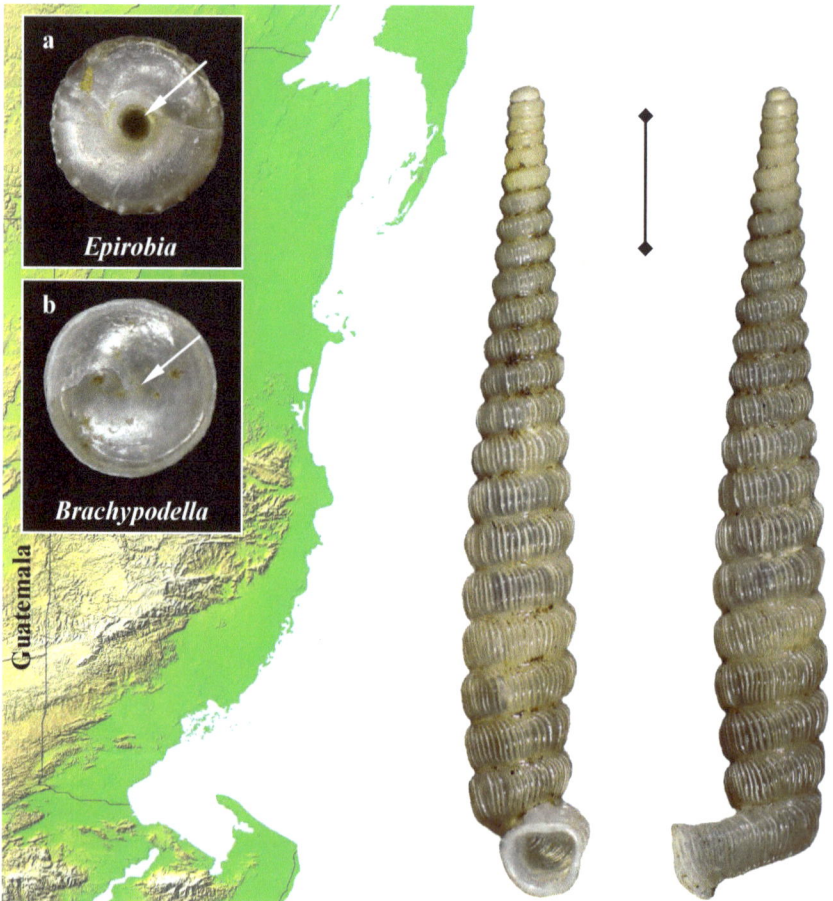

**Painted Quill,** *Epirobia longior* **Thompson, 2012**
**Honduras, north slope of Cerro Santa Barbara Mountain, 1700**
**meters, Holotype UF 242712**

# Limestone Drum                    UROCOPTIDAE

## *Coelocentrum fistulare* (Morelet, 1849)

**Height**: 28 mm (without top), Diameter: 7-10 mm

**Description**: Cylinder-shape; <u>shell widest in the middle</u>; lip reflected; shell without top, 11-12 whorls; top (when present) has around 15 whorls; minutely perforate (opposite page); shell thin but solid; <u>internal axis (column) is hollow (a) and ribbed (opposite page)</u>, a feature used to separate it from *Eucalodium,* which has a solid column (b); suture moderately impressed; shell with a fine gloss and bronze color; well developed transverse striae on all whorls; without any notable spiral striae; the last whorl roundish.

**Similar Species**: Blue Creek Drum, *Coelocentrum* species, is smaller but is widest at its base not in the center of the shell as in *C. fistulare.*

**Habitat**: Found sparsely under leaf litter on limestone hillsides becoming more frequent on talus slopes of limestone.

**Status**: Common, a relatively frequent species of southern Belize.

**Specimen**: Belize, Toledo District, Bladen Nature Reserve (Dourson collection).

**Type Locality**: Petén, Guatemala.

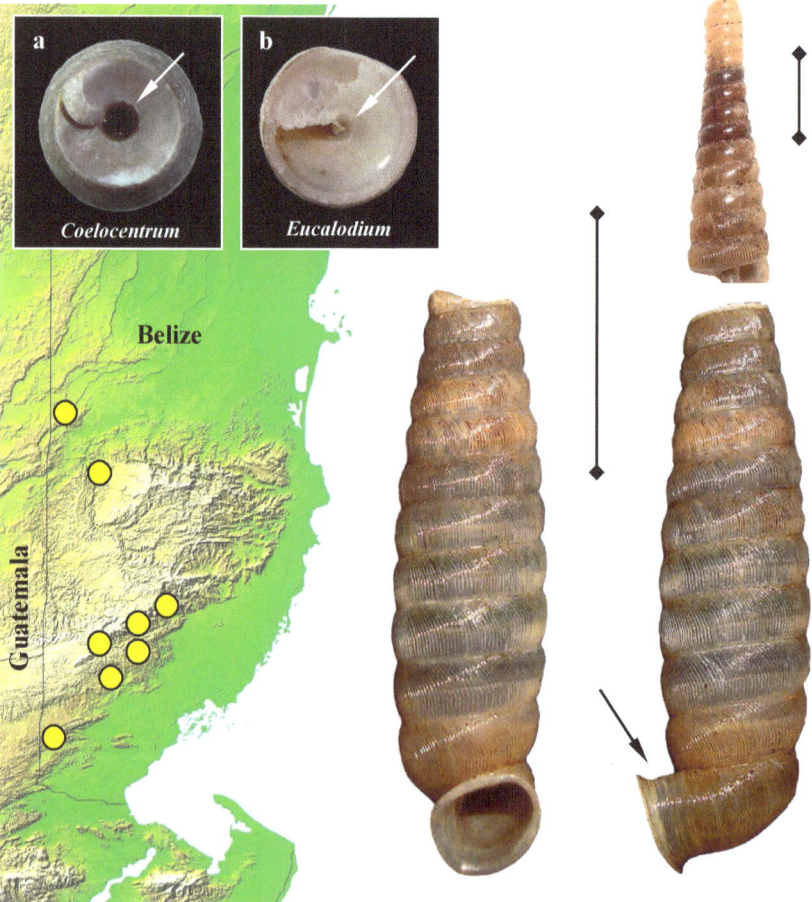

234

*Coelocentrum fistulare* (Morelet, 1849) Syntype NHMUK 1893.2.4.326

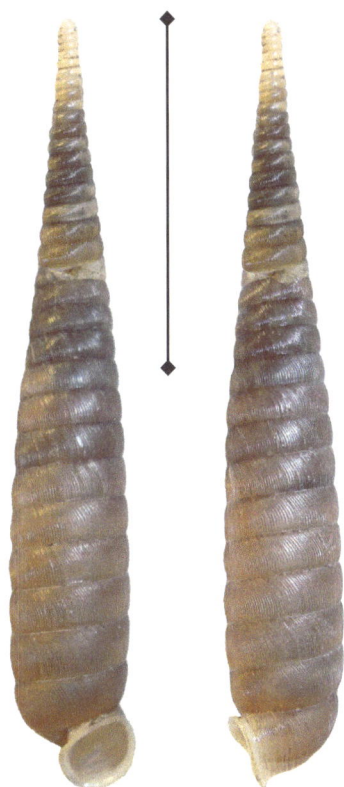

Decollation of shell
enlarged and showing
the mostly smooth tip

**Fine transverse striae**

**Umbilicus perforate**

**Internal column ribbed**

*Coelocentrum fistulare* showing details of various internal and exterior shell features. Rio Frio Cave, Cayo District, Belize.

# Blue Creek Drum                    UROCOPTIDAE

### *Coelocentrum* species (undetermined)

**Height**: 24 mm (without top), Diameter: 8-10 mm

**Description**: Cylinder-shape; <u>shell widest at the bottom</u>; lip reflected; shell without top, 10-12 whorls; top (when present) has around 18 whorls; perforate (a); shell thin but solid; internal axis is hollow and ribbed as in *C. fistulare*; sutures moderately impressed; shell with a fine gloss and bronze color; well developed transverse striae on all whorls; without any notable spiral striae; the last whorl roundish.

**Similar Species**: Similar to *C. fistulare* but has a wider umbilicus, less protruding lip and is widest at it base, not the center of the shell as in *C. fistulare*.

**Habitat**: Found sparsely on a talus slope of limestone behind the cabins and dining room at International Zoological Expeditions (IZE) at Blue Creek Cave where it occurs with *C. fistulare*.

**Status**: Rare, currently known only from a hillside directly behind the cabins at Blue Creek Cave in southern Belize. <u>Additional specimens are needed to make a final determination</u>.

**Specimen**: Belize, Toledo District, Blue Creek Cave area (Dourson collection).

# UROCOPTIDAE Shells Compared (Proportionate)

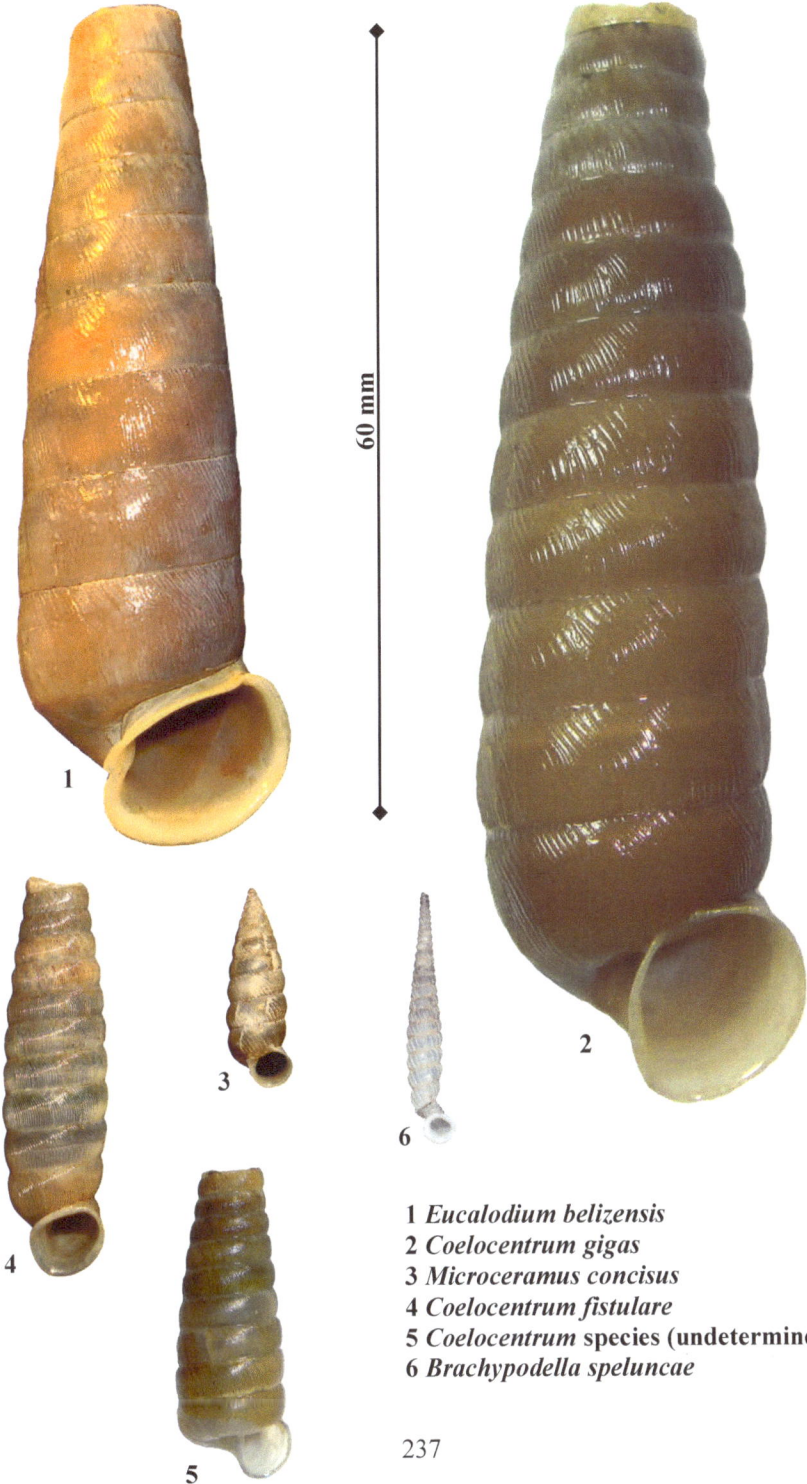

60 mm

1 *Eucalodium belizensis*
2 *Coelocentrum gigas*
3 *Microceramus concisus*
4 *Coelocentrum fistulare*
5 *Coelocentrum* species (undetermined)
6 *Brachypodella speluncae*

237

# Colossal Drum                    UROCOPTIDAE

*Coelocentrum gigas* **Martens, 1897**

**Height**: 65-95 mm (without top), Diameter: 15 mm

**Description**: Cylinder-shape; shell widest near the bottom; lip slightly reflected; shell without top having 13-15 whorls; the detached top (a) having around 15 whorls; the entire shell reaching a length of more than 115 mm; rimate shell well constructed and heavy-duty; the shell's internal axis is hollow (b); sutures moderately impressed; shell with a fine gloss and copper color; well developed transverse striae on all whorls; without any notable spiral striae; the last whorl roundish.

**Similar Species**: Similar to *C. fistulare* but much larger; *Eucalodium* species have a hollow axis, (b) (not solid as in *Coelocentrum*, (c).

**Habitat**: Found under leaf litter wet limestone of hillsides sheltered in tropical jungle.

**Status**: Not yet reported from Belize but known at the border with Guatemala.

**Specimen**: Guatemala, 1 mile west of Lagunita Creek Village, UF 463282.

**Type Locality**: Livingston, Guatemala.

# Obese Drum                                    UROCOPTIDAE

*Coelocentrum badium* Pilsbry (date unknown)

**Height**: 70-75 mm (without top), Diameter: 15-18 mm

**Description**: Cylinder-shape; shell widest near the bottom; lip slightly reflected; shell without top having 8-10 whorls; rimate; shell well constructed and thick; the shell's internal axis is hollow (a); sutures moderately impressed; shell with a fine gloss and copper color; well developed transverse striae on all whorls; without any notable spiral striae; the last whorl roundish.

**Similar Species**: Similar to *C. gigas* but shorter with fewer whorls; differs from *Eucalodium* species by having a hollow axis.

**Habitat**: Found under leaf litter covering wet limestone of hillsides sheltered in tropical jungle.

**Status**: Not yet reported from Belize but close to the country's southern border around Livingston, Guatemala.

**Specimen**: Guatemala, Izabal Dept. near the town of Livingston, mountains near Cavech, UF 34329.

**Type Locality**: Livingston, Guatemala.

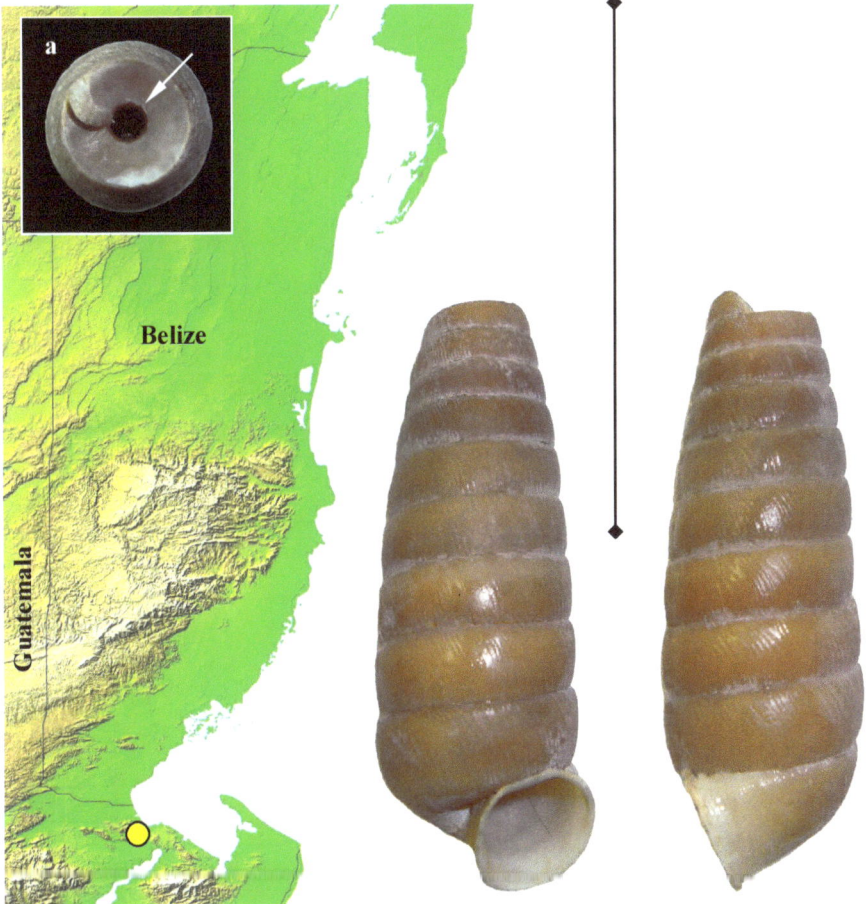

# Mayan Drum                                    UROCOPTIDAE

*Eucalodium belizensis* **Thompson & Dourson, 2013**

**Height**: >75 mm entire shell, 50-60 mm without top, Diameter: 15 mm

**Description**: Cylinder-shape; shell widest at the bottom; lip reflected; shell without top having around 10 whorls; the lost top (a) having around 15 whorls; the entire shell reaching a length of more than 75 mm; rimate; shell well constructed and thick; the shells internal axis is solid (b); suture shallow; shell with a dull gloss and rust to rosy color; live animal white to flesh; well developed transverse striae on all whorls; last whorl roundish with a slight angle (c).

**Similar Species**: Similar to *Coelocentrum gigas* but smaller, lighter in color and most importantly has a solid axis not hollow as in *Coelocentrum*.

**Habitat**: Found in tropical rainforests, under leaf litter covering karst foothills.

**Status**: Rare, ENDEMIC to BELIZE known only from the type locality; less than a dozen shells have been found and one live specimen. *Eucalodium belizensis* was formally described based on shell morphology.

**Specimen**: Belize, Toledo District, 5 miles west of San Jose, UF 449720.

**Type Locality**: San Jose, Belize.

*Eucalodium belizensis,* Holotype UF 44972

Known only from the type locality in southern Belize, the Mayan Drum, *Eucalodium belizensis*, is one of the rarest land snails on Earth. Holotype specimen illustrated above, below image of a live individual on limestone rock.

# Land Dart                                    UROCOPTIDAE

## *Microceramus concisus* (Morelet, 1849)

**Height**: 10-15 mm, Diameter: 6-7 mm

**Description**: Cylinder-shape; shell widest at the bottom; lip slightly reflected; mature shell have around 12 whorls; rimate; aperture roundish; shell thin but not fragile; height of the shell varies greatly as can be seen below; sutures moderately impressed; shell with a dull gloss, frequently with contrasting lighter and darker streaks; well developed transverse striae on all whorls; without any notable spiral striae; last whorl roundish.

**Similar Species**: Most similar to *Microceramus kieneri* in form but slightly smaller with a lighter color and rounder aperture.

**Habitat**: Found at the base of limestone rock ledges and under leaf litter covering karst foothills.

**Status**: Common, occurs throughout Belize.

**Specimen**: Belize, Toledo District, Bladen Nature Reserve (Dourson collection).

**Type Locality**: Yucatan, Mexico.

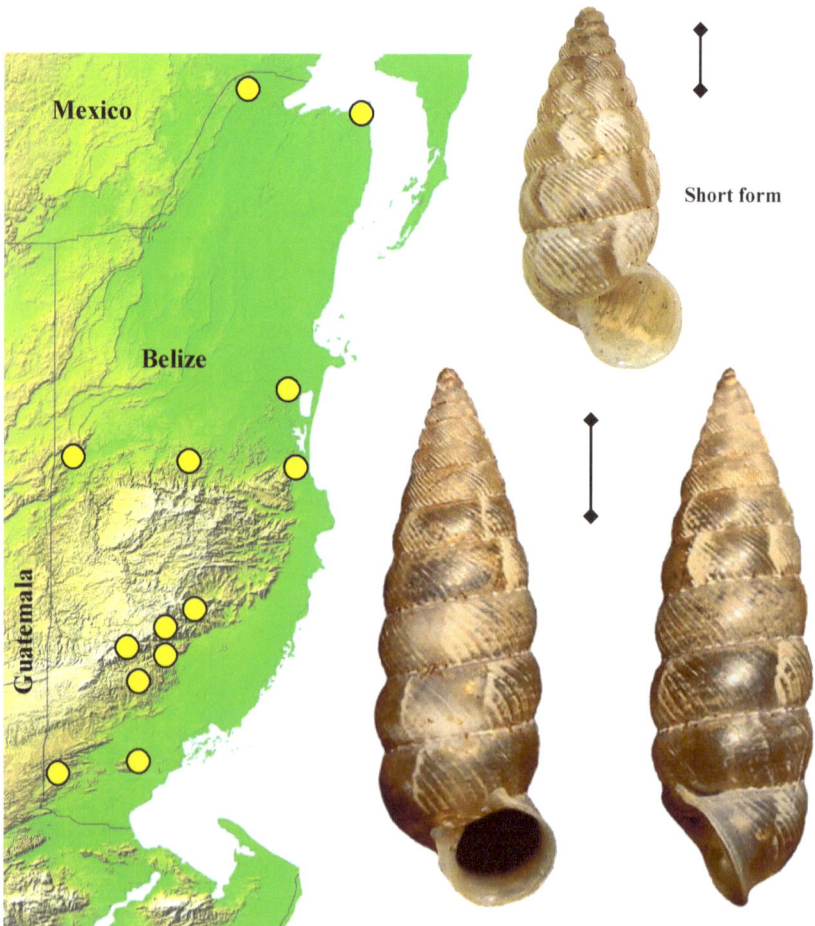

Short form

242

# Foothills Land Dart                    UROCOPTIDAE

*Microceramus kieneri* **(Pfeiffer, 1846)**

**Height**: 18 mm, **Diameter**: 6 mm

**Description**: Cylinder-shape; shell widest at the bottom; lip slightly reflected; mature shells have around 13 whorls; rimate; <u>aperture squared</u>; shell thin but not fragile; height of the shell varies greatly; sutures moderately impressed; shell with a dull gloss, marbled brown and white in color; <u>apex is darkened (a)</u>; well developed transverse striae on all whorls; without any notable spiral striae; the last whorl contains a revolving colored line that may appear raised a little from the surface and sharp like a delicate carina (c) (Bland 1883).

**Similar Species**: Similar to *M. concisus* but slightly larger, a different color and darkened apex (a) and a straight ledge on upper lip of aperture (b) illustration (Bland 1883).

**Habitat**: Under leaf litter of rocky limestone foothills.

**Status**: Uncommon, **ENDEMIC to BELIZE** reported from Gracie Rock, 10 miles southwest of Belize City off the Western Highway.

**Specimen**: Belize, Belize District, top of Gracie Rock, 10 mi SW of Belize, UF 260497.

**Type Locality**: "Honduras" Belize.

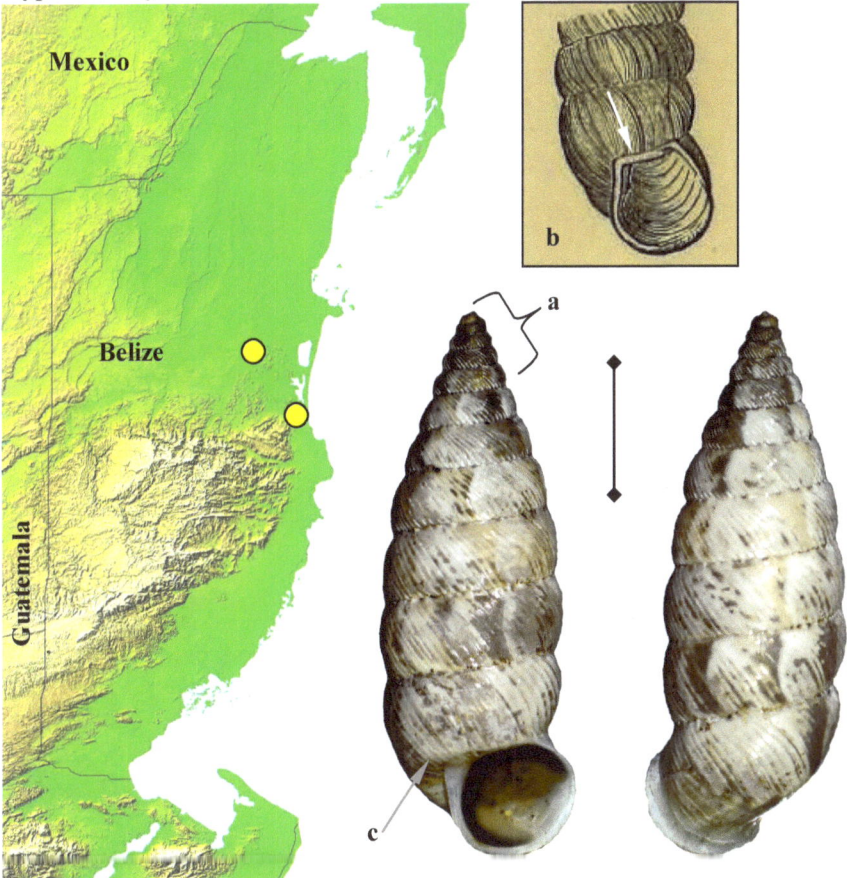

243

One of the more spectacular species in the family UROCOPTIDAE is the Saw-toothed Quill, *Callonia ellioti* (Poey, 1857). These ornate gastropods live on limestone cliffs and boulders, the specimen above from Prov. Pinar del Río, Sierra de Guane, Cuba.

# 11 Shells Wider Than Tall, Less Than 5 mm in Diameter

This section includes minute land snails that are wider than tall although some shells included here will be nearly equal in height. A few species have shells slightly larger than 5 mm. These tiny gems are usually found in leaf litter. One representative species from each family is illustrated to the right below.

## Families and Genera Included:

**Not proportionate**

# Fringed Phora                    THYSANOPHORIDAE
## *Thysanophora plagioptycha* (Shuttleworth, 1854)
**Diameter**: 2.5 mm, Height: 2.25 mm

**Description**: Heliciform; lip simple, aperture roundish to squared; shell with 4.5 whorls; perforate to umbilicate; shell surface dull, thin and fragile; transverse striae a weak feature with crossing diagonal fringes (a), scope required to see this fine detail; periphery rounded.

**Similar Species**: Most like *Thysanophora caecoides* in build and size but has a larger umbilicus and slightly smaller aperture; *T. villosus* has a lower shell profile, a wider umbilicus and long hairs instead of fringes.

**Habitat**: Karst hills and ravines where it is found under leaf litter and among rocky terrain.

**Status**: Common; currently known from the northern half of Belize especially in the lower elevation forests.

**Specimen**: Bahamas, Nassau, UF 251585.

**Type Locality**: Humacoa, Puerto Rico.

# Rimate Phora                    THYSANOPHORIDAE

## *Thysanophora caecoides* (Tate, 1870)

**Diameter**: 2 mm, Height: 2.5 mm

**Description**: Heliciform; lip simple, aperture roundish; shell with 4.5 whorls; perforate; shell surface rather glossy, corneous, thin and fragile; transverse striae a weak feature with crossing diagonal fringes (a) scope required to see this fine detail; periphery rounded.

**Similar Species**: *Thysanophora caecoides* varies little from *T. plagioptycha* in size, general build and sculpture but is readily distinguished by the smaller umbilicus; the umbilicus of *T. caecoides* is also notably smaller than other *Thysanophora* of Belize.

**Habitat**: Found in leaf litter of karst foothills rocky terrain.

**Status**: Common, found across Belize where habitat is present.

**Specimen**: Belize, Toledo District, Bladen Nature Reserve (Dourson collection).

**Type Locality**: Chontales, Nicaragua.

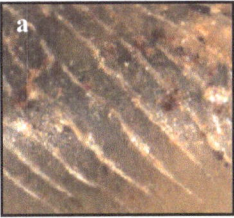

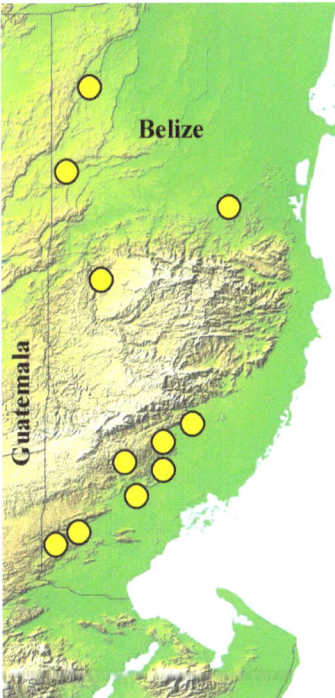

# Tall Phora                    THYSANOPHORIDAE

*Thysanophora rhoadsi* Pilsbry, 1919

**Diameter**: 3 mm, Height: 2.5-3 mm

**Description**: Heliciform, <u>shell as tall or taller than it is wide</u>; lip simple; aperture roundish, and in mature shells, with a slight reflection; shell with 5-6 whorls; perforate to umbilicate; shell surface dull, thin but solid; transverse striae a weak but a constant feature, may be smooth or nearly so on aged shells; periphery well rounded.

**Similar Species**: Most similar to *Thysanophora caecoides* in build but is slightly larger, has a higher shell and wider umbilicus.

**Habitat**: Limestone slopes.

**Status**: Rare, currently known only from the Bladen Nature Reserve, but also expected to occur elsewhere in Belize.

**Specimen**: Belize, Toledo District, a limestone slope above Quebrada de Oro Creek, GPS location: 0315327E, 1830584N. Bladen Nature Reserve (Dourson collection).

**Type Locality**: Gualan, Guatemala.

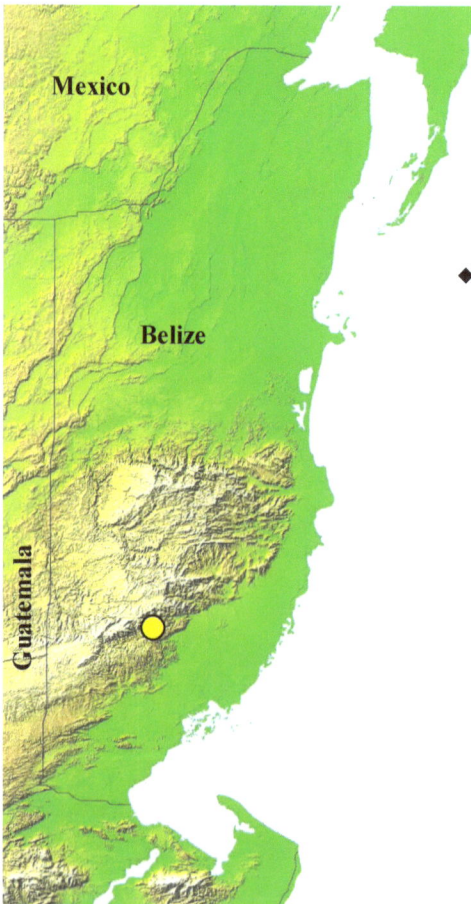

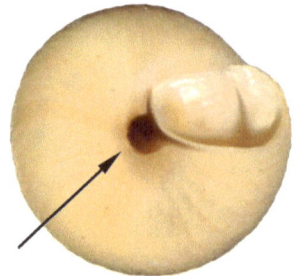

# Hairy Phora                    THYSANOPHORIDAE

*Thysanophora meermani* (new species)

**Diameter**: 2 mm, **Height**: 1.8 mm

**Description**: Heliciform; lip simple, aperture oval; shell with 4.5-5 whorls; umbilicate; shell surface dull, thin and fragile; transverse striae relativity well developed, distinctive spiral striae on the embryonic whorl; long hairs cover the entire shell surface; periphery rounded.

**Similar Species**: *Thysanophora plagioptycha* (page 246) has a higher shell profile, a narrower umbilicus and diagonal fringes; *T. caecoides* (page 247) also has a higher build, is without hairs and perforate.

**Habitat**: Found under leaf litter at the base of hills above the flood zone in well developed limestone talus.

**Status**: Rare, ENDEMIC to BELIZE currently known only from a few scattered locations across Belize.

**Specimen**: Belize, Belize District, Peccary Hills, just past mile marker 20 along Coastal Highway in limestone outcrop, Holotype UF 505445.

**Type Locality**: Peccary Hills, Belize (17°10'15"N, 88°22'57W).

**Etymology**: Named in honor of Jan Meerman, an extraordinary Belizean biologist whose work to create a database for wildlife in Belize has been an invaluable resource.

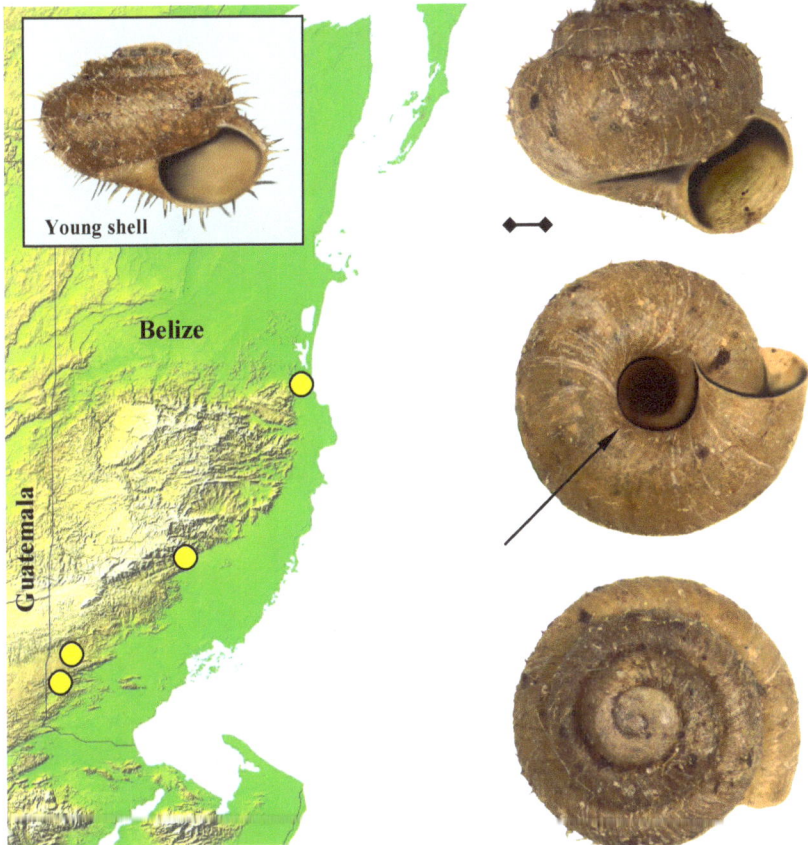

Young shell

Belize

Guatemala

249

# Flat Phora

# THYSANOPHORIDAE

## *Thysanophora conspurcatella* (Morelet, 1851)

**Diameter**: 2.5-3.5 mm, Height: 2 mm

**Description**: Depressed heliciform; lip simple or slightly reflected; aperture oval; shell with 4.5 whorls; umbilicate, wide for a *Thysanophora*; embryonic whorls with fine spiral striae (a); shell surface dull, thin and fragile; transverse striae a relatively strong feature on entire shell surface (a); upper periphery angular or strongly shouldered (b).

**Similar Species**: *Thysanophora plagioptycha* has a higher shell profile, is perforate and has diagonal fringes not hairs; *T. villosus* has long hairs and a more narrow umbilicus.

**Habitat**: Found in karst hills under leaf litter.

**Status**: Not yet reported from Belize; found at Tikal Maya Archaeological Site Petén, Guatemala near Belize border; also Chiapas, Mexico; likely to occur at Shipstern Nature Preserve in northern Belize.

**Specimen**: Mexico, Chiapas, 8.6 mi E Chiapa de Corzo, 3100 feet in elevation, UF 00019214.

**Type Locality**: Merida, Mexico.

Umbilicus wide

# Talus Phora          THYSANOPHORIDAE

### *Thysanophora impura* (Pfeiffer, 1866)

**Diameter**: 3–4.6 mm, Height: 2.6 mm

**Description**: Depressed heliciform; lip simple or slightly reflected, aperture roundish or squared; shell with 4.5 whorls; umbilicate; thin, fragile shell surface dull-glossy usually covered in soil; transverse striae a weak feature with crossing diagonal fringes, scope required to see this fine detail; upper periphery roundish and strongly shouldered (a).

**Similar Species**: *Thysanophora conspurcatella* has a similar build but smaller umbilicus; its aperture is more oval shaped.

**Habitat**: Karst foothills where it is found under leaf litter and rocky terrain or near the entrances of caves and lowlands; *Thysanophora impura* has the practice of covering its shell with dirt (Goodrich and Van Der Schalie 1937).

**Status**: Rare; reported from two locations in Belize, St. Herman's Cave, Cayo District and around the entrance of Cerrito Cave, near Trio Village, Toledo District (Thompson 1998).

**Specimen**: Mexico, Campeche 11.4 mi E Cayal, UF 00019210.

**Type Locality**: Mirador, Mexico.

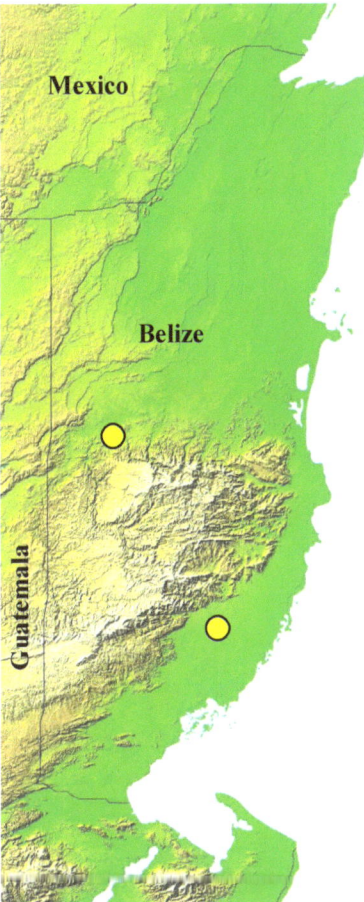

# Giant Phora                        THYSANOPHORIDAE

*Thysanophora cf. canalis* Pilsbry, 1910

**Diameter**: 4.5-5 mm, Height: 4 mm

**Description**: Heliciform; lip simple or slightly reflected, <u>aperture roundish</u>; shell with 4.5 whorls; <u>species large for a *Thysanophora*</u>; umbilicate, wide for a *Thysanophora*; embryonic whorls with fine spiral striae; shell surface dull, thin and fragile; transverse striae a weak feature, the crossing diagonal fringes stand out more boldly; periphery rounded.

**Similar Species**: Most similar to *Thysanophora plagioptycha* but fully 2-3 mm larger with a wider umbilicus; this is the largest *Thysanophora* in Belize.

**Habitat**: Found in karst regions of its range on limestone hills and in ravines under leaf litter.

**Status**: Rare, known only from Caracol Maya Archaeological Site based on collections made by Fred Thompson. Since the type locality for *Thysanophora canalis* is in Panama, specimens from Belize deserve further scrutiny to confirm identity.

**Specimen**: Belize, Cayo District, Caracol Archaeological Site, UF 207412.

**Type Locality**: Las Cascades, Canal Zone, Panama.

# Rufus Speck　　　　　THYSANOPHORIDAE

*Microconus rufus* Thompson, 1958

**Diameter**: 3.5 mm, Height: 2.8 mm

**Description**: Heliciform; lip simple; dull glossy shell that is thin and translucent; aperture roundish; shell with 4-5 whorls; umbilicate; <u>embryonic whorl smooth and protruding with minute but distinct, unequal, microscopic growth-wrinkles which are cut by weakly incised spiral lines; color horn yellow;</u> columellar margin slightly reflected; upper periphery rounded with a shoulder.

**Similar Species**: *Thysanophora plagioptycha* has a higher shell profile, stronger shell surface ornamentation and is perforate; *Microconus pilsbryi* is more compact, has a smaller umbilicus (a) and differs in color.

**Habitat**: Limestone hills of the Petén region.

**Status**: Rare in Belize, reported from Orange Walk and Cayo Districts but without specific location information; map locations are estimates of probable location; reported from the Petén region of Guatemala.

**Specimen**: Guatemala Dept. Izabal, 8 mi SW Puerta Matias de Galvez, also known from the Tikal Ancient Maya Archaeological Site, UF 00132525.

**Type Locality**: Pueblo Nueva, Guatemala.

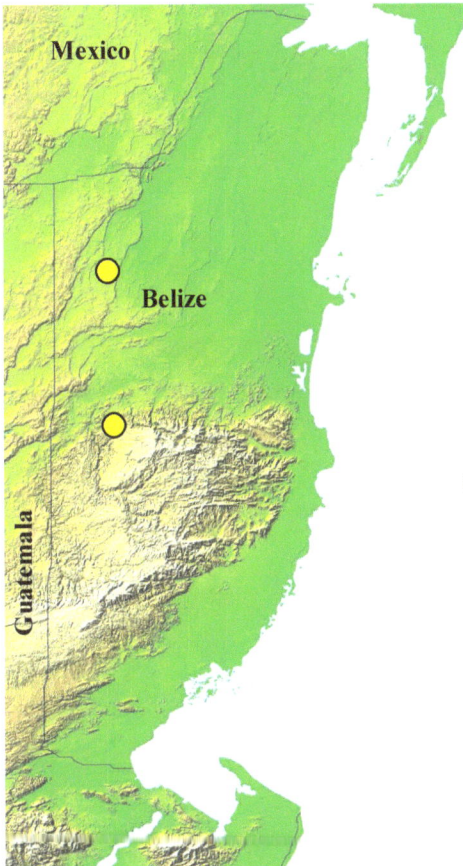

# Pilsbry's Speck                    THYSANOPHORIDAE

*Microconus pilsbryi* **Thompson, 1958**

**Diameter**: 3 mm, Height: 1.5 mm

**Description**: Heliciform; lip simple, aperture roundish; shell with 4 whorls; umbilicate; <u>shell surface dull with a grainy texture</u>, thin and fragile; transverse striae are spaced rather sparsely, crossed by fine spiral striae; upper periphery rounded with a shoulder.

**Similar Species**: *Thysanophora plagioptycha* has a higher shell profile, stronger shell surface ornamentation and is perforate; *Microconus rufus* is only slightly larger, a different color and the umbilicus (a) may be minutely wider but this feature varies among different populations.

**Habitat**: Karst foothills of the Maya Mountains where it is found under leaf litter and rocky terrain.

**Status**: Common, where it occurs in the Maya Mountains.

**Specimen**: Belize, Toledo District, Bladen Nature Reserve (Dourson collection).

**Type Locality**: Matagalpa, Nicaragua.

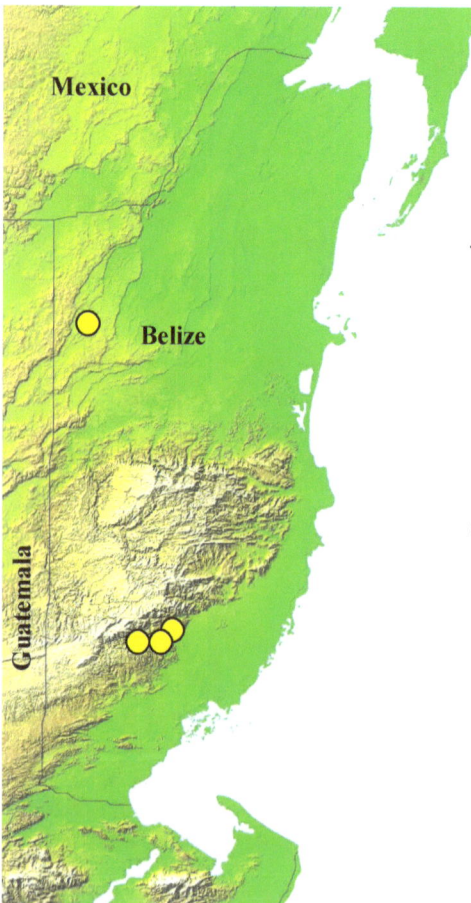

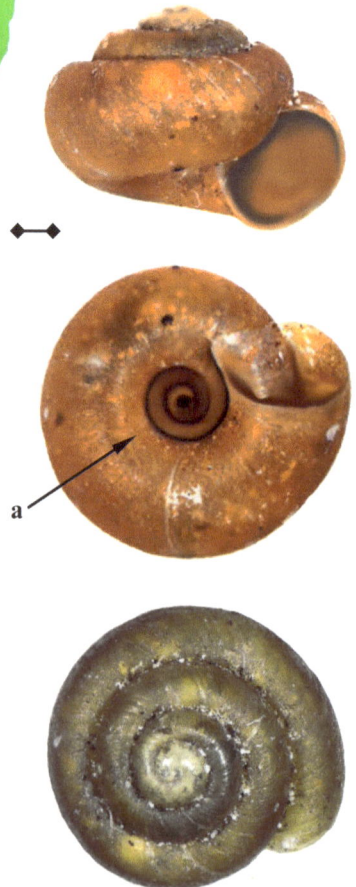

# Morelet's Habroconus      EUCONULIDAE

*Habroconus trochulinus* **(Morelet, 1851)**

**Diameter**: 3-4 mm, Height: 4 mm

**Description**: Heliciform; shell as tall as it is wide; lip simple; shell with 5-6 whorls; minutely perforate, opening only visible under favorable light (a); shell surface with a silky luster, translucent, thin and fragile; without notable transverse striae resulting in a nearly smooth shell; under the scope there are very faint irregular spiral striae not close together but with large intervals (Von Martens 1890); periphery rounded or very slightly angular.

**Similar Species**: *Thysanophora plagioptycha* has more rounded whorls and stronger shell surface ornamentation; *Microconus pilsbryi* has a lower shell, a rounded aperture and wider umbilicus.

**Habitat**: Found in a variety of calcareous habitats including dry limestone outcrops and under leaf litter.

**Status**: Rare, new record for Belize reported only from August Pine Ridge, Orange Walk District.

**Specimen**: Belize, August Pine Ridge, Orange Walk District, UF 0134973.

**Type Locality**: San Luis, Petén, Guatemala.

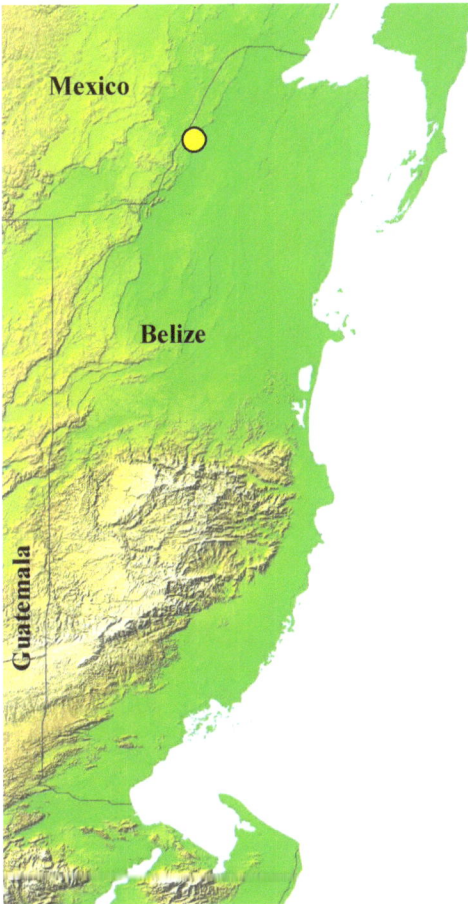

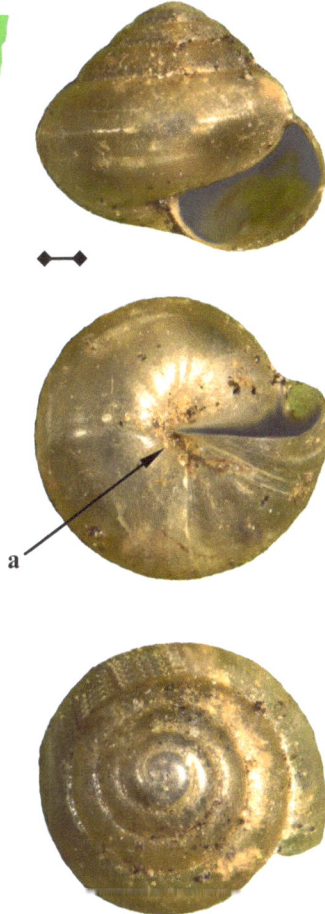

255

# Pittier's Habroconus                    EUCONULIDAE

*Habroconus pittieri* (Von Martens, 1892)

**Diameter**: 3 mm, Height: 2.5 mm

**Description**: Heliciform; lip simple; shell with 5-6 whorls; perforate/rimate; shell surface with a silky luster, translucent, thin and fragile; without notable transverse striae resulting in a nearly smooth shell; under the scope there are very faint and very close spiral striae found chiefly at the base of shell (Von Martens 1892); periphery rounded or slightly angular, the angle highlighted below in the frontal view of the shell (a).

**Similar Species**: Differs from *H. trochulinus* primarily by the narrower and more gradually increasing whorls, the last of which occupies a smaller part of the whole height (Von Martens 1892).

**Habitat**: Found in a variety of calcareous habitats under leaf litter.

**Status**: Common, one of the most common *Habroconus* species in Maya Mountains; also occurs in scattered locations throughout Belize.

**Specimen**: Belize, Forest Hill, Bladen Nature Reserve (Dourson collection).

**Type Locality**: Near San Jose, Costa Rica.

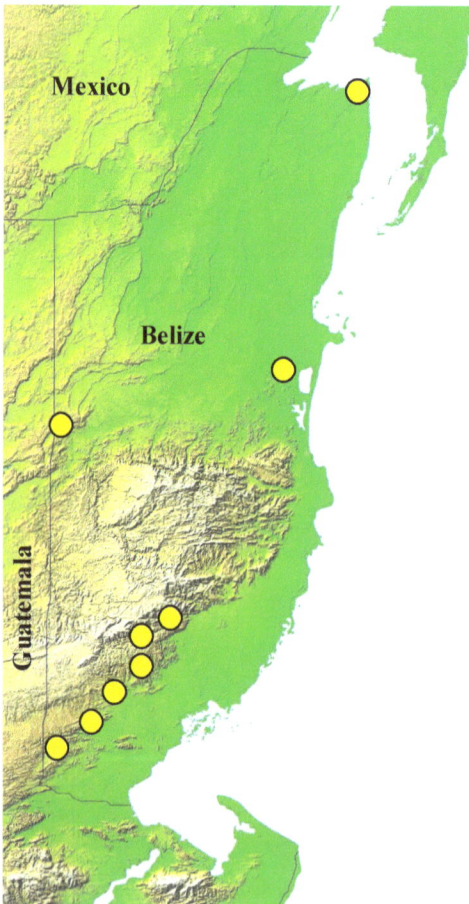

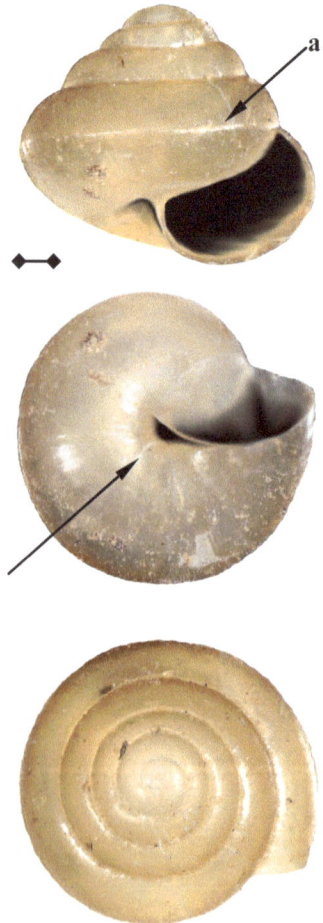

# Petén Habroconus                    EUCONULIDAE

*Habroconus elegantulus* **(Pilsbry, 1919)**

**Diameter**: 3.3 mm, Height: 3.2 mm

**Description**: Shell heliciform or dome-shape; lip simple; shell with 6 whorls; perforate/rimate (a); shell thin, glossy; surface above periphery has microscopic sculpture of fine, close, nearly vertical striae cut by equally close spiral lines (not always present); base glossy, engraved with spiral striae far more widely spaced than upper surface; last whorl rounded.

**Similar Species**: *Strobilops salvini* is around the same size and build but with well developed ribs on all its whorls and two parietal lamellae; *Guppya gundlachi* is shorter in build and displays fine spiral striae.

**Habitat**: A species of karst foothills usually found in low numbers under leaf litter and around rock structure covered in tropical rain forest.

**Status**: Uncommon, found in only two locations in Belize, Rio Bravo Conservation Management area, Orange Walk and Bladen Nature Reserve, Maya Mountains, Toledo; more common throughout the Petén region of Guatemala.

**Specimen**: Mexico, San Luis Potosi, UF 479651.

**Type Locality**: San Luis Potosi, Mexico.

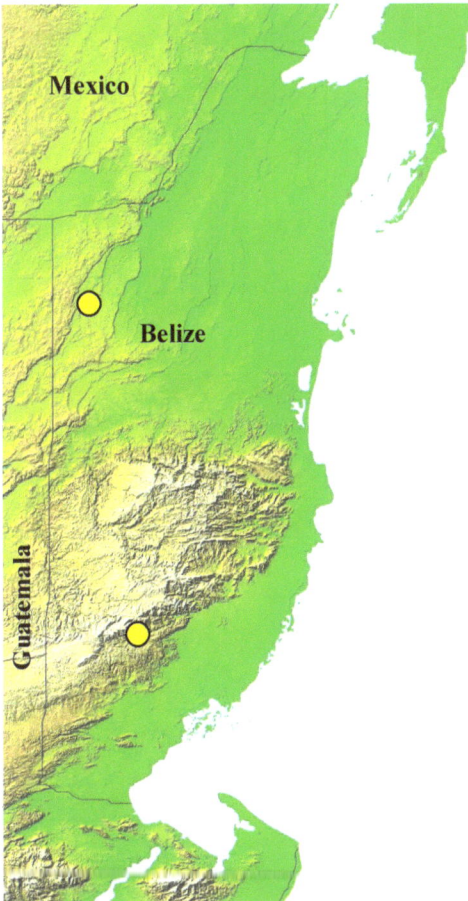

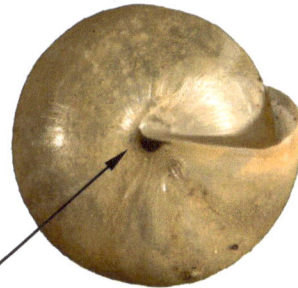

# Common Granule
# EUCONULIDAE

## *Guppya gundlachi gundlachi* (Pfeiffer, 1840)

**Diameter**: 2 mm, Height: 1.5 mm

**Description**: Depressed heliciform; lip simple, aperture roundish; shell with 5-6 whorls; perforate; shell surface glossy, thin and fragile; without notable transverse striae; the ever present spiral striae (a) are a constant and reliable diagnostic feature and should not be ignored, however, a strong scope is required to see this fine detail; old and weathered shells will generally lack this fine feature (wetting the surface will sometimes expose the striae); periphery rounded (b); dark specks seen on the shells below are actually dirt particles.

**Similar Species**: *Guppya gundlachi orosciana* has an angular not rounded periphery.

**Habitat**: Karst foothills scattered across Belize.

**Status**: Common across Belize in scattered locations

**Specimen**: Belize, Toledo District, base of Forest Hill, Bladen Nature Reserve (Dourson collection).

**Type Locality**: Cuba.

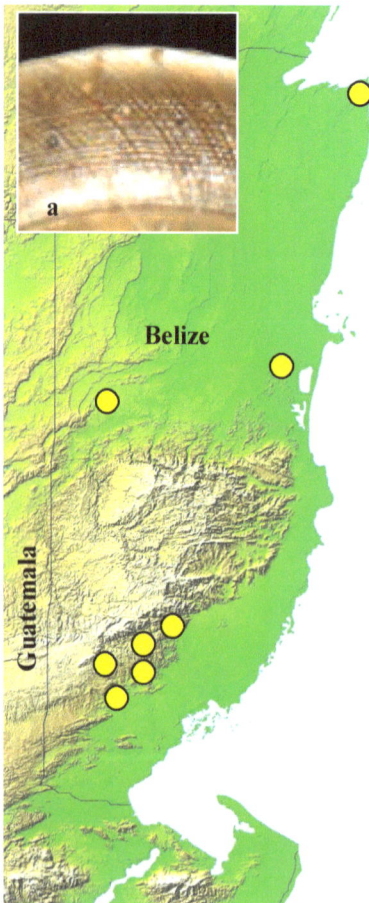

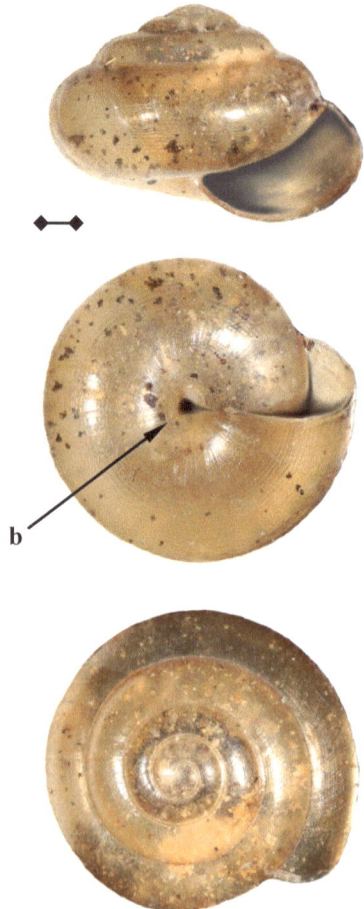

# Angled Granule                    EUCONULIDAE

*Guppya gundlachi orosciana* **Von Martens, 1892**

**Diameter**: 2.75 mm, Height: 1.75 mm

**Description**: Depressed heliciform; lip simple; aperture roundish; shell with 5-6 whorls; perforate; shell surface glossy, thin and fragile; without notable transverse striae; spiral striae (a) are a constant and reliable diagnostic feature and should not be disregarded, however a strong scope required to see this fine detail; periphery distinctly angular (b).

**Similar Species**: *G. gundlachi gundlachi* has a rounded periphery and is more glossy; *Habroconus* species have more elevated shells.

**Habitat**: Karst foothills of the Maya Mountains under leaf litter and among limestone rock talus.

**Status**: Rare, new record for Belize, known only from one location around Xunantunich Archaeological Site.

**Specimen**: Belize, road to Xunantunich Archaeological Site near San Ignacio, UF 146335.

**Type Locality**: Calera de San, Ramon, Costa Rica.

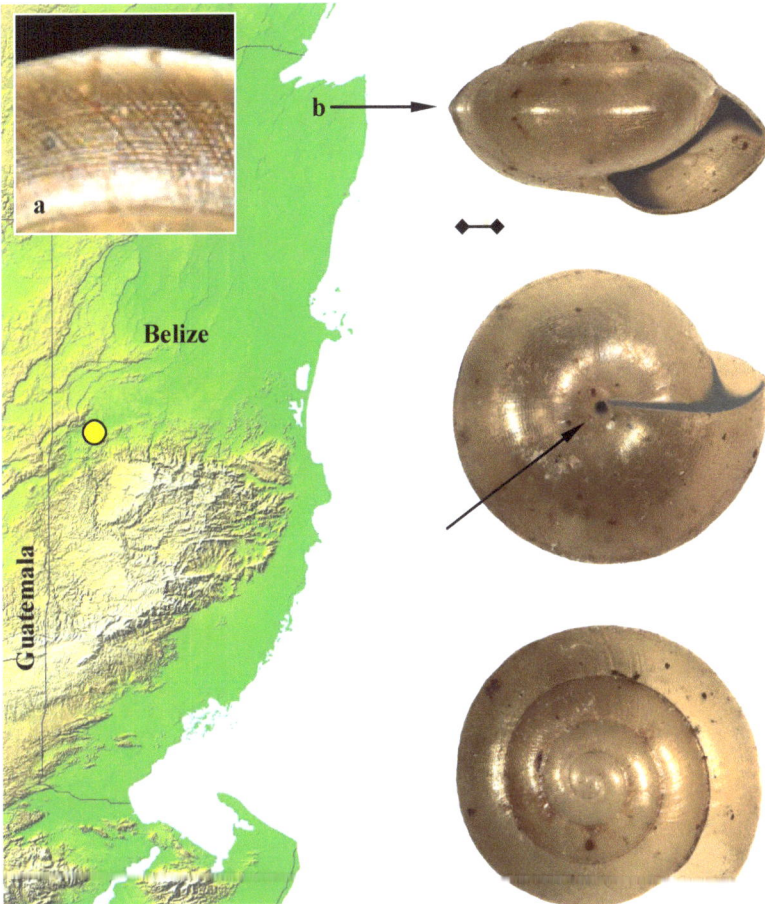

259

# Limestone Pinecone        STROBILOPSIDAE

## *Strobilops strebeli guatemalensis* (Tristram, 1863)

**Diameter**: 2.5 mm, Height: 1.5 mm

**Description**: Heliciform to <u>depressed heliciform</u>; lip reflected; aperture squared and small; inside the aperture there are two entering lamellae reaching deep into the shell for several whorls where they terminate; shell with 8 whorls; <u>perforate</u>; shell surface dull glossy, cinnamon-brown; shell solid for its small size; low riblets across the entire surface; without notable spiral striae; <u>periphery angular (a)</u>.

**Similar Species**: Similar to *Strobilops salvini* but smaller, flatter with a more angular periphery and a much smaller umbilicus (b).

**Habitat**: Found in calcareous soils, around limestone outcrops and on top of large boulders with ample detritus cover.

**Status**: Rare; previously known only from the type locality in Guatemala, the species was recently discovered at the Farm Inn in southern Belize.

**Specimen**: Guatemala, Puerto Matias De Galvez, UF 210193.

**Type Locality**: Alta Verapaz, Guatemala.

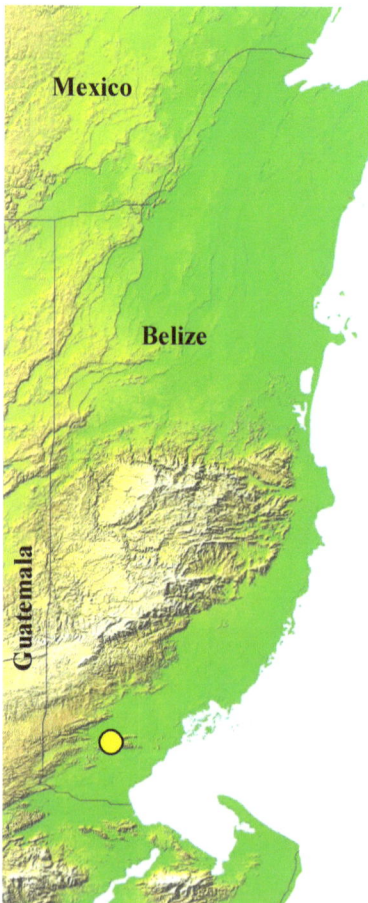

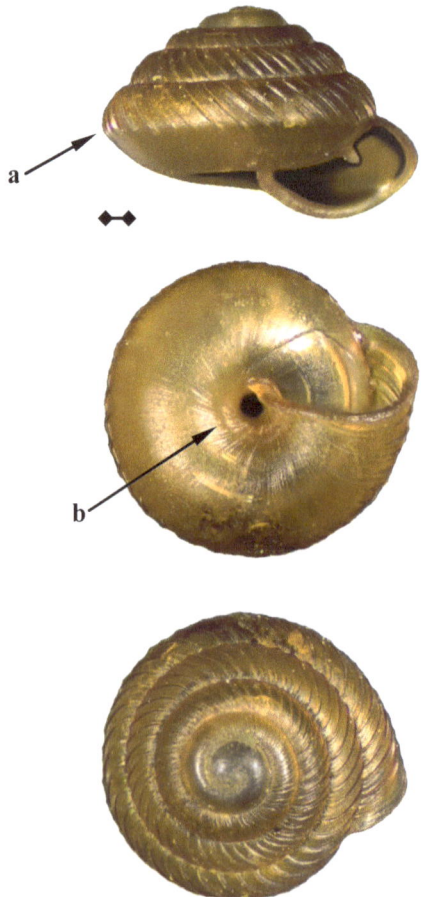

# Pyramid Pinecone       STROBILOPSIDAE

*Strobilops salvini* **(Tristram, 1863)**

**Diameter**: 3-4 mm, Height: 2.5-3 mm

**Description**: Heliciform; lip reflected, aperture squared and small, inside the aperture are two entering lamellae (a) reaching deep into the shell for several whorls where they terminate; shell with 8 whorls; <u>umbilicate</u>; shell surface dull glossy, shell solid for its small size; low <u>riblets </u>across the entire surface; without notable spiral striae; <u>periphery bluntly angular (b)</u>.

**Similar Species**: Most similar to *Strobilops strebeli guatemalensis* but larger and has a notably wider umbilicus (c); *Habroconus* and *Guppya* species have the same general shell shape but are smoother, without any internal armature and have a nearly closed umbilicus.

**Habitat**: Found in calcareous soils and around limestone outcrops.

**Status**: Uncommon, found mostly in the southern Maya Mountains but also found in the Peccary Hills, Belize District.

**Specimen**: Belize, Toledo District, Bladen Nature Reserve (Dourson collection).

**Type Locality**: Alta Verapaz, Guatemala.

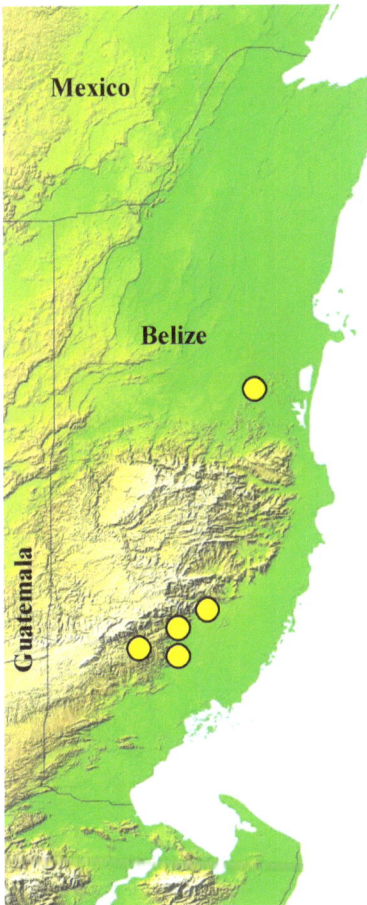

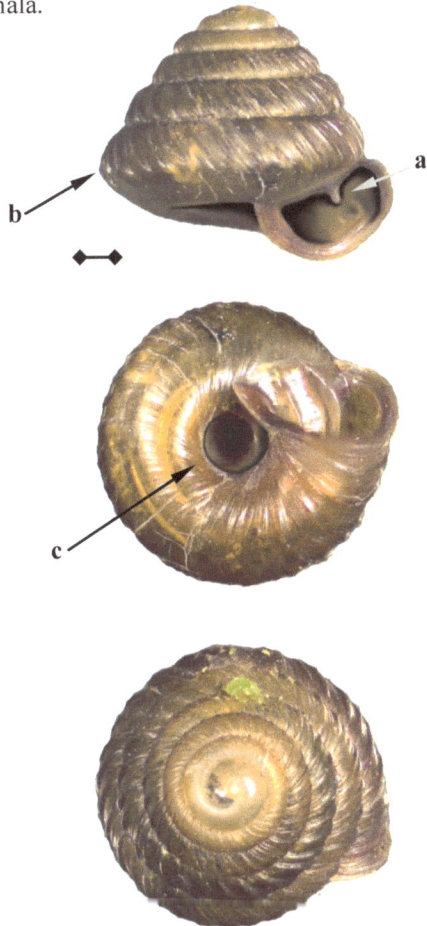

# Toothed Gem                                      SAGDIDAE

*Xenodiscula taintori* **Goodrich & Van der Schalie, 1937**

**Diameter**: 1.3 mm, Height: 0.5 mm

**Description**: Depressed heliciform; lip simple; aperture roundish; <u>shell armed with one short, entering lamellae (a)</u>; shell with 3 whorls; <u>widely umbilicate</u>; surface of shell with a glass-like gloss, translucent and exceedingly delicate; widely-spaced, engraved transverse striae on all whorls; without spiral striae; periphery roundish; the deceased and dried animal can be seen through the clear shell in all three views; this is Belize's smallest land snail.

**Similar Species**: No other species of land snail of similar size in Belize has the clear shell and wide umbilicus of *Xenodiscula taintori.*

**Habitat**: Associated with moist leaf litter at the base of limestone hills and also higher volcanic regions of montane oak forests.

**Status**: Common; found at scattered locations throughout Belize.

**Specimen**: Belize, Toledo District, base of Forest Hill, Bladen Nature Reserve (Dourson collection).

**Type Locality**: El Paso de los Caballos, Guatemala.

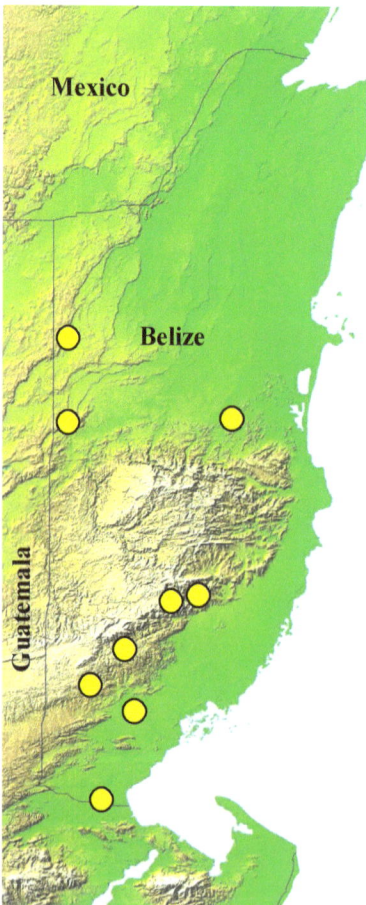

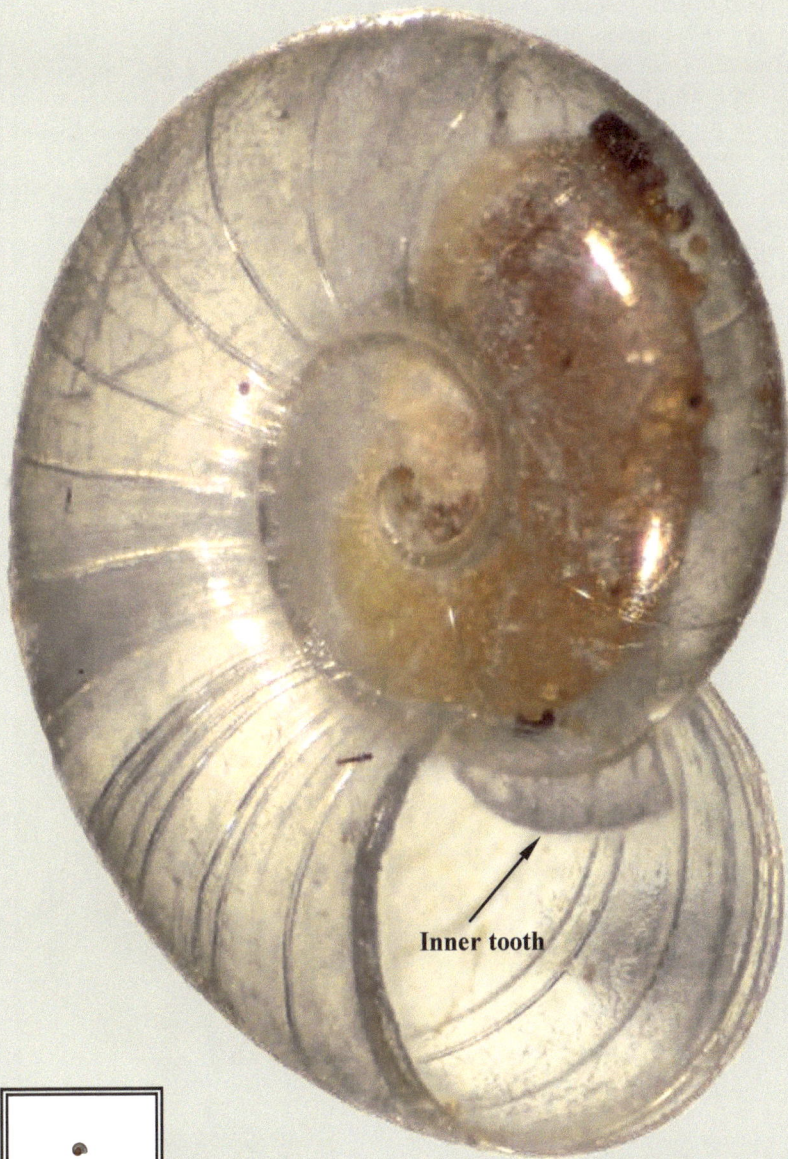

Inner tooth

Actual size

**Toothed Crystal,** *Xenodiscula taintori*
**Belize's Smallest Land Snail**

# Tall Turbinella                                        SAGDIDAE

## *Hyalosagda turbinella* (Morelet, 1851)

**Diameter**: 6-7 mm, Height: 5 mm

**Description**: Heliciform; <u>spire pointed, having a pyramid shape</u>; lip simple; shell with 5 whorls; perforate; without color bands; shell with a smooth translucent surface with widely spaced (low) transverse fringes; periphery compressed-roundish to sub-angular.

**Similar Species**: Snails in the HELICINIDAE family have much thicker shells and are usually with color bands; *Hyalosagda turbinella* is most similar to *Pyrgodomus microdinus* in shape but is a larger, thinner, more delicate shell, a different color and an open umbilicus, (a), not closed like *P. microdinus*; it differs from *Lacteoluna selenina* by its taller shape and smaller umbilicus.

**Habitat**: A rare to uncommon gastropod of limestone hills, usually around outcrops of limestone cliffs and talus slopes.

**Status:** Uncommon in Belize, further surveys are expected to produce additional populations.

**Specimen**: Belize, Cayo District, mouth of Rio Frio Cave (Dourson collection).

**Type Locality**: Petén, Guatemala.

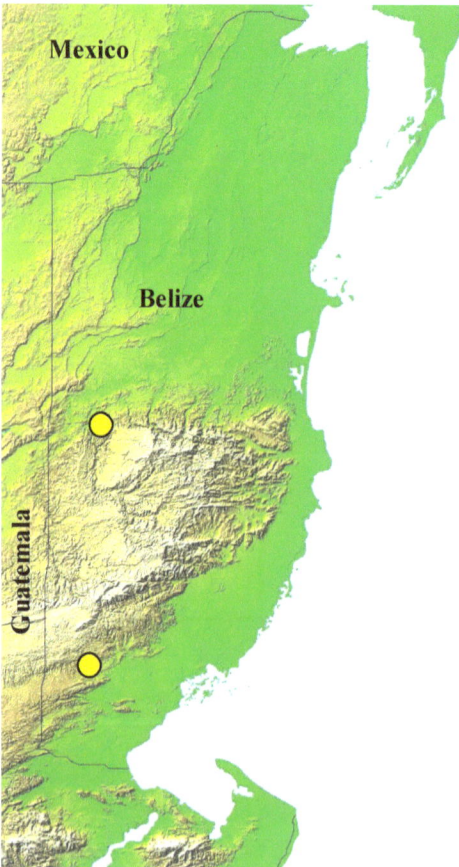

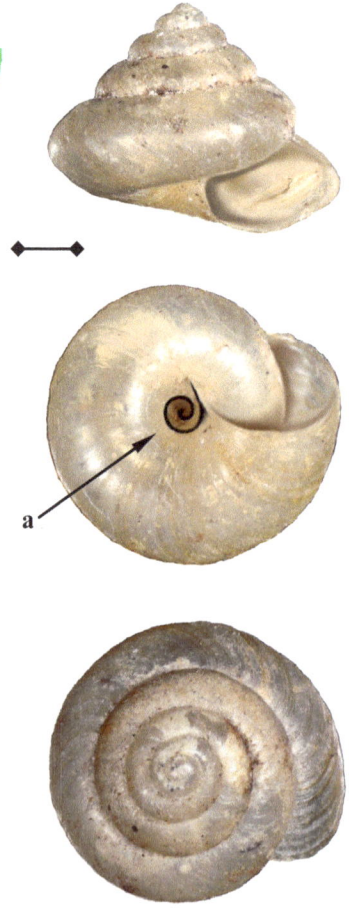

264

# Polished Selenina                    SAGDIDAE

## *Lacteoluna selenina* (Gould, 1848)

**Diameter**: 5 mm, Height: 3 mm

**Description**: Depressed heliciform; spire flattened; lip simple; aperture oval; shell with 5 whorls; umbilicate; color pale opaline; without teeth; shells are thin and fragile; shell nearly smooth or with weakly defined transverse striae; live specimens possess a very coarse and ragged periostracum which is quickly lost when the snail dies; shells are usually heavily coated with fine dead plant material arranged to give shell a carinate appearance (Hubricht 1985); without notable spiral striae; periphery bluntly angular .

**Similar Species**: *Miradiscops maya* and *M. bladenensis* do not have flat spirals; *Chanomphalus* and *Radiodiscus* species have distinctive ribs.

**Habitat**: A calciphile found under rocks, logs and old palm fronds.

**Status**: Uncommon, Las Milpas Ancient Maya Archaeological Site; Rockville Quarry, Gracie Rock Hill (Thompson 1993).

**Specimen**: Belize, Orange Walk District, Las Milpas Ancient Maya Archaeological Site at 150 m, UF 207264.

**Type Locality**: Unknown.

# Minute Gem                                        ZONITIDAE

*Hawaiia minuscula* (A. Binney, 1840)

**Diameter**: 2-3 mm, Height: 1.5 mm

**Description**: Depressed heliciform; lip simple, aperture roundish; shell with 4.5 whorls; widely umbilicate; shell glossy, translucent and fragile; transverse striae on all whorls; <u>faint spiral striae (a) are always present </u>(a strong lens required to see this feature); periphery rounded.

**Similar Species**: *Zonitoides* species are larger having a glossier, nearly smooth surface; *Striatura meridionalis* is notably smaller with a more oval aperture (not rounded as in *H. minuscula*).

**Habitat**: Found living in low places under and among moist leaf litter in calcareous soils and rocky places.

**Status**: Common, found in scattered locations throughout Belize in a variety of habitats; a common micro-snail in eastern North America.

**Specimen**: Belize, Toledo District, Forest Hill, Bladen Nature Reserve (Dourson collection).

**Type Locality**: Ohio, USA.

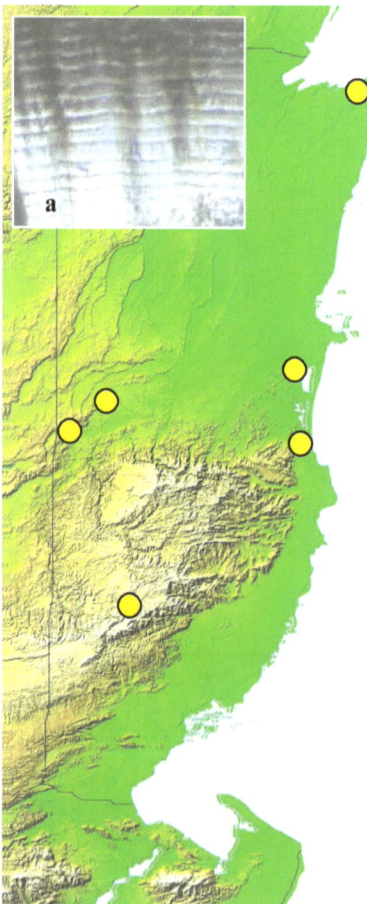

# Ground Creeper                    ZONITIDAE

## *Zonitoides cf. hoffmanni* (Von Martens, 1892)

**Diameter**: 6-7 mm, Height: 2 mm

**Description**: Depressed heliciform; lip only slightly reflected; aperture roundish and without an operculum; shell with 5 whorls; umbilicate; shell surface glossy, corneous, thin and fragile; transverse striae poorly developed and nearly smooth; without any color bands or notable spiral striae; periphery rounded.

**Similar Species**: *Zonitoides arboreus* is fully 2-3 mm smaller and a different color.

**Habitat**: Under leaf litter of limestone foothills of the Maya Mountains; it was found in Blue Creek Cave and at the entrance of Rio Frio Cave, likely carried into the caves during rain events.

**Status**: Rare; as leaf litter surveys for small species increase in Belize additional sites should be discovered.

**Specimen**: Belize, Toledo District, 5 miles west of San Jose (Dourson collection).

**Type Locality**: Quebrada Honda, Guanacaste District, Costa Rica.

# Tree Creeper                                    ZONITIDAE

## *Zonitoides arboreus* (Say, 1816)

**Diameter**: 5 mm, Height: 2 mm

**Description**: Depressed heliciform; lip simple not reflected; shell with 4.5–5 whorls; umbilicate; olive buff and glossy; transverse striae are poorly developed; spiral striae always present but a weak feature; without teeth; periphery rounded.

**Similar Species**: *Zonitoides cf. hoffmanni* is larger, found on the ground (although sometimes on vegetation) and is a different color.

**Habitat**: Found on or under exfoliating bark of standing or down rotting trees in advanced stages of decay; usually found in small colonies in higher elevation jungle.

**Status**: Uncommon; in Belize appears restricted to the higher elevation montane forests of the Maya Mountains; a widespread snail from eastern North America to Panama.

**Specimen**: USA, Kentucky, Powell County, Furnace Mountain (Dourson collection).

**Type Locality**: Philadelphia, Pennsylvania, USA.

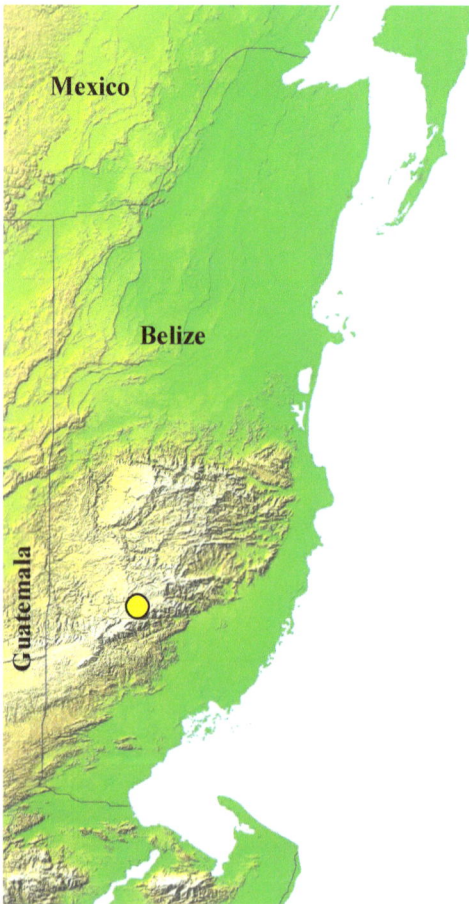

268

# Tiny Spot                                    PUNCTIDAE

## *Punctum cf. coloba* (Pilsbry, 1894)

**Diameter**: 1.1-1.3 mm, Height: 0.5 mm

**Description**: Depressed heliciform; lip simple; aperture roundish, not re-flected; shell with 3.5-4.5 whorls; umbilicate; pale brown to corneous; no teeth present; transverse striae closely spaced (a) and crisscross each other (b), this amazing micro-feature is so small it can only be seen with the use of a strong lens; fresh and live shells are thin and translucent; periphery rounded.

**Similar Species**: *Punctum vitreum* is similar in size and build but differs pri-marily by having more widely spaced ribs especially on the final whorls; *Punctum coloba* has closely placed ribs.

**Habitat**: Found in deep moist pockets of leaf litter in depressions or around logs; between the layers of moist leaves at higher elevations.

**Status**: Rare; restricted to highest elevations along the Maya Mountain Divide.

**Specimen**: Belize, Toledo District, Bladen Nature Reserve along the divide (Dourson collection).

**Type Locality**: Polvon, Nicaragua.

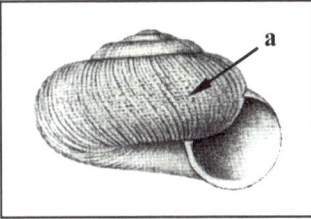

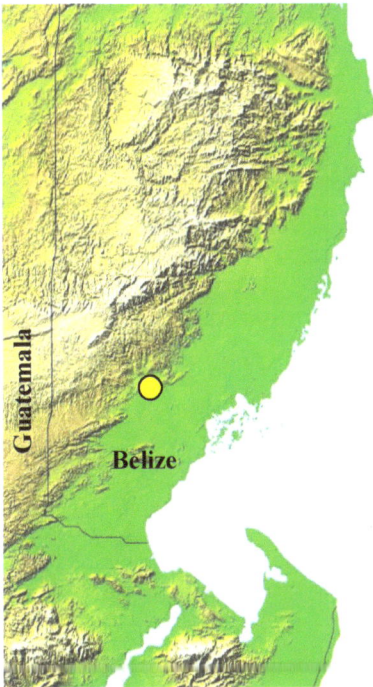

269

# Glass Spot                                     PUNCTIDAE
## *Punctum vitreum* (H. B. Baker, 1930)

**Diameter**: 1.2-1.4 mm, Height: 0.5 mm

**Description**: Depressed heliciform; lip simple; aperture roundish; shell with 4-4.5 whorls; umbilicate; corneous to colorless; no teeth present; transverse striae form low riblets and are sparsely spaced (a), these riblets crossed by lower spiral striae forming a netted appearance (b); shells thin, fragile and more or less translucent; periphery rounded.

**Similar Species**: Both *Hawaiia minuscula* and *Striatura meridionalis* have a notably wider umbilicus.

**Habitat**: Found between moist, matted leaves around small seeps in higher elevation jungle.

**Status**: Uncommon; in Belize it appears restricted to higher elevations; a widespread land snail reported from eastern North America to Panama;

**Specimen**: North Carolina, Haywood County, Purchase Knob (GSMNP collection).

**Type Locality**: New Braunfels, Texas, USA.

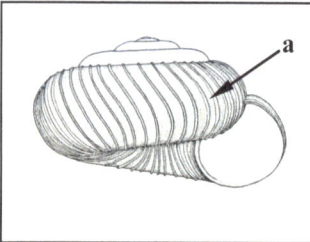

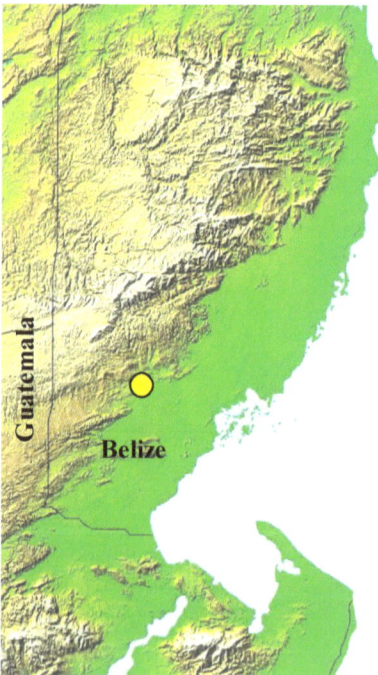

# Median Striate                    PUNCTIDAE

*Striatura meridionalis* **Pilsbry & Ferriss, 1906**

**Diameter**: 1.7-1.8 mm, Height: 0.5 mm

**Description**: Depressed heliciform; aperture roundish; lip simple; shell with 3-3.5 whorls; widely umbilicate; shell translucent; embryonic whorl (a) with spiral striae but not transverse striae; transverse striae become well developed on later whorls forming minute riblets that are not parallel with the peristome and thus the growth lines cross them at an angle, this character alone distinguishes *S. meridionalis* from *Punctum* species; periphery rounded.

**Similar Species**: *Punctum* species are smaller with a narrower umbilicus.

**Habitat**: Found in mixed hardwood forests on hillsides and ravines living between the layers of moist leaves, in higher elevation jungle.

**Status**: Uncommon; a wide ranging gastropod from eastern North America to Panama; in Belize, appears to be restricted to the higher elevation forests.

**Specimen**: USA, North Carolina, Macon County, Nantahala National Forest (Dourson collection).

**Type Locality**: Barranca, Colorado, Mexico.

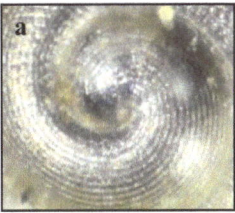

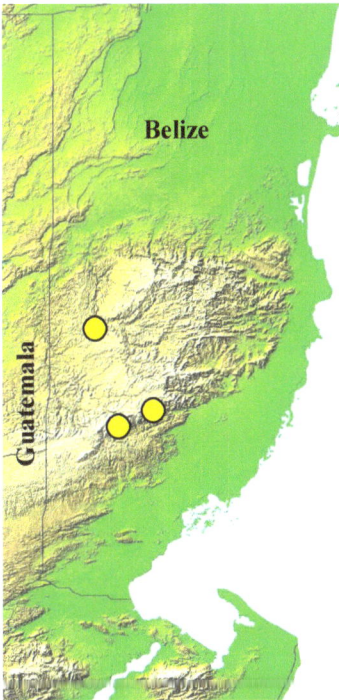

# Mountain Gloss                    SCOLODONTIDAE

## *Miradiscops striatae* (new species)

**Diameter**: 2 mm, **Height**: 1 mm

**Description**: Heliciform; lip simple; aperture roundish; shell with 4 whorls; narrowly umbilicate; shell glossy, horn to yellowish color; translucent and fragile; very fine, closely-spaced transverse striae crossed by even finer spiral striae (a) which may be absent in some specimens; periphery rounded.

**Similar Species**: *Miradiscops maya* (page 274) has the same build but is more translucent and has a smooth shell without striae; *M. youngii* (page 273) is flatter with a wider umbilicus; *Hawaiia minuscula* is larger and has a much wider umbilicus.

**Habitat**: Found living in low places under and among moist leaf litter in calcareous soils and karst hills.

**Status**: Common, ENDEMIC to BELIZE but expected to occur throughout the Maya Mountains of Belize.

**Specimen**: Belize, Forest Hill, Toledo District (Dourson collection). Holotype UF 505449, (not pictured). Paratype UF 505450, (not pictured).

**Type Locality**: Forest Hill, Toledo District, Belize (16°33'21"N, 88°42'45"W)

**Etymology**: Striatae means striate in Latin.

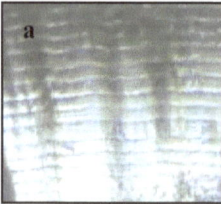

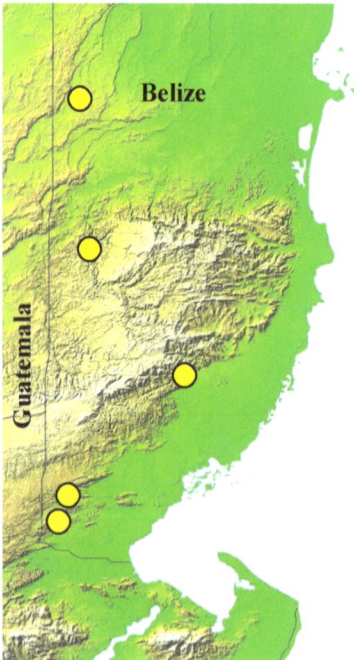

# Hillside Gloss                                    SCOLODONTIDAE

## *Miradiscops youngii* (new species)

**Diameter**: 1.8 mm, Height: 1.6-1.8 mm

**Description**: Depressed heliciform; lip simple; aperture roundish; shell with 4-4.5 whorls; <u>widely umbilicate</u>; shell whitish, translucent and fragile; very fine, closely-spaced transverse striae; without any discernible spiral striae; <u>shell surface covered in minute pits giving it a grainy texture (a); periphery rounded</u>.

**Similar Species**: *Miradiscops maya* (page 274) has the same build but is more translucent and has a smooth shell without pitting; *M. striatae* (page 272) is slightly larger, has a more elevated shell, a notably narrower umbilicus and has spiral striae (not pitting as in *M. youngii*).

**Habitat**: Found living in low places under and among moist leaf litter in calcareous soils and karst hills.

**Status**: Uncommon, **ENDEMIC to BELIZE**.

**Specimen**: Belize, Toledo District, road to San Jose, Holotype UF 505451, Paratype UF 505452 (not pictured).

**Type Locality**: San Jose, Toledo District, Belize (16°17'5"N, 89°3'20"W).

**Etymology**: Named in honor of Colin Young in recognition of his outstanding work in the field of conservation and biology in Belize.

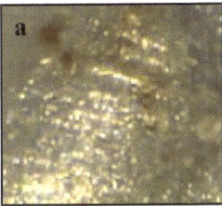

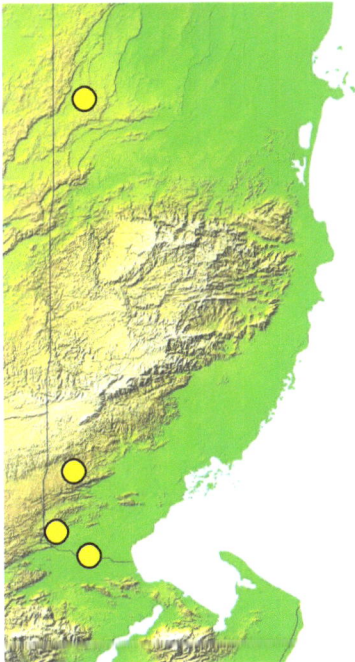

# Mayan Gloss                           SCOLODONTIDAE

*Miradiscops maya*  **Pilsbry, 1920**
**Diameter**: 1.5 mm, Height: 0.5 mm
**Description**: Depressed heliciform; lip simple; aperture roundish; shell with 4.5 whorls; umbilicate; color of shell opaline; no teeth present; shells are thin, fragile, translucent and smooth (a) or with very faint transverse striae; without spiral striae; periphery rounded.
**Similar Species**: *Punctum* species are similar but have notable shell ornamentation whereas the shell surface of *Miradiscops maya* is smooth; *Hawaiia* species are much larger.
**Habitat**: Found between the moist, matted leaves at the base of foothills of the Maya Mountains upriver from Blue Pool, Bladen Nature Reserve.
**Status**: Uncommon; the status of this species in Belize remains unknown but it likely occurs in many more locations than indicated; the type locality is reported in Guatemala.
**Specimen**: Belize, Toledo District, Bladen Nature Reserve near Blue Pool up river from BFREE crossing (Dourson collection).
**Type Locality**: Quirigua, Guatemala.

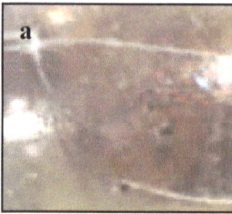

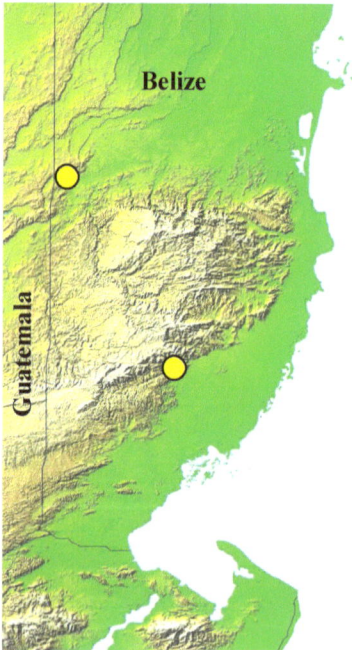

# Bladen Gloss                    SCOLODONTIDAE

*Miradiscops bladenensis* **(new species)**

**Diameter**: 2 mm, Height: 0.5 mm

**Description**: Depressed heliciform; lip simple; aperture roundish to oval; shell with 4.5 whorls; umbilicate; color of shell a pale-beige; no teeth present; shells are thin, fragile, translucent and smooth (a), without any notable transverse or spiral striae; periphery rounded.

**Similar Species**: *Miradiscops maya* (page 274) has a higher shell profile and wider umbilicus; *M. puncticipitis* (page 276) has a notably wider umbilicus and spiral papillae.

**Habitat**: Found in deep moist leaf litter of old cohune palm cavities along the floodplains of the Bladen River.

**Status**: Rare; ENDEMIC to BELIZE known only from the Bladen Nature Reserve.

**Specimen**: Belize, Toledo District, Bladen Nature Reserve, Forest Hill, cavity created by rotting cohune palm near trail. Holotype UF 505447, Paratype UF 505448, (not pictured).

**Type Locality**: Bladen Nature Reserve, Toledo District, Belize (16°33'21"N, 88° 42'45"W).

**Etymology**: Named in honor of the Bladen Nature Reserve, the "crown jewel" of Belize's numerous protected areas.

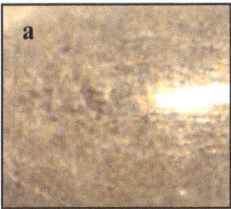

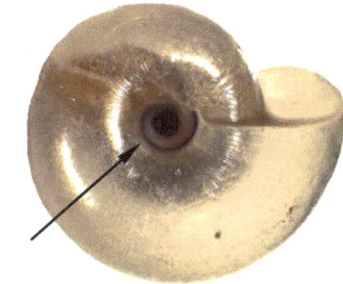

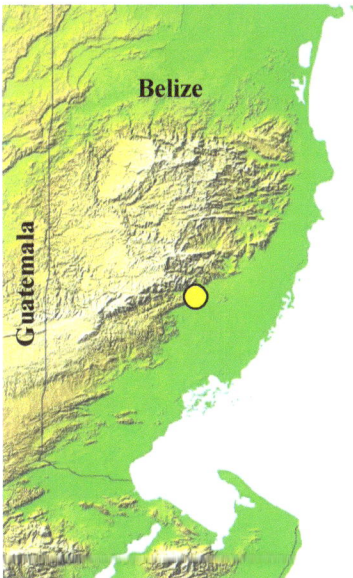

# Textured Gloss                    SCOLODONTIDAE

## *Miradiscops puncticipitis* (Pilsbry, 1926)

**Diameter**: 1.5-2 mm, Height: 0.5 mm

**Description**: Depressed heliciform; lip simple; aperture roundish; shell with 4.5 whorls; widely umbilicate; shell color corneous; shells are thin, fragile and translucent; closely and regularly spaced transverse striae on all whorls; spiral papillae always present but may be a hard detail to see; periphery rounded.

**Similar Species**: *Miradiscops maya* and *Miradiscops bladenensis* have shells that are smooth (without any delicate ornamentation) and a more narrow umbilicus.

**Habitat**: Found in deep moist leaf litter in both limestone foothills and mountain valleys.

**Status**: Relatively common, found in scattered locations throughout Belize

**Specimen**: Belize, Toledo District, Oak Ridge, 1100 m in elevation, Bladen Nature Reserve (Dourson collection).

**Type Locality**: Chama, Guatemala.

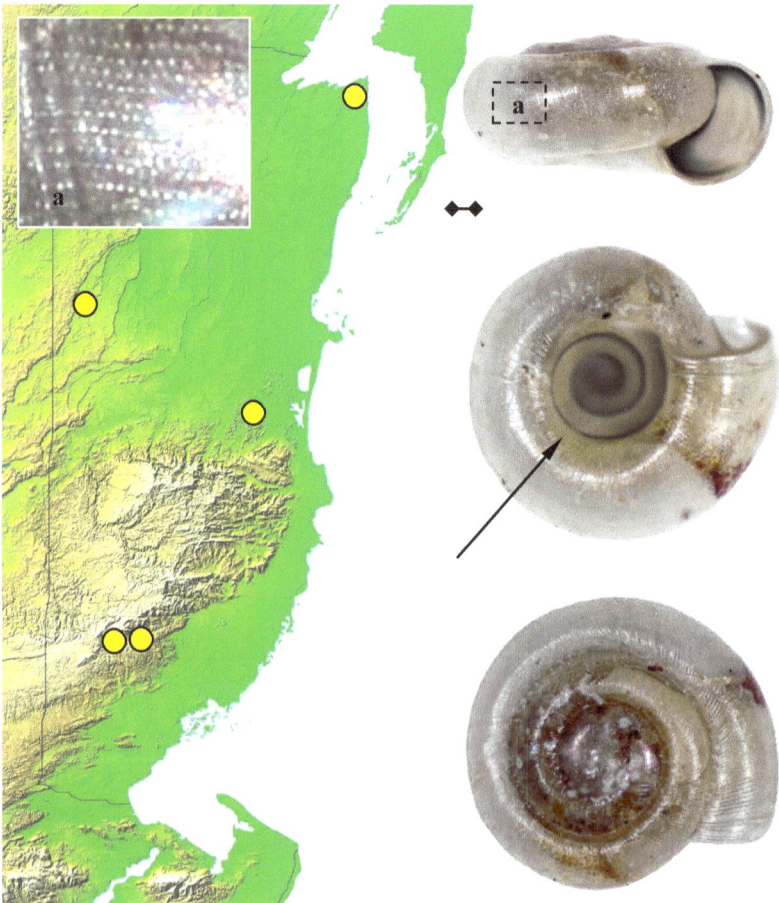

# Common Radiodiscus                    CHAROPIDAE

*Radiodiscus millicostatus* Pilsbry & Ferriss, 1906

**Diameter**: 2 mm, Height: 1.1 mm
**Description**: Depressed heliciform; lip simple; shell with 5 whorls; first whorls depressed and bluish-white are minutely engraved spirally, the rest of the shell chestnut-brown; umbilicate; without teeth; shells are thin and fragile; embryonic whorls with fine spiral striae (a); closely and regularly spaced, rib-like transverse striae on all whorls; periphery rounded.
**Similar Species**: *R. proameri* is much smaller, has more closely spaced riblets and more compressed whorls than *R. millicostatus.*
**Habitat**: Found in moist leaf litter covered by forests.
**Status**: Widely distributed from Arizona south to Nicaragua; although not yet recorded from Belize the species is expected to occur based on its wide ranging distribution in Mexico and Northern Central America.
**Specimen**: Arizona, USA Tucson, UF 295330.
**Type Locality**: Carr Canyon, Arizona, USA.

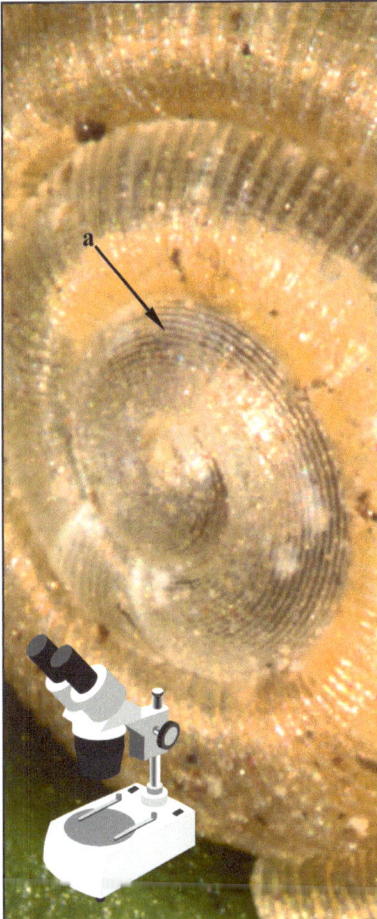

# Maya Mountain Rotadiscus          CHAROPIDAE

*Rotadiscus saqui* (new species)
**Diameter**: 1.5 mm, Height: 0.5 mm
**Description**: Depressed heliciform; lip simple and squared; shell with 4-5 shouldered whorls; sutures deep; embryonic whorl smooth, without sculpture; widely umbilicate; bronze color; closely and regularly spaced riblets on all whorls (a); just inside shell there is a long, thick palatal lamellae located on the outer wall (b); no spiral striae; periphery boxy in form, each whorl conspicuously shouldered.
**Similar Species**: *Rotadiscus hermanni* (page 279) is slightly larger, has more rounded whorls and is without the internal lamella of *R. saqui.*
**Habitat**: A mountain species of moist leaf litter in montane oak forests.
**Status**: Rare; ENDEMIC to BELIZE the species is expected to occur in other locations along the Maya Mountain Divide.
**Specimen**: Belize, Toledo District, Oak Ridge, 1000 m elevation, Bladen Nature Reserve, Holotype UF 505457.
**Type Locality**: Oak Ridge, Bladen Nature Reserve, Belize (16°31'1"N, 88°55'43"W).
**Etymology**: Named in honor of Ernesto and Aurora Garcia Saqui for their extraordinary contributions to conservation, Mayan cultural preservation, art and alternative healing in Belize.

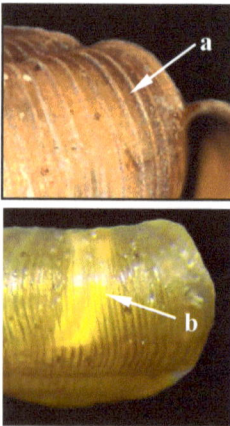

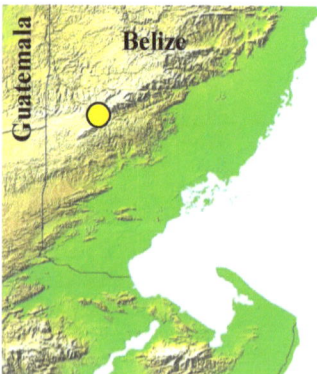

# Herman's Rotadiscus

## CHAROPIDAE

*Rotadiscus hermanni* **(Pfeiffer, 1866)**

**Diameter**: 2 mm, Height: 1 mm

**Description**: Depressed heliciform; spire flattened; lip simple; aperture roundish-lunate; shell with 5 whorls; umbilicate; without teeth; shells are thin and fragile; shell quite shiny, translucent, white turning to tawny; embryonic whorls smooth, without sculpture (a); closely and regularly spaced rib-like transverse striae on all whorls; periphery rounded.

**Similar Species**: The smooth embryonic whorls alone separate *Rotadiscus* from *Radiodiscus;* other species of similar size have more elevated shells not discoidal (flat). *R. saqui* (page 278) is slightly smaller, has whorls that are not as rounded and has a thick, long internal lamella.

**Habitat**: A calciphile found under rocks, logs and in moss in limestone hills and ravines.

**Status**: Appears to be rare in Belize, reported from around Cuevos Gemelas, Augustine, Cayo District and one site in the Bladen Nature Reserve.

**Specimen**: Belize, Toledo District, limestone outcrops above Blue Pool, Bladen Nature Reserve (Dourson collection).

**Type Locality**: Mirador, Veracruz, Mexico.

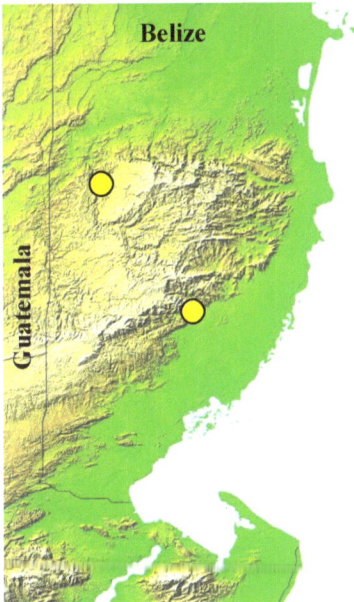

279

# Common Crystal                    CHAROPIDAE

*Chanomphalus pilsbryi* **(Baker, 1927)**

**Diameter**: 1.3 mm, Height: 0.5 mm

**Description**: Depressed heliciform; lip simple; shell with 3-4 whorls; widely umbilicate; corneous; without teeth; shells are thin and fragile; embryonic whorl nearly smooth, remaining whorls with <u>closely but regularly spaced riblets</u>; without notable spiral striae; periphery shouldered.

**Similar Species**: *Miradiscops maya* and *M. bladenensis* are around the same size but have shells that are smooth, without notable ornamentation and a more narrow umbilicus; *Chanomphalus angelae* is larger, has higher more widely spaced riblets, deeper sutures, a higher shell profile and a more oval aperture.

**Habitat**: Found in deep moist leaf litter at all elevations in both acidic and limestone soils.

**Status**: Common; found in scattered locations throughout Belize and expected to be found in additional locations as surveys continue especially if micro species are targeted.

**Specimen**: Belize, Toledo District, base of Forest Hill, Bladen Nature Reserve (Dourson collection).

**Type Locality**: Veracruz, Mexico.

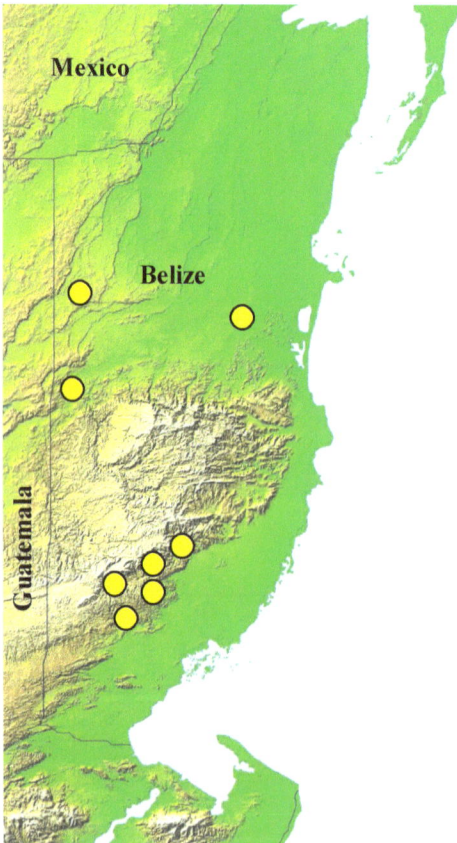

280

# Ornate Crystal

CHAROPIDAE

*Chanomphalus angelae* **(new species)**

**Diameter**: 2 mm, Height: 0.5 mm

**Description**: Depressed heliciform; lip simple; shell with 3-4 whorls; widely umbilicate; sutures deeply impressed; corneous; without teeth; shells are thin and fragile; embryonic whorl nearly smooth, remaining whorls with <u>sparsely but regularly spaced riblets</u>; spiral striae present but can only be viewed under a very strong lens; periphery is shouldered.

**Similar Species**: *Chanomphalus pilsbryi* (page 280) is smaller, flatter, with more crowded riblets and less deeply impressed sutures; *Miradiscops* species have shells that are without riblets (page 272-276).

**Habitat**: Found in deep moist leaf litter located in a sinkhole in karst foothills near San Benito Poite in southern Belize.

**Status**: Rare, **ENDEMIC to BELIZE** known only from the type locality.

**Specimen**: Belize, Toledo District, 1.27 miles east of San Benito Poite, Holotype UF 505548.

**Type Locality**: A sinkhole 2 miles east of San Benito Poite, Belize (16° 6'49"N, 89°12'41"W).

**Etymology**: Named in honor of my daughter, Angela Dourson Christensen, a woman of uncommon courage and genuine honesty of whom I am deeply proud.

**Ornate Crystal, *Chanomphalus angelae* (new species)**

# 12 Shells Taller Than Wide, Less Than 5 mm

This section includes minute land snails that are taller than wide and less than 5 mm tall. These snails are usually found in leaf litter but a few species are found living in bromeliads and other epiphytes growing on trees. These tiny land snails are best viewed under a strong dissecting scope so that micro-features of the snail can be used to determine species. One representative species from each family is illustrated to the right below.

## The Families and Genera Included in Chapter 12

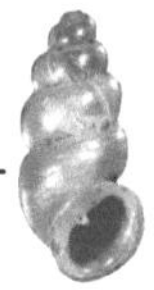

# Belize Thorn                    CARYCHIIDAE

*Carychium belizeense* Jochum & Weigand, 2017

**Height**: 1.5 mm, Diameter: 0.5 mm

**Description**: Pupa-shape; <u>lip reflected and thickened</u>; shell with 5-6 whorls; sutures deep (a); <u>one horizontal entering lamella on the left side of the aperture (b)</u>; transverse striae are poorly defined; very faint spiral papillae on first 3 whorls vanishing and replaced by a smooth surface on remaining whorls; shells are translucent; periphery well rounded. DNA sequencing by Adrienne Jochum *et al.* (2017) has shown this species to be genetically distinct from *Carychium exile.*

**Similar Species**: No other small pupa-shape snail has the deep sutures (a) and lamellae; *Gastrocopta* species are larger with more distinctly defined teeth in apertures.

**Habitat**: In pockets of moist leaves that accumulate in low depressions and around small seeps; live animals live between layers of wet leaves.

**Status**: Common; throughout the Maya Mountain at all elevations.

**Specimen**: Belize, Toledo District, Forest Hill, Bladen Nature Reserve (Dourson collection).

**Type Locality**: Forest Hill, Bladen Nature Reserve, Toledo District, Belize.

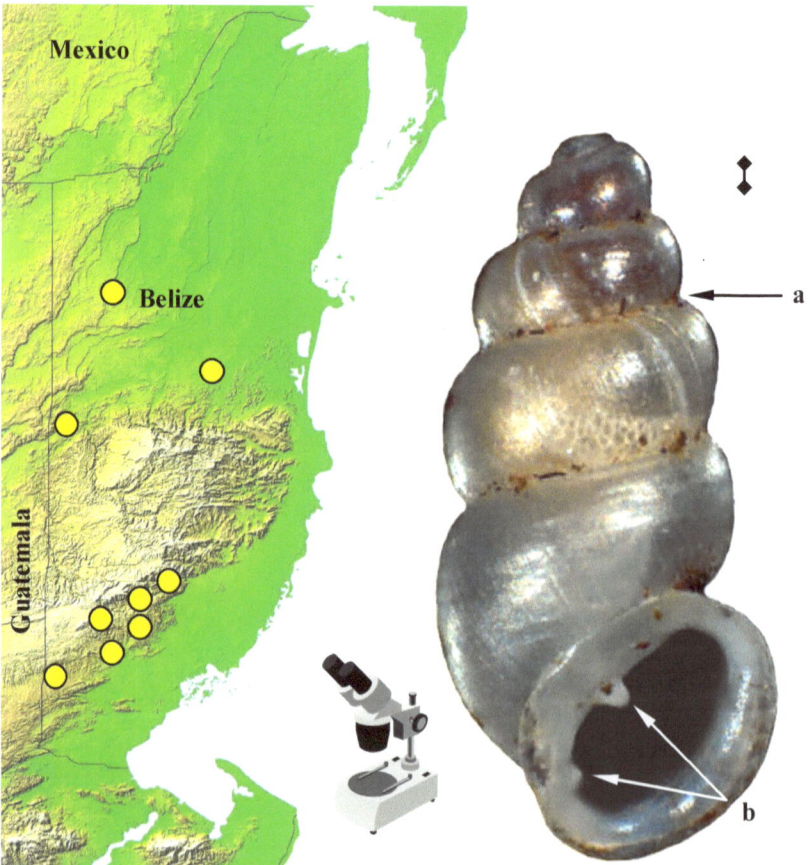

## *Myxastyla* (Thompson, 1995)　　　　**SPIRAXIDAE**

A relatively new genus described by Thompson in 1995, *Myxastyla* are minute land snails 2-4 mm tall; cylindrical-ovate or pupa-form in shape with the spire comprising about half the length of the shell of around 4-5 whorls. Sculpture is simple consisting of nearly uniformly spaced impressed growth varices. Aperture pinched inward along the outer lip (a). A strongly twisted columellar plait forms a narrow, deep channel between the parietal wall and the edge of the plate (b). This genus is placed provisionally in the subfamily STREP-TOSTYLINAE. It differs from other streptostylid genera by its sculpture. It resembles some species of *Streptostyla* in shape but mature *Streptostyla* are much larger without regularly spaced impressed growth varices. To a lesser extent, *Myxastyla* resembles some *Spiraxis* (subgenus *Volutaxis*) because of the strongly twisted columellar plait. However, various species of *Volutaxis* have raised axial sculpture, are much more attenuate in shape, and have a greater number of whorls (Thompson 1995). At present, there are three described *Myxastyla* species from central Guatemala and Roatan Island, Honduras. Recently, two of the named species of this genus, *Myxastyla hyalina* and *Myxastyla pycnota* were documented in Belize during our surveys. Illustrations and information courtesy of the Florida Museum of Natural History, USA.

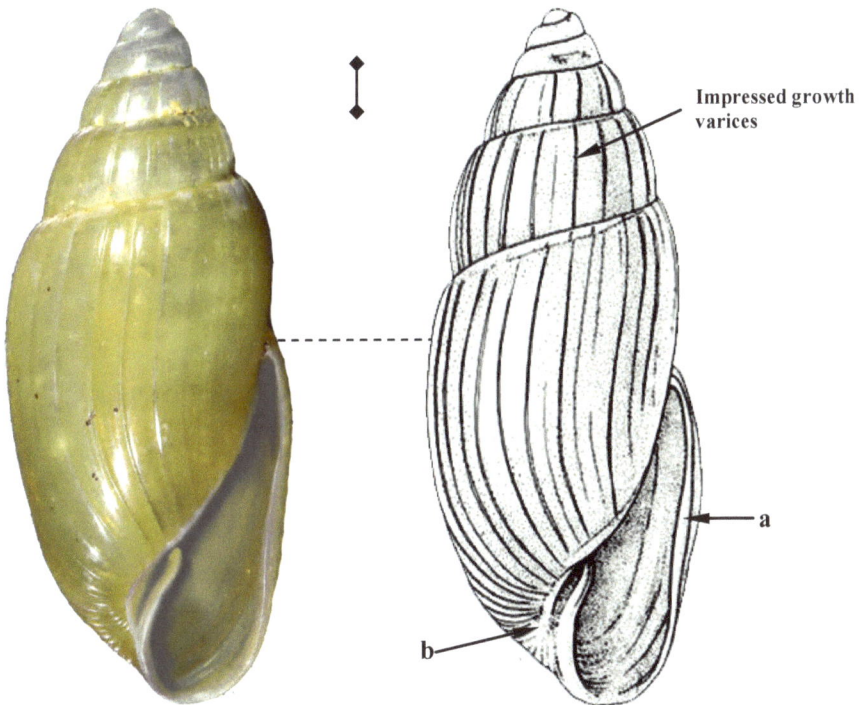

*Myxastyla coxeni* (Richards, 1938) from Honduras, Roatan Island

# Short Myxastyla                    SPIRAXIDAE

*Myxastyla hyalina* Thompson, 1995

**Height**: 2.5 mm, Diameter: 0.5 mm

**Description**: Pupa-shape; lip simple; shell with 5 whorls; sutures shallow; sculpture simple consisting of nearly uniformly spaced impressed growth varices or indented sculpture that is characteristic of the genus (Thompson 1995); columellar plait strongly twisted (a); shells are translucent; live animal pale-yellow; periphery well rounded.

**Similar Species**: *Cecilioides* species have a simple not a strongly twisted columellar plait as seen in *Myxastyla* species and are typically smaller in build; *Carychium* species have a much different build and deep sutures.

**Habitat**: In pockets of moist leaves that accumulate in low depressions along floodplains.

**Status**: Common, found at scattered locations throughout Belize.

**Specimen**: Figure (b & c) from base of Forest Hill, Bladen Nature Reserve (Dourson collection); *M. hyalina* illustrations (Thompson 1995).

**Type Locality**: Tactic, Alta Vera Paz Dept., Guatemala.

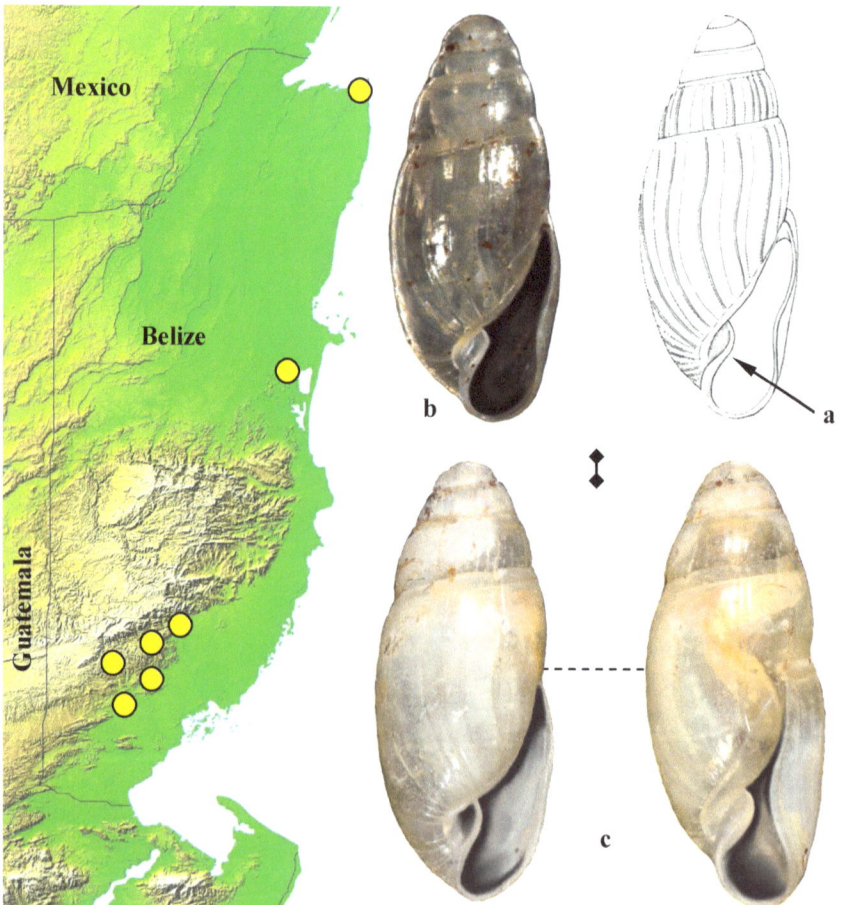

286

# Tall Myxastyla                                    SPIRAXIDAE

*Myxastyla pycnota* Thompson, 1995

**Height**: 3 mm, Diameter: 1 mm

**Description**: Pupa-shape; lip simple; shell with 5 whorls; sutures shallow; sculpture simple consisting of nearly uniformly spaced impressed growth varices or indented sculpture that is characteristic of the genus (Thompson 1995); columellar plait strongly twisted (a); shells are translucent.

**Similar Species**: *Cecilioides* species have a simple not strongly twisted columellar plait as seen in *Myxastyla* species and are generally smaller.

**Habitat**: In pockets of moist leaves that accumulate in low depressions in mid-elevation karst hills.

**Status**: Rare; new country record for Belize; until recently known only from the type locality but was discovered in 2016 on a limestone hillside above Blue Pool, Bladen Nature Reserve.

**Specimen**: Guatemala, Huehuetenango Prov. 12 km. SSE of La Democracia, 950 m. Paratype UF 193057, *M. pycnota* illustrations (Thompson 1995).

**Type Locality**: La Democracia, Huehuetenango Dept., Guatemala.

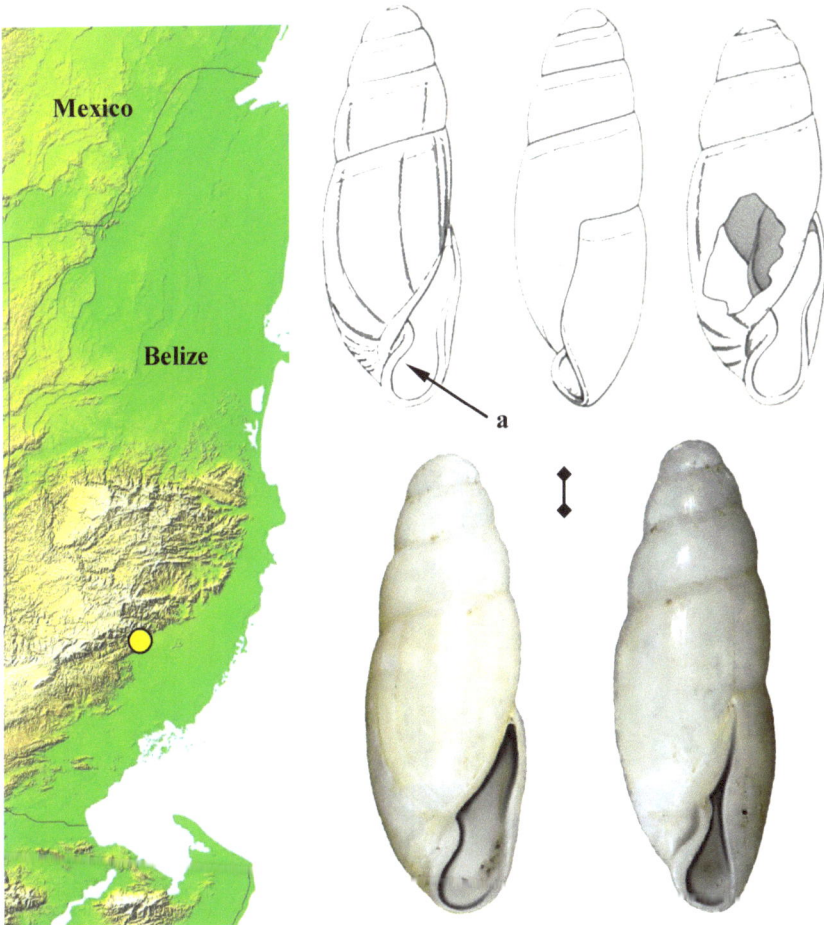

# Common Teardrop                    FERUSSACIDAE
*Cecilioides consobrinus primus* (De Folin, 1870*)*
**Height**: 1.9 mm, Diameter: 1 mm wide
**Description**: Pupa-shape, narrow; lip simple; shell with 5 whorls; imperforate; sutures shallow; shell surface smooth, thin and translucent, axis visible through shell; color pellucid; live animal pale-yellow; in the first few whorls the deceased and dried animal can be seen through the clear shell; weakly developed spiral striae; the columellar plait not twisted (a), as it is in *Myxastyla* species; periphery broadly rounded.
**Similar Species**: *Myxastyla* species have a strongly twisted columellar plait; *Cecilioides dicaprio* has an obese build and is found only in montane forests; *Cecilioides consobrinus veracruzensis* has a trimmer physique.
**Habitat**: In pockets of moist leaves that accumulate in low depressions, and rock talus slopes of limestone foothills.
**Status**: Common, scattered across Belize.
**Specimen**: Belize, Toledo District, base of Forest Hill, Bladen Nature Reserve. (Dourson collection).
**Type Locality**: Veracruz, Mexico.

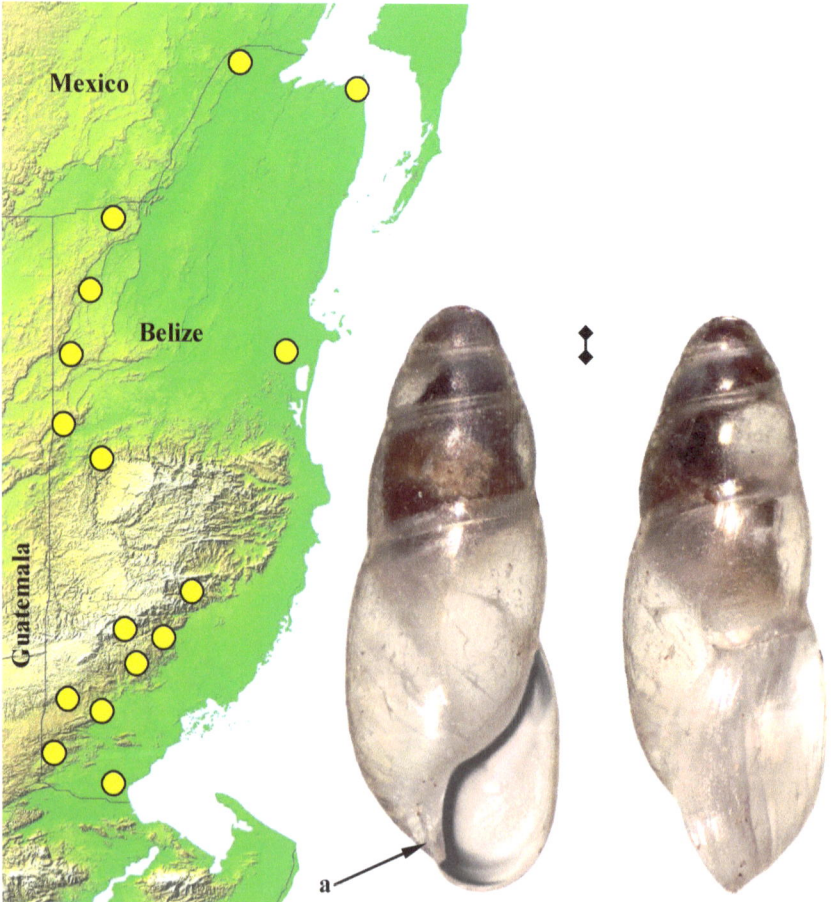

288

# Slender Teardrop ' FERUSSACIDAE

## *Cecilioides consobrinus veracruzensis* (Crosse & Fischer, 1870)

**Height**: 2.1 mm, Diameter: 0.75 mm

**Description**: <u>Pupa-shape, narrow</u>; lip simple; shell with 4-5 whorls; imperforate; sutures shallow facing downward, the last one at around a 20° angle (see below); shell surface smooth, thin and translucent, the axis visible through shell; color pellucid; spiral striae more distinctly developed than other *Cecilioides*; <u>columellar plait not twisted</u>.

**Similar Species**: *Myxastyla* species have a strongly twisted columellar plait; other *Cecilioides* species reported in Belize have wider builds and much weaker spiral striae and sutures that slant at around 10°; *Carychium belizeense* has a reflected lip.

**Habitat**: In pockets of moist leaves that accumulate in depressions around the base of limestone foothills and montane forests at higher elevations.

**Status**: Common in the Maya Mountains of southern Belize.

**Specimen**: Belize, Toledo District, Oak Ridge, Bladen Nature Reserve (Dourson collection).

**Type Locality**: Rio Antigua, Mexico.

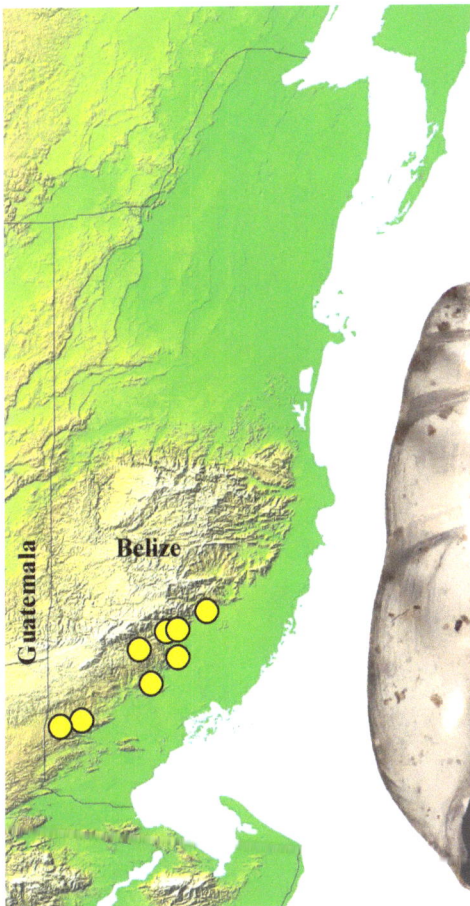

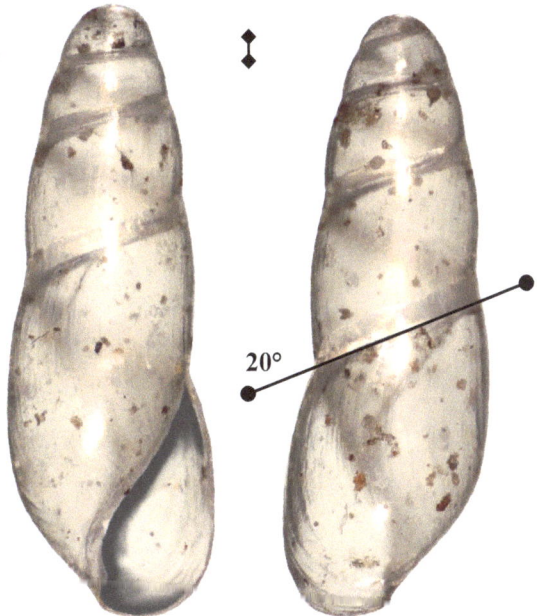

# Oak Ridge Teardrop         FERUSSACIDAE

*Cecilioides dicaprio* (new species)

**Height**: 2 mm, Diameter: 1.2 mm

**Description**: Pupa-shape, wide; lip simple; shell with 5 whorls; imperforate; sutures shallow; shell surface smooth, thin and translucent, axis visible through shell; color pellucid; very faint spiral striae; in the first few whorls the deceased and dried animal can be seen through the clear shell; columellar plait not twisted as in *Myxastyla.*

**Similar Species**: Both *C. consobrinus primus* (page 288) and *C. consobrinus veracruzensis* (page 289) have thinner builds and are lower elevation species; *Myxastyla* species have a strongly twisted columellar plait.

**Habitat**: In pockets of moist leaves that accumulate in depressions around large oak trees at 1000 meters;

**Status**: Rare, **ENDEMIC to BELIZE**; likely occurs in other high elevation montane oak forests along the Maya Mountain Divide but this remains to be more fully investigated in this largely unsampled region of Belize.

**Specimen**: Belize, Toledo District, Oak Ridge, Bladen Nature Reserve, elevation 1000 m Holotype UF 505459.

**Type Locality**: Belize, Toledo District, Oak Ridge, Bladen Nature Reserve (16°31'1"N, 88°55'43"W).

**Etymology**: Named in honor of American actor, Leonardo Dicaprio for his interest in creating a sustainable, green eco-resort in Belize and his use of film to bring attention to the challenges facing our natural world and planet.

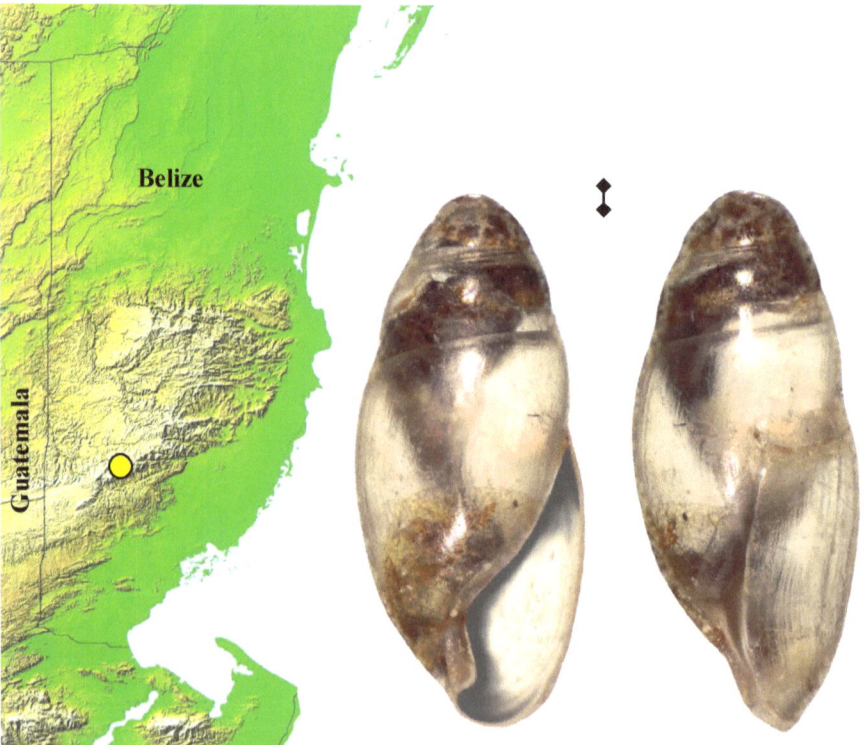

Similar Size Shells Compared (proportionate)
*All specimens are from the Bladen Nature Reserve*

1.5 mm

*Myxastyla hyalina*

*Carychium belizeense*

*Cecilioides dicaprio*

*Cecilioides consobrinus primus*

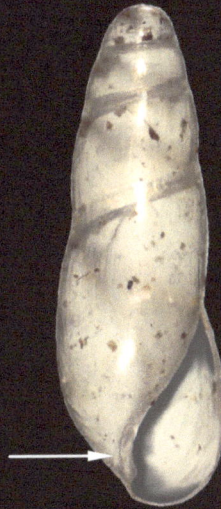

*Cecilioides consobrinus veracruzensis*

# Comb Snaggletooth                                  VERTIGINIDAE

*Gastrocopta pentodon* **(Say, 1821)**

**Height**: 1.5-1.8 mm, Diameter: 0.5 mm

**Description**: Pupa-shape; lip narrowly reflected; shell with 5-6 whorls; parietal tooth is largest; outer lip with a distinct palatal callus or ridge usually <u>containing 5 or 6 teeth but sometimes up to 9 teeth</u>; shells are somewhat translucent in live individuals and fresh shells are typically covered in soil.

**Similar Species**: *Gastrocopta contracta* is larger in size with fewer larger teeth in its aperture and is notably more tapering toward apex of shell.

**Habitat**: A species that is found in a variety of habitats including dry upland forests, acidic forests such as pine savannas, around limestone outcrops but also occasionally found in low wet places.

**Status**: Uncommon in Belize but found from eastern North America to Panama; current records in Belize are scarce but as surveys continue new sites are expected to be added.

**Specimen**: Belize, Toledo District, base of Forest Hill, Bladen Nature Reserve (Dourson collection).

**Type Locality**: Pennsylvania, USA.

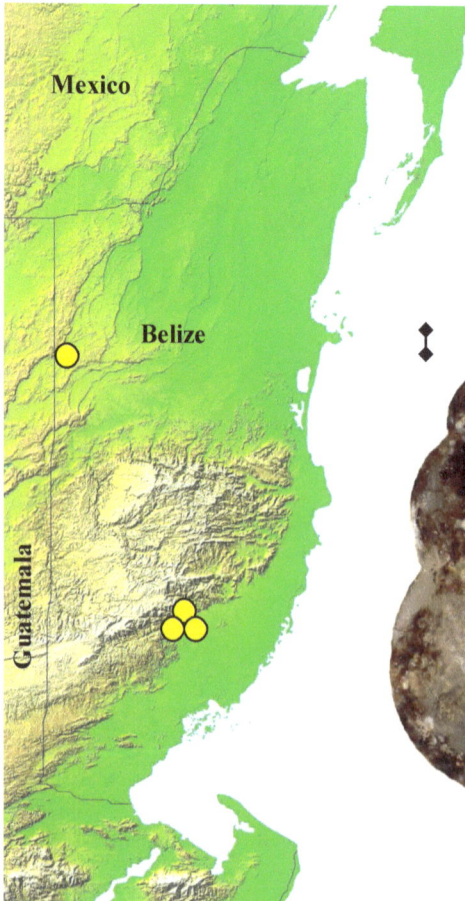

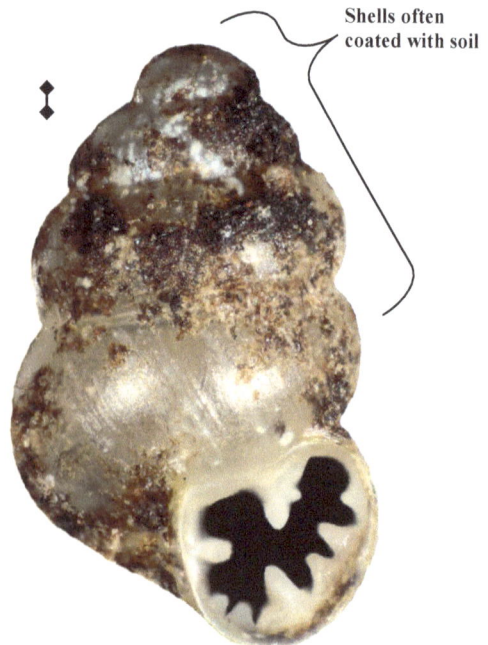

Mexico

Belize

Guatemala

Shells often coated with soil

292

# Bottleneck Snaggletooth            VERTIGINIDAE

*Gastrocopta contracta* **(Say, 1822)**

**Height**: 2.2-2.5 mm, Diameter: 0.5 mm

**Description**: Pupa-shape; lip widely reflected; shell with 5.5 whorls; 3-4 large teeth that crowd the aperture; parietal tooth forked (a); spire of shell tapering quickly; aperture more or less triangular in shape; transverse striae a weak but present feature; shells are somewhat translucent in live individuals and fresh shells are frequently covered with soil; old shells quickly become bleached.

**Similar Species**: *Gastrocopta servilis* is around the same size but with a more narrow build and less crowded teeth in the aperture.

**Habitat**: This species is found in nearly all terrestrial habitats including low wet places to dry mountainsides and acidic forests.

**Status**: Uncommon in Belize, current records are scarce but as surveys continue new sites are expected to be added; a widespread land snail from eastern North America to Panama.

**Specimen**: Belize, Toledo District, base of Forest Hill, Bladen Nature Reserve (Dourson collection).

**Type Locality**: Occoquan, Virginia, USA.

# Wandering Snaggletooth          VERTIGINIDAE

*Gastrocopta servilis* (Gould, 1843)

**Height**: 2.5 mm, Diameter: 1.1 mm

**Description**: Pupa-shape; lip widely reflected; shell with 6 whorls; with 4-5 short but prominent teeth that do not crowd the aperture; parietal tooth forked; spire of shell tapering little; aperture more or less oval in shape; shells are somewhat translucent in live individuals and fresh shells are frequently covered with soil; old shells quickly become bleached.

**Similar Species**: *Gastrocopta pentodon* is shorter in build with teeth that crowd the aperture, containing a fewer number of teeth than *G. servilis.*

**Habitat**: This species is found in nearly all terrestrial habitats from low wet places to dry mountainsides and acidic forests.

**Status** Rare in Belize, found only in one location in the southern Maya Mountains but additional locations are expected to be found especially if snail searches include leaf litters that harbor smaller species.

**Specimen**: Belize, Toledo District, 5 miles west of San Jose (Dourson collection).

**Type Locality**: Matanzas, Cuba.

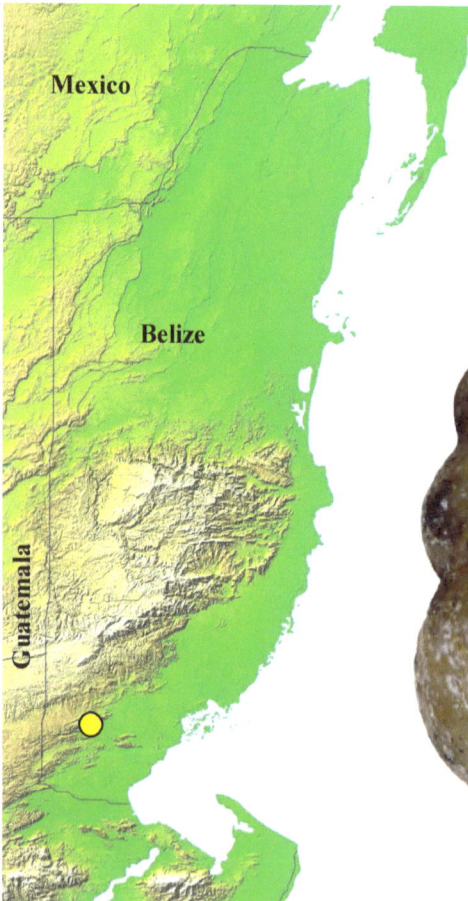

# Caribbean Birddrop          VERTIGINIDAE

*Sterkia eyriesii* (Drouet, 1859)
**Height**: 2 mm, Diameter: 0.5 mm
**Description**: Pupa-shape; lip widely reflected; shell with 5 whorls; with 5 rather large teeth that crowd the aperture; spire of shell tapering little; barrel shape; aperture more or less square; transverse striae a weak but present feature; shells are somewhat translucent in live individuals and fresh shells; old shells quickly become bleached.
**Similar Species**: The most barrel shaped of the VERTIGINIDAE species in Belize, all other species having at least some taper to the shells.
**Habitat**: This species is arboreal, living in trees and shrubs, clinging to the undersides of leaves (pers. comm. Fred Thompson, 2012); possibly found on epiphytes but this remains to be studied.
**Status**: Common where it occurs in Belize; this species is also common ranging from eastern North America to Panama.
**Specimen**: Belize, Toledo District, base of Forest Hill, Bladen Nature Reserve (Dourson collection).
**Type Locality**: Ilet-la-Mere, Guyana.

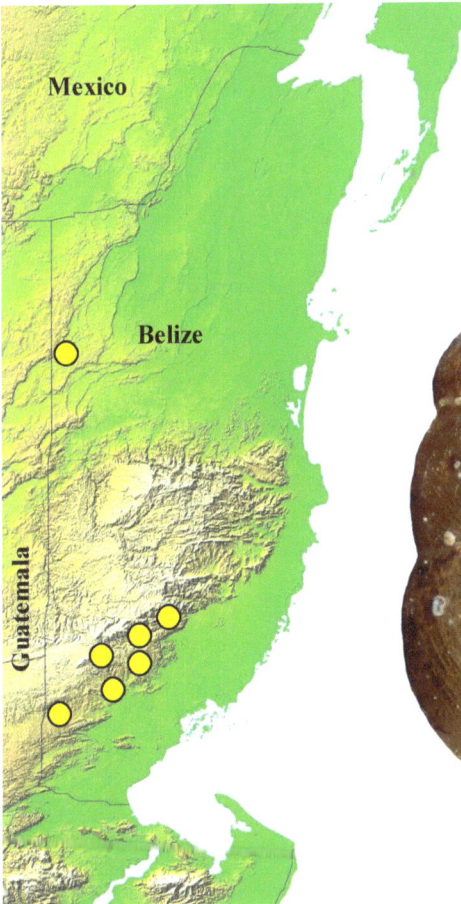

295

# Yam Babybody                    VERTIGINIDAE

*Pupisoma dioscoricola insigne* Pilsbry, 1920

**Height**: 1.95 mm, Diameter: 1.85 mm

**Description**: Pupa-shape; lip simple; shell with 3-4 whorls; perforate; without teeth; spire of shell tapering greatly; shell color cinnamon and glossy; aperture more or less roundish; transverse striae or low riblets are a strong feature with distinctive spiral striae (a); shells are somewhat translucent in live individuals and fresh shells; old shells quickly become bleached.

**Similar Species**: Most similar to *Pupisoma mediamericanum* but has a shorter shell form and has spiral striae instead of pitting.

**Habitat**: Limestone regions of Belize covered in tropical rainforest.

**Status**: Uncommon; known from three locations in Belize but additional locations are expected to be found especially if snail searches include leaf litter that harbor smaller species.

**Specimen**: Belize, Toledo District, Bladen Nature Reserve (Dourson collection).

**Type Locality**: Brownsville, Texas.

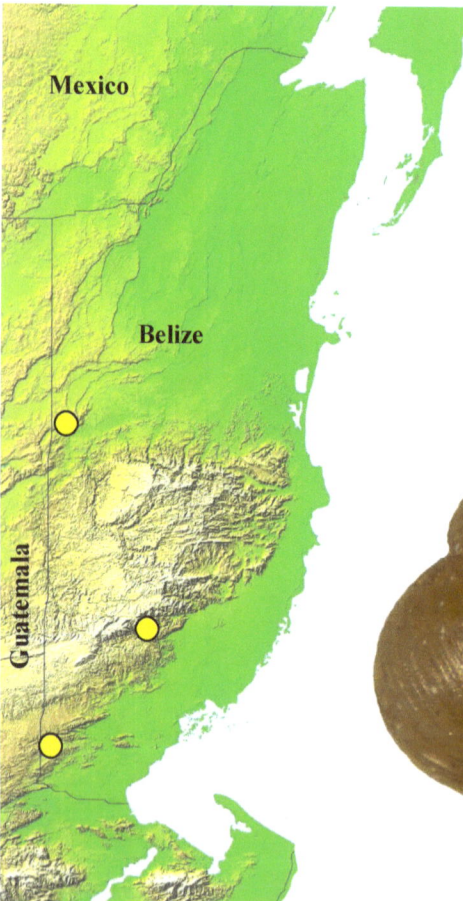

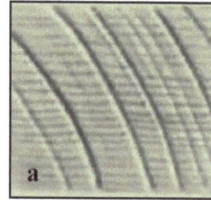

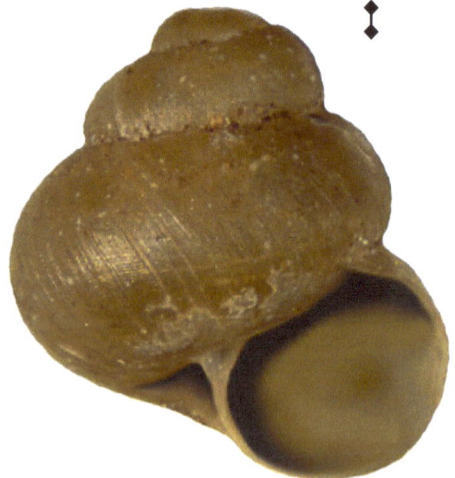

# Pitted Babybody                    VERTIGINIDAE

*Pupisoma mediamericanum* **Pilsbry, 1920**
**Height**: 1.6 mm, Diameter: 1.3-1.46 mm
**Description**: Pupa-shape; lip simple; shell with 4-4.5 whorls; perforate; without teeth; spire of shell tapering greatly; shell color olive-buff and somewhat glossy; aperture more or less roundish to oval; transverse striae or low riblets are a strong feature; without spiral striae but shell covered with shallow pitting; shells are somewhat translucent in live individuals and fresh shells; old shells quickly become bleached.
**Similar Species**: The high form, distinct riblets and greater number of whorls separates this from *Pupisoma dioscoricola;*
**Habitat**: Limestone regions of Belize covered in tropical rainforest.
**Status**: Rare in Belize, known only from the Bladen Nature Reserve but expected to be discovered elsewhere as sampling efforts for smaller species increases countrywide.
**Specimen**: Belize, Toledo District, hillside above Bladen River near Quebrada de Oro, Bladen Nature Reserve (Dourson collection).
**Type Locality**: Brownsville, Texas.

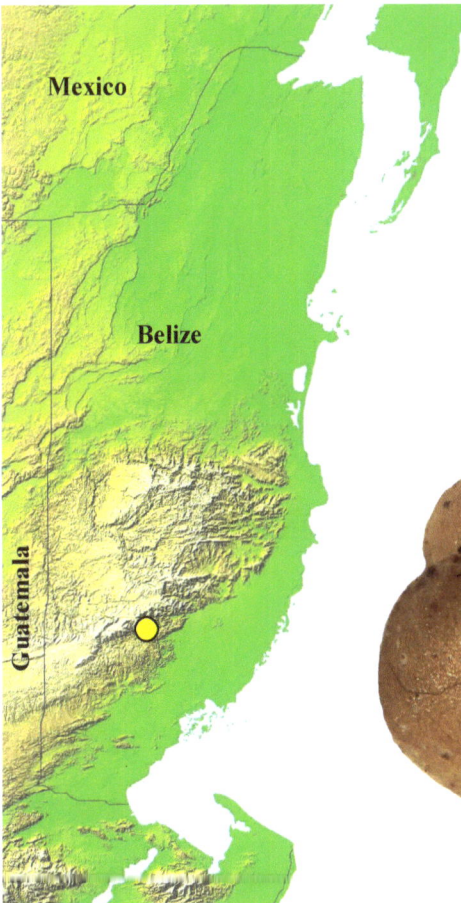

297

# Three-tooth Snaggletooth        VERTIGINIDAE

*Bothriopupa breviconus* **(Gould, 1848)**

**Height**: 2 mm, Diameter: 0.75-1 mm

**Description**: Pupa-shape, short and compact; lip widely reflected; perforate; shell with 6 whorls; with 3-4 small aperture teeth that do not crowd the aperture; color of shell bronze; spire of shell tapering considerably; aperture more or less square; transverse striae a weak but present feature.

**Similar Species**: Most similar to *Pupisoma dioscoricola insigne* and *P. mediamericanum* but containing 3 or 4 aperture teeth.

**Habitat**: Found in leaf litter on exposed limestone rocks and anthropogenic areas such as roadsides, yards and vacant lots.

**Status**: Rare in Belize, found only at one location; current records in Belize are scarce but as surveys continue new sites are expected to be added especially in southern portions of the country.

**Specimen**: Belize, Toledo District, base of rock wall; The Farm Inn, located off the Southern Highway (Dourson collection).

**Type Locality**: Livingston, Guatemala.

Mexico

Belize

Guatemala

Tooth may
be absent

298

# VERTIGINIDAE Shells Compared (proportionate)

1 mm

*Gastrocopta servilis*

*Gastrocopta pellucida hordeacella*
Tikal, Guatemala, H 1.8-2.5 mm
W.76-1 mm); Not yet reported from
Belize

*Gastrocopta pentodon*

*Gastrocopta contracta*

*Sterki eyriesii eyriesii*

*Bothriopupa breviconus*

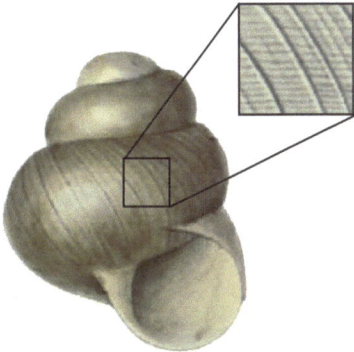

*Pupisoma dioscoricola insigne*

*Pupisoma mediamericanum*

(Illustrations above from Manual of Conchology, Vol. 26)

# Can Tiny Land Snails Really Fly?

Web image

Tiny land snails are easily overlooked and therefore rarely included in collections. However, in any given location, species under 5 mm will usually make up around half or more of the total land snail fauna. The majority of tiny species live among or under leaf litter of the forest floor, feeding on detritus, mycelium, molds and minuscule mushrooms. These miniatures of the land snail world are best viewed under a strong dissecting scope where micro-features of the shell surface can be examined in detail.

Believe it not, tiny land snails such as *Punctum minutissimum* and *Hawaiia minuscula*, although microscopic in size, are among the widest-ranging land snails in the Americas; found from eastern North America southward into South America. It is unknown precisely how these minute species have dispersed so well but has been speculated that tiny snails can be carried by high altitude winds, floating debris from flood events and birds (Abbott 1989, Fred Thompson pers. comm. 2010). Charles Darwin was perhaps the first one to propose the idea that snails might be dispersing by rafting or hitchhiking on migratory birds. Some land snails were thought to have been carried 5,500 miles from Europe to Tristan de Cunha Island in the South Atlantic Ocean and back (Gittenberger *et al.* 2006; Miura *et al.* 2012).

A recent study by Shinichiro Wada of the University of Tohoku in Japan showed that 15 percent of *Tornatellides boeningi* (a tiny land snail around 2.5 mm) consumed by birds can survive the bird's gut with smaller snails appearing to have a survival advantage over larger ones. This is the first study that demonstrates that birds can indeed transport a substantial number of micro land snails in their gut alive. Neotropical birds moving back and forth between the eastern US and Central America could be the conduit that help spread tiny species such as *Punctum vitreum*, this theory awaiting further investigation.

# 13 Slugs of Belize

This section includes slugs which are land snails that never developed a hard external shell for protection. Native slugs are represented by only two native species in Belize and one introduced. While exotic slugs may become quite prolific, native slugs are usually uncommon to rare. Interestingly, several species of slugs carry love darts used to stimulate copulation (mating). One representative species from each family is illustrated to the right below.

## The Families and Genera Included in Chapter 13

**VERONICELLIDAE**

**PHILOMYCIDAE**

# Morelet Slug                               VERONICELLIDAE

*Leidyula moreleti* **(Fischer, 1871)**

**Length**: 40-50 mm while crawling

**Description**: A thick tongue-like slug, brownish in color, sometimes with one light mid-dorsal stripe with a pair of dark bands (a) on the foot of the slug.

**Similar Species**: *Leidyula floridana* is larger without the dark bands on its underside and has two dorsal lines running the length of the animal.

**Habitat**: Slugs are generally rare in the tropics and only two native species have been found in the Maya Mountains; sensitive to drying, slugs are usually only found out crawling during wet weather; during dry periods they can be found under or inside rotting logs.

**Status**: Uncommon, currently known from only one location in Belize but likely occurs throughout the country.

**Specimen**: Belize, Toledo District, Bladen Nature Reserve (authors photo collection).

**Type Locality**: Palenque, Mexico.

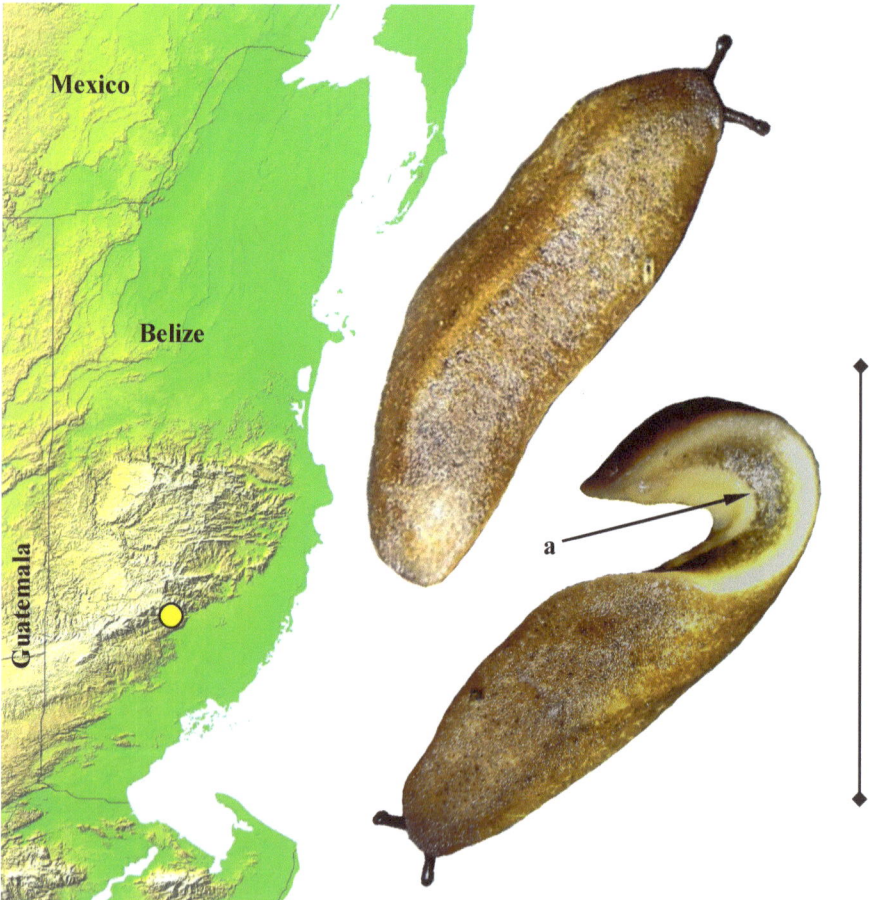

# Caribbean Slug (exotic)         VERONICELLIDAE

*Leidyula floridana* **Leidy, 1851**

**Length**: 50-70 mm while crawling

**Description**: A thick tongue-like slug; brownish in color, with or without two thin, dark dorsal stripe, a pair of dark bands (a) are found on the foot of the slug.

**Similar Species**: *Leidyula moreleti* is similar but without the two dorsal top lines seen in *L. floridana*; the Montane Mantleslug, *Pallifera* species (undetermined) is much smaller, has a thinner build and has only been found in higher elevation montane forests of Belize.

**Habitat**: A slug of urban areas and highly degraded forests; it has been found in large numbers at night during rainy weather on the road to the Cockscomb Basin Wildlife Sanctuary next to the citrus farms (Dourson pers. obs.); during the day slugs can be found under old refuse (tires, boards, etc.)

**Status**: Common in highly disturbed areas only.

**Specimen**: Belize, Toledo District, road to the Cockscomb (Dourson photo collection).

**Type Locality**: Punta Rasa, Florida.

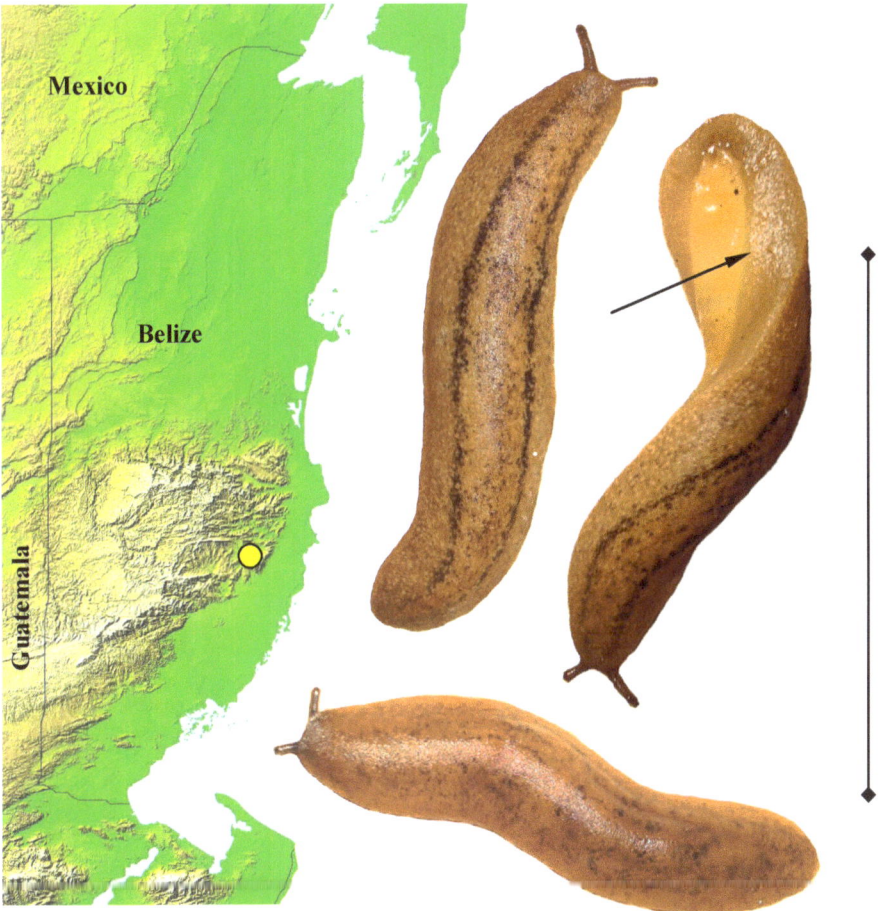

# Montane Mantleslug                    PHILOMYCIDAE

*Pallifera* species (undetermined)

**Length**: 15-25 mm while crawling

**Description**: A thin bodied slug with numerous scattered black spots on an otherwise charcoal-gray animal, anterior end of the animal orange, foot of the species whitish with a black tip (a).

**Similar Species**: *Leidyula moreleti* and *L. floridana* are much larger in all dimensions and both are found in lower elevation habitats of Belize.

**Habitat**: A species found in upper oak montane forests under leaf litter at an elevation of around 1000 meters.

**Status**: Rare; the status of this species in Belize remains unknown; currently reported from only one location (Oak Ridge, BNR), an area located along the Maya Mountain Divide, the species however is expected to occur across other portions of the divide.

**Specimen**: Belize, Toledo District, Oak Ridge at 1000 meters, Bladen Nature Reserve (BFREE collection, Belize).

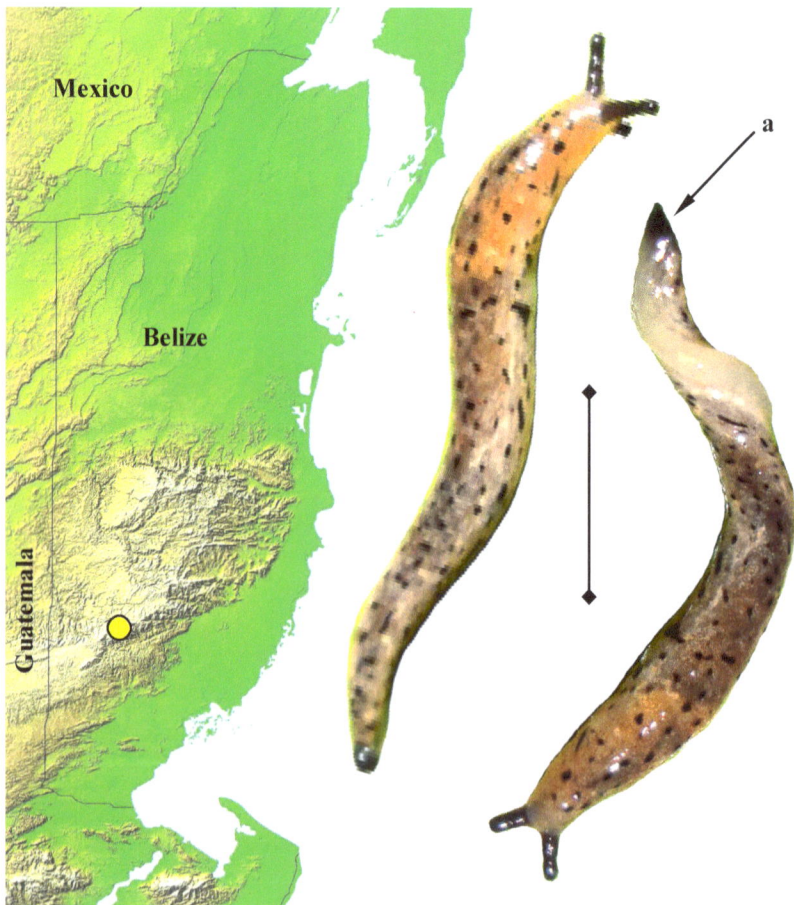

# 14 Freshwater Snails of Belize

This section includes aquatic snails reported from Belize or close to its borders. Although not nearly as diverse as the land snails, aquatic snails are extremely important components of healthy functioning ecosystems and should not be disregarded. This chapter is provided to assist biologists and archaeologists who often document snail shells during excavation of Ancient Maya sites and other researchers with some basic aquatic species found in the country. It is by no means complete as little literature exists for this group of organisms in Central America and much field work remains to be done.

# Freshwater Snails of Belize

Freshwater snails are gastropods found in rivers, streams, springs, stillwater lagoons and vernal pools. Although the diversity of aquatic snails is substantially less than that of land snails, they make up in numbers what they lack in diversity and are considered key organisms in aquatic systems. Aquatic snails are primary decomposers or "shredders" that consume large quantities of abscissed leaves (i.e. *Ficus* species) that fall within their reach (see images below). The left picture is of a horde of River Darts, *Pachychilus largillierti,* feeding on a fig leaf and the right image of the same leaf one hour later. As the snails feed on the leaf tissues, waste products of the gastropods are released into the water column spawning a new flush of nutrients. These nutrients become food for a whole host of aquatic micro-organisms which are then fed on by fish. Aquatic gastropods also feed on algae growing on submerged logs and rocks. Clearly, the loss of aquatic gastropods due to water pollution, dredging, or other forms of habitat degradation would threaten a multitude of wildlife species.

River snails were also utilized extensively by early Maya civilizations as a reliable food source, as evidenced by the presence of shells in so many ancient Maya archaeological sites. Even today, aquatic snails are used by Belizeans to treat eye problems such as glaucoma. Live snails are gathered from the river, broken in half, exposing a watery slime which is then applied to the affected eye (pers. comm. William Garcia 2009, Trio Village, Belize).

Aquatic snails have much thicker shells, as much as ten times more dense than land snails. This is an evolutionary response to living in an environment of ever-shifting material such as sand particles, rolling rocks and boulders which hammer the shells during flood events. Most land snails are relatively thin-shelled gastropods therefore would be pulverized in such a place with the exception of land snail species like *Helicina, Amphicyclotus, Orthalicus* and *Eucalodium*. These land snail genera are fairly difficult to break between your fingers unless you have an iron grip.

While it is generally uncommon to find land snails along gravel bars of rivers and creeks, it is not uncommon to find aquatic snails on hilltops or along overhanging clifflines and dry caves in Belize. Early Maya gathered large quantities of gastropods from rivers and carried them to camps where they were consumed and the empty shells were simply discarded on site.

A key character for separating aquatic snails from land snails is shell thickness. In general, if you can break the shell between your fingers, it is likely a land snail. If not, its likely an aquatic species.

The face of a River Dart, *Pachychilus largillierti,* looks more like a creature from a Star Wars movie than a snail.

# The River Darts, *Pachychilus* of Belize (shells proportionate)

100 mm

***Pachychilus glaphyrus immanis,*** height 100-105 mm Guatemala, Lake Isabel (UF 82846)

***Pachychilus glaphyrus obeliscus,*** height 57-75 mm Guatemala (UF 230026)

***Pachychilus lacustris lacustris***, height 52-6 mm Belize, Sibun River (UF 207511)

***Pachychilus glaphyrus,*** height 50-100 mm, Belize, Toledo District, Rio Grande at Big Falls (UF 207631)

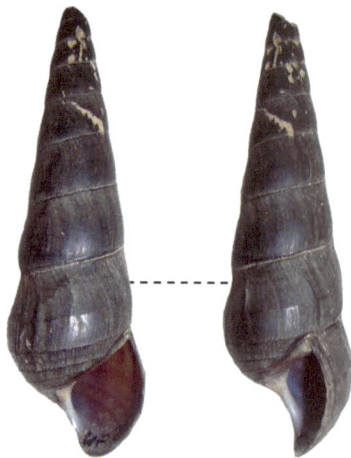

***Pachychilus corvinus***, height 30-40 mm, Guatemala, Alta Verapaz Prov. (UF 82839)

# The River Darts, *Pachychilus* of Belize (shells proportionate)

*Pachychilus largillierti*, height 50 mm
Belize, Punta Gorda (UF 167070)

*Pachychilus glaphyrus pyramidalis*,
height 75 mm, Rio Blanco River (BFREE
collection).

*Pachychilus indiorum*, height 60-65 mm Belize,
small creek near San Antonio (UF 135228)

*Pachychilus* species (undetermined), height
50-55 mm found only in springs above Bladen
River, BNR.

*Pachychilus indiorum*, height 45-61 mm,
Belize, St Margaret's Creek (UF 135114)

Enlarged

*Melanoides tuberculata* height 10-20 mm
(exotic), Belize, Toledo District, Bladen

309

# The Apple Snails, *Pomacea* of Belize (shells proportionate)

100 mm

*Pomacea belizensis,* Belize, Orange Walk District, New River (UF 41473)

*Pomacea ghiesbreghti*, Belize, Belize District, Big Snail (UF 254445)

*Pomacea ghiesbreghti,* Belize, Toledo District, Bladen Nature Reserve (BFREE collection)

# The Apple Snails, *Pomacea* of Belize (shells proportionate)

3 color morphs

*Pomacea flagellata flagellata*, Belize Orange Walk District, Doubloon Bank Lagoon (UF 41475)

*Pomacea flagellata erogata*, Mexico, Oaxaca River (UF 22559)

The protective operculum
all apple snails possess

*Pomacea flagellata livescens*, Mexico, Veracruz (UF 35031)

311

# 15Exotic Land Snails

**Giant African land snail**
**web image**

Land snails including slugs can be agricultural pests. Most often, snails and slugs that are problematic in gardens are non-native species that have been inadvertently introduced from other countries. Most have been accidently released into Belize by way of imported plants, potting soils or shipping crates. However, these introduced gastropods can naturalize quickly and multiply.

The degradation of native habitats in Belize through development only makes things worse providing the conduit for dispersal and movement of these foreign pests into new, unaffected areas of the country. These exotics often carry molluscan diseases and problematic parasites that can effect domestic animals, livestock, native wildlife and even humans. Exotic slugs can be especially damaging pests in greenhouses and agricultural lands, costing millions of dollars worth of damage.

In contrast, native snails and slugs are rarely trouble and most species actually become scarce or disappear entirely in areas where the natural vegetation has been eliminated. The Giant African Land Snail (pages 313-314) is a major pest in many places across the globe on agriculture crops including bananas causing millions of dollars of damage every year. This major crop nuisance that can grow to the size of an orange continues to spread and is now reported from several Caribbean islands including Barbados. If seen in Belize, these invasive pests should be quickly eliminated but care should be taken to ensure that the species is not the smaller, similar looking native Belizean snail, the Princess Cone, *Orthalicus princeps* (page 314).

# Giant African Land Snail

Multiple species; shells enormous 152-204 mm tall; 9 whorls; shell brownish with color streaks; lip not reflected at any stage of growth. Giant African Land Snails that occur outside of the continent of Africa can cause destruction on agricultural crops costing millions of dollars. They also harm and displace native wildlife. Now listed as one of the top 100 invasive species in the world, the Giant African Land Snail is reported to consume at least 500 different types of plants (many of which are food crops) and can even cause structural damage to plaster and stucco.

**Shells are enormous**

## ~Warning~

Giant African snails are known to carry a parasitic nematode that can lead to serious health issues like meningitis in humans. Reported cases of meningitis usually result when a person eats the raw or undercooked snail flesh sometimes found on leafy vegetables or by handling live wild snails. Depending on the species, snails can live nine years and are prolific breeders laying over 1200 eggs each year. The genus, *Achatina,* has been reported from Hawaii, USA, Caribbean islands of Martinique and Guadeloupe and has been recently detected in Saint Lucia, Cuba and Barbados.

**If these snails reach Belize, the environmental effect on the country will be disastrous!**

**Any Giant African Snails found should be caught, exterminated and reported to the Belize Agricultural Department promptly!**

# Giant African Land Snails are Bad News Exotics
(all images of specimens from the Florida Museum of Natural History, USA)

Juvenile

*Achatina zebra*

*Achatina panthera*

*Orthalicus princeps*
(native to Belize)

*Achatina weynsi*

*Achatina marginta*

# Asian Tramp Snail

*Bradybaena similis* **(Ferussac, 1821)** Shell 9-12 mm, 5 whorls; without teeth; umbilicus rimate (a); shell whitish, with or without light color bands; lip slightly reflected in adult shells; an exotic gastropod from Asia, found mostly around long established towns and cities. Specimen: Belize, Cayo District, town of San Ignacio.

a

# Tooted Gulella

*Gulella bicolor* **(Hutton, 1834)** Shell 7 mm tall; 8 whorls; large teeth that crowd the aperture; an exotic from Asia found around disturbed areas such as banana and citrus farms; well-established communities; the species is carnivorous, feeding on other land snails and is likely harmful to native land snail populations. Specimen: Belize, Toledo District, Trio Village.

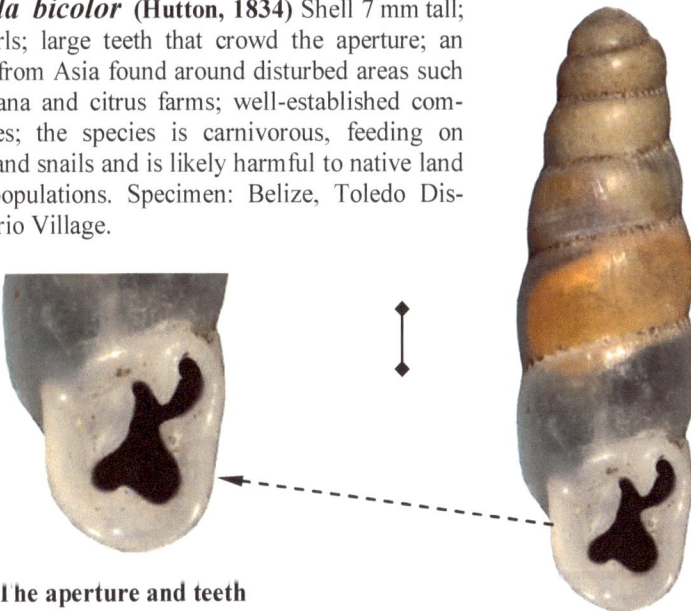

The aperture and teeth

315

# Caribbean Slug

*Leidyula floridana* Leidy, 1851

While crawling 50-70 mm; no hard shell; with or without two lateral stripes down the back (a), no color bands on bottom of foot as seen in *V. moreleti;* overall color may vary slightly; *V. floridana* is native to Jamaica and Florida, USA, and in general, not found in continuous forests; it has however been found in large numbers on the road to the Cockscomb Basin Wildlife Sanctuary next to the citrus grove. Specimens: Belize, Toledo District, Maya Center.

a

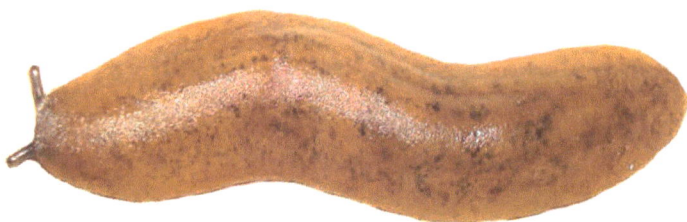

**Plain form**

# Glossary of Terms

**Abiotic:** absence of living organisms.

**Aestivate:** a form of hibernation; a cellophane-like covering formed over the aperture to prevent desiccation during periods of drought.

**Amphibious:** suited for both land and water.

**Angular periphery:** shell with an angular rather than a round shape or contour.

**Arcuate:** bow-shaped; curved.

**Aperture**: opening or mouth of the snail shell.

**Apex**: the top end of a shell where the embryonic whorl begins.

**Apex whorl:** initial whorl, the oldest part of the shell.

**Apical plug:** the sealed portion of a shell that is decollate.

**Axial ribs:** shell sculpture that may be axial or longitudinal, running parallel to the axis of the body from the anterior to the posterior.

**Axis:** the imaginary line around which the whorls of a coiled shell are formed.

**Basal tooth:** the tooth on the basal (lower) lip of a snail shell.

**Biogeographical:** geographical distribution of living organisms.

**Calcareous:** composed of calcium carbonate.

**Callus:** a deposit of lime or shell often seen as a thickening near the umbilicus.

**Capacious:** roomy.

**Carinate:** the periphery of the shell appears compressed creating a sharper angular appearance with a ridge-like rim.

**Cations:** minerals contained in the soil like magnesium, aluminum and calcium.

**Chemoreceptor:** a sense organ that responds to chemical stimuli.

**Circumneutral:** water with a pH of 5.5 (acidic) to 7.4 (alkaline).

**Columella:** the internal column of a spiral shell around which the whorls revolve.

**Columellar insertion:** the convergence of the peristome and the body whorl.

**Contiguous:** touching.

**Corneous:** horn-like in color.

**Crypsis:** the ability of an animal to avoid observation or detection by other animals. It may be a predation strategy or an anti-predator adaptation. Methods include camouflage and mimicry.

**Decollate, decollation:** the process of discarding the top 3-4 whorls including the apex of a snail's shell in some species.

**Depressed heliciform:** shell with flattened spire.

**Dessication**: process of drying out.

**Dextral aperture:** gastropods with apertures that open on the right.

**Dioecious:** having male reproductive organs in one individual and female in another of the same species.

**Discoidal:** having a flat spired shell.

**Embryonic whorl:** the earliest whorls of a snail shell that are formed in the egg.

**Endemic:** restricted to a specific locality or region.

**Epiphragm:** a calcified membrane produced by some land snails during hibernation that covers the aperture and prevents desiccation.

**Evapotranspiration**: the transport of water into the atmosphere from the earth's surface; the water cycle.

**Foot:** the locomotion organ of snails that is often modified for digging or grasping prey.

**Fulvous:** reddish-yellow, tawny.

**Geomorphology:** the study of the characteristics, origin, and development of land forms.

**Globose:** formed like a globe, spherical.

**Growth varices:** lines on the shell indicating a rest period during growth.

**Heliciform:** shell that has an elevated spire or globose form

**Hermaphrodite:** possessing both male and female reproductive organs in the same individual.

**Imperforate:** lacking an opening on the ventral or bottom end of the snail shell.

**Impressed lines:** marked by a furrow.

**Instar:** a stage in the life of an arthropod between two successive molts.

**Lamella, lamellae:** a fold, elongated "tooth" or raised callus in the aperture of a shell.

**Lip:** the edge of the aperture; also called the peristome.

**Loosely coiled:** having just a few, widely expanding whorls on a shell

**Mandible:** strong biting jaws of an arthropod.

**Mantle:** organ responsible for building the shell.

**Montane:** higher elevation forests where the forest line or timberline is often marked by a change to hardier species that occur in less dense stands.

**Operculum:** a circular calcium-based trapdoor structure on the top rear part of the fleshy foot that serves to thwart predators and prevent desiccation in some snails.

**Palatal tooth:** the tooth located on the outer lip.

**Papillae:** small calcium deposits that appear as minute bumps on the surface of the shell.

**Parietal tooth:** the tooth located on inner wall of the aperture or body of the snail shell.

**Pellucid:** translucent, clear.

**Penult whorl:** the next to the last whorl in an adult shell.

**Perforate:** a minute opening in the umbilicus.

**Periostome:** the opening of the snail shell.

**Periostricum:** a thin protein-based skin that covers the shell of the gastropod.

**Periphery:** edge of whorl farthest from the central axis of the shell.

**Phytogeography:** the science of geographical relationships of plants.

**Proboscis:** the sucking or feeding organ of an insect.

**Protoconch:** the larval shell of a gastropod.

**Pupate:** the time between larva and adult in some insects; usually in a cocoon.

**Radula:** a ribbon-like organ with many fine teeth used in rasping food located on the foot of a snail.

**Sinistral:** an aperture that occurs on the left side of the shell instead of the right side; opposite of dextral.

**Speciose:** rich in species.

**Spiral papillae:** raised bumps or papillae arranged spirally on a shell.

**Spiral striae:** surface features of a shell that are indented or raised that run parallel with the whorls.

**Spire:** all the whorls above the aperture.

**Striae:** surface features of a shell that are either indented or raised.

**Succiniform:** shell that is higher than wide with a very large aperture; spire is generally brief and the body whorl is greatly expanded.

**Suture:** indentation of the shell surface where one whorl of the shell is in contact with another.

**Sympatric:** occurring in the same geographical area; overlapping in distribution.

**Teleconch:** shell minus the protoconch.

**Transverse striae:** surface features indented or raised in the shell surface running perpendicular with the whorls.

**Truncate:** terminating abruptly, ending in a transverse line.

**Umbilicus:** an opening in the center of the axis of the shell bottom that is rather wide; umbilicate.

**Whorl:** one complete turn of a gastropod shell.

# Gastropods of Belize
# Species List
*(arranged by family)*
*Species Not Yet Documented in Belize or provided for comparison)

VESTIGASTROPODA– The OPERCULATES
HELICINIDAE
    *Helicina amoena* Pfeiffer,1849
    *Helicina arenicola* Morelet, 1849
    *Helicina bocourti* Crosse & Fischer, 1869
    *Helicina durangoana* Mousson, 1883
    *Helicina flavida* Menke, 1828
    *Helicina fragilis* Morelet, 1851
    *Helicina ghiesbreghti* Pfeiffer, 1856*
    *Helicina notata* Pfeiffer, 1856*
    *Helicina pterophora* Sykes, 1901*
    *Helicina oweniana* (Pfeiffer,1849)
    *Helicina rostrata* Morelet,1849
    *Helicina tenuis* Pfeiffer, 1849*
    ***Lucidella caldwelli* new species**
    *Lucidella lirata* (Pfeiffer,1847)
    *Pyrgodomus microdinus* (Morelet,1851)
    *Pyrgodomus simpsoni* (Ancey, 1886)
    *Schasicheila* species (undetermined)
AMPULLARIIDAE
    *Pomacea belizensis* (Crosse & Fischer, 1888)
    *Pomacea flagellata flagellata* (Say, 1827)
    *Pomacea flagellata erogata* (Fischer & Crosse, 1890)
    *Pomacea flagellata livescens* (Reeve, 1856)
    *Pomacea ghiesbreghti* (Reeve, 1856)
NEOCYCLOTIDAE
    *Amphicyclotus ponderosus* (Pfeiffer, 1851)
    *Neocyclotus dysoni berendti* (Pfeiffer, 1861)
    *Neocyclotus dysoni cookei* (Bartsch, & Morrison, 1942)
    *Neocyclotus dysoni dysoni* (Pfeiffer,1851)
    *Neocyclotus dysoni dyeri* (Bartsch & Morrison, 1942)
    *Neocyclotus dysoni hinkleyi* (Bartsch & Morrison, 1942)*
    *Tomocyclus fistularus* Thompson, 1963
    *Tomocyclus simulacrum* (Morelet, 1849)
PACHYCHILIDAE
    *Pachychilus corvinus* (Morelet,1849)
    *Pachychilus glaphyrus glaphyrus* (Morelet,1849)
    *Pachychilus glaphyrus immanis* (Morelet, 1851)
    *Pachychilus lacustris lacustris* (Morelet, 1849)
    *Pachychilus glaphyrus obeliscus* (Reeve, 1861)
    *Pachychilus glaphyrus pyramidalis* (Morelet, 1851)
    *Pachychilus indiorum* (Morelet, 1849)
    *Pachychilus largillierti largillierti* (Philippi,1843)

*Pachychilus largillierti pyramidalis* (Morelet, 1851)
*Pachychilus* species (undetermined)
THIARIDAE
*Melanoides tuberculata* (O. F. Muller,1774)
ANNULARIIDAE
*Choanopomops largillierti* (Pfeiffer, 1846)
*Diplopoma rigidulum* (Morelet, 1851)
*Halotudora gaigei* (Bequaert & Clench, 1931)
*Halotudora gruneri* (Pfeiffer, 1846)
*Halotudora kuesteri* (Pfeiffer, 1852)
*Parachondria rubicundis* (Morelet, 1849)
*Paradoxipoma enigmaticum* Watters, 2014
*Tudorisca andrewsae* (Ancey, 1886)
PULMONATA– The PULMONATES
CARYCHIIDAE
*Carychium belizeense* Jochum & Weigand, 2017
VERONICELLIDAE
*Leidyula floridana* Leidy,1851
*Leidyula moreleti* (Fischer,1871)
SUCCINEIDAE
*Succinea guatemalensis* Morelet, 1849
*Succinea luteola luteola* Gould, 1849
STROBILOPSIDAE
*Strobilops salvini* (Tristram,1863)
*Strobilops strebeli guatemalensis* (Tristam, 1863)
VERTIGINIDAE
*Bothriopupa breviconus* (Gould, 1848)
*Gastrocopta contracta* (Say,1822)
*Gastrocopta pentodon* (Say, 1821)
*Gastrocopta servilis* (Gould, 1843)
*Pupisoma dioscoricola insigne* Pilsbry, 1920
*Pupisoma mediamericanum* Pilsbry, 1920
*Sterkia eyriesii* (Drouet, 1859)
ORTHALICIDAE
*Bulimulus corneus* (Sowerby, 1833)
*Bulimulus coriaceus* (Pfeiffer, 1856)
*Bulimulus dysoni* (Pfeiffer, 1846)
*Bulimulus unicolor* (Sowerby, 1833)
*Bulimulus species (undetermined)*
*Bulimulus species (undetermined)*
*Bulimulus species (undetermined)*
*Drymaeus attenuatus* (Pfeiffer, 1851)
*Drymaeus dominicus* (Reeve, 1850)
*Drymaeus emeus* (Say, 1829)
*Drymaeus cf. hondurasanus* (Pfeiffer, 1846)
*Drymaeus serperastrus* (Say,1829)
*Drymaeus shattucki* Bequaert & Clench, 1931
*Drymaeus* species (undetermined)

*Drymaeus sulphureus* (Pfeiffer,1857)
*Drymaeus translucens alternans,* (Beck, 1837)
*Drymaeus cf. tropicalis* (Morelet, 1849)
**Drymaeus tzubi new species**
*Orthalicus cf. livens* Shuttleworth, 1856
*Orthalicus princeps princeps* (Broderip,1833)
*Orthalicus princeps crossei* (Von Martens, 1893)
*Orthalicus princeps deceptor* (Pilsbry, 1899)
UROCOPTIDAE
*Brachypodella dubia (*Pilsbry, 1891)
**Brachypodella levisa new species**
*Brachypodella morini* (Morelet, 1849)
*Brachypodella speluncae* (Morelet, 1852)
*Brachypodella subtilis* (Morelet, 1849)
*Coelocentrum badium* Pilsbry
*Coelocentrum fistulare* (Morelet,1849)
*Coelocentrum gigas* Martens, 1897
*Coelocentrum* species (undetermined)
*Epirobia polygyrella* (Von Martens, 1863)
*Eucalodium belizensis* Thompson & Dourson, 2013
*Microceramus concisus* (Morel,1849)
*Microceramus kieneri* (Pfeiffer, 1846)
FERUSSACIDAE
*Cecilioides consobrinus primus* (De Folin, 1870)
*Cecilioides consobrinus veracruzensis* (Crosse & Fischer, 1870)
**Cecilioides dicaprio new species**
SUBLINIDAE
*Allopeas gracilis (*Hutton, 1934)
*Beckianum beckianum* (Pfeiffer,1846)
*Lamellaxis cf. fordianus* (Ancey, 1886)
**Lamellaxis matola new species**
**Leptinaria doddi new species**
*Leptinaria lamellata* (Potiez & Michaud, 1838)
*Leptinaria livingstonensis* Hinkley, 1920
**Leptopeas corwinii new species**
*Leptopeas cf. guatemalense* (Strebel, 1882)
*Leptopeas micra* (Orbigny, 1835)
*Leptopeas yucatanense* (Pilsbry, 1906)
**Opeas marlini new species**
*Opeas pumilum* (Pfeiffer, 1840)
*Subulina octona* (Brugiere, 1789)
SPIRAXIDAE
*Euglandina cylindracea* (Phillips, 1846)
*Euglandina cumingi* (Beck, 1827)
*Euglandina ghiesbreghti* (Pfeiffer,1856)
**Euglandina fosteri new species**
*Euglandina* species (undetermined)
*Euglandina titan* Thompson, 1827*

321

*Pseudosubulina juancho* **new species**
*Mayaxis martensiana* (Pilsbry, 1920)
*Myxastyla hyalina* Thompson, 1995
*Myxastyla pycnota* Thompson, 1995
*Rectaxis alvaradoi* (Goodrich & van der Schalie, 1937)
*Rectaxis funibus* (Goodrich & van der Schalie, 1937)
*Rectaxis breweri* **new species**
*Salasiella guatemalensis* Pilsbry, 1920
*Salasiella modesta* (Pfeiffer, 1862)
*Salasiella cf. pulchella* (Pfeiffer, 1856)
*Streptostyla delibuta* (Morelet,1851)
*Streptostyla dysoni* (Pfeiffer, 1846)
*Streptostyla cf. labida* (Morelet, 1851)
*Streptostyla lattrei* Pfeiffer,1845
*Streptostyla ligulata* (Morelet, 1849)
*Streptostyla meridana* (Morelet, 1849)
*Streptostyla nigricans* (Pfeiffer,1845)
*Streptostyla cf. thomsoni* (Ancey, 1888)
*Streptostyla ventricosula* (Morelet, 1849)
*Streptostyla* species (undetermined)
*Varicoglandina monilifera* (Pfeiffer, 1845)
*Volutaxis livingstonensis* (Pilsbry, 1920)
*Volutaxis longior* (Pilsbry, 1920)
*Volutaxis similaris* (Strebel, 1882)
*Volutaxis sulciferus* (Morelet, 1851)
SCOLODONTIDAE
*Miradiscops bladenensis* **new species**
*Miradiscops maya* Pilsbry, 1920
*Miradiscops puncticipitis* (Pilsbry, 1926)
*Miradiscops striatae* **new species**
*Miradiscops youngii* **new species**
PUNCTIDAE
*Punctum cf. coloba* (Pilsbry, 1894)
*Punctum vitreum* (H. B. Baker,
*Striatura meridionalis* Pilsbry & Ferriss, 1906
CHAROPIDAE
*Chanomphalus angelae* **new species**
*Chanomphalus pilsbryi* (H. B. Baker,1927)
*Radiodiscus millicostatus* Pilsbry & Ferriss, 1906
*Rotadiscus hermanni* (Pfeiffer, 1866)
*Rotadiscus saqui* **new species**
SAGDIDAE
*Hyalosagda turbinella* (Morelet, 1851)
*Lacteoluna selenina* (Gould, 1848)
*Xenodiscula taintori* Goodrich & Van der Schalie,1937
EUCONULIDAE
*Guppya gundlachi* (Pfeiffer,1840)
*Guppya gundlachi orosciana* Von Martens, 1892

*Habroconus elegantulus* (Pilsbry, 1919)
*Habroconus pittieri* (Von Martens, 1892)
*Habroconus trochulinus* (Morelet, 1851)
ZONITIDAE
*Hawaiia minuscula* (A. Binney,1840)
*Zonitoides arboreus* (Say,1821)
*Zonitoides cf. hoffmanni* (Von Martens, 1892)
PHILOMYCIDAE
*Pallifera species* (undetermined)
XANTHONICHIDAE
*Trichodiscina coactiliata* Férussac,1838
*Trichodiscina hinkleyi* Pilsbry, 1919
*Trichodiscina suturalis* (Pfeiffer, 1846)
*Leptarionta aff. trigonostoma* (Pfeiffer,1844)
*Leptarionta trigonostoma* Pfeiffer, 1844)*
*Leptarionta trigonostoma stolliana* (Von Martens, 1892)*
POLYGYRIDAE
*Polygyra dysoni* (Shuttleworth, 1852)
*Polygyra yucatanea* (Morelet, 1849)
*Praticolella griseola* (Pfeiffer, 1841)
THYSANOPHORIDAE
*Itzamna sigmoides*
*Microconus rufus* Thompson, 1958
*Microconus pilsbryi* Thompson, 1958
*Thysanophora caecoides* (Tate, 1870)
*Thysanophora cf. canalis* Pilsbry, 1910
*Thysanophora conspurcatella* (Morelet, 1851)
*Thysanophora impura* (Pfeiffer, 1866)
*Thysanophora plagioptycha* (Shuttleworth,1854)
**Thysanophora meermani new species**
*Thysanophora rhoadsi* Pilsbry, 1919

# Bibliography

ABBOTT, R. T. 1989. *Compendium of Landshells*. American Malacologists, Inc. Melbourne, Florida. 220 pp.

ARTHUR, M. A., L. M. TRITTON, & T. J. FAHEY. 1993. Dead bole mass and nutrients remaining 23 years after clear-cutting of a northern hardwood forest. Canadian Journal of Forest Reserves. 23:1298-1305.

BAALBERGEN E, R. HELWERDA, R. SCHELFHORST, R. F. CASTILLO CAJAS, C.H.M. VAN MOORSEL & R. KUNDRATA. 2014. Predator-prey interactions between shell-boring beetle larvae and rock-dwelling land snails. PLoS ONE. 9(6): e100366. https://doi.org/10.1371/journal.pone.0100366

BAKER, H. B. 1922a. Notes on the radula of the Helicinidae. Proceedings of the Academy of Natural Sciences of Philadelphia, 74: 29-67.

BAKER, H. B. 1922b. The Mollusca collected by the University of Michigan-Walker Expedition in southern Vera Cruz, Mexico, IIII. Occasional papers of the Museum of Zoology, University of Michigan, (106): 1-60.

BAKER, H. B. 1923a. The Mollusca collected by the University of Michigan-Walker Expedition in southern Vera Cruz, Mexico, IV. Occasional papers of the Museum of Zoology, University of Michigan, (135): 1-19.

BAKER, H. B. 1923b. The Mollusca collected by the University of Michigan-Williamson Expedition in Venezuela, I. Occasional Papers of the Museum of Zoology, University of Michigan, (137): 1-59.

BAKER, H. B. 1925a. The Mollusca collected by the University of Michigan-Williamson Expedition in Venezuela, III. Occasional Papers of the Museum of Zoology, University of Michigan,(156): 1-57.

BARTSCH, P. &

BARTSCH, P. 1959. Land Mollusca of the Tikal National Park, Guatemala. Occasional Papers of the Museum of Zoology, University of Michigan. 612: 1-16.

BLAND, T. 1883. Notes on *Macroceramus kieneri* and *M. pontificus* with wood cuts. Annals of New York Academy of Sciences. 2:127-128.

BONATO, V., K. G. FACURE, & W. UIEDA. 2004. Food habits of bats of subfamily VAMPYRINIAE in Brazil. Journal of Mammalogy. 85 (4):708-713.

BOYCOTT, A. E. 1934. The habitats of land mollusca in Britain. Journal of Ecology. 22:1-38.

BRAUN, E. L. 1940. An ecological transect of Black Mountain, Kentucky. Ecological Monographs. 10:193-241.

BREURE, A. S. H. 1978. Notes and descriptions of Bulimulidae. Zoologische Verhandlungen, Uitgegeven door het Rijkmuseum van Natuurlijke Historie te Leiden (164): 1-255.

BREURE, A. S. H 1979. Systematics, phylogeny and zoogeography of Bulimulinae (Mollusca). Zoologische Verhandlungen, Uitgegeven door het Rijkmuseum van Natuurlijke Historie te Leiden (168): 1-215.

BREURE, A.S.H. 2011. Dangling snails and dangerous spiders: malacophagy and mimicry in terrestrial gastropods. Folia conchyliologica. 7: 7–13.

BREURE, A.S.H. & J. D. ABLETT. 2014. Annotated type catalogue of the Bulimulidae (Mollusca, Gastropoda, Orthalicoidea) in the Natural History-

Museum, London.— *ZooKeys* 392: 1–367, figs 1–75, L1–L67. [PDF: doi: 10.3897/zookeys.392.6328 | 77]

BREURE, A. S. H. & A. S. C. ESKENS. 1981. Notes on and descriptions of Bulimulidae (Mollusca, Gastropoda), II. Zoologische Verhandelingen, Uitgegeven door het Rijkmuseum van Natuurlijke Historie te Leiden, (186): 1-111.

BREURE, A.S.H. & A.A.C. ESKENS. 1981. Notes on and descriptions of Bulimulidae (Mollusca, Gastropoda), 2. *Zoologische Verhandelingen (Leiden)* 186: 1-111. Breure, A.S.H. & J. R. Schouten. Notes on and descriptions of Bulimulidae (Mollusca, Gastropoda), 3. *Zoologische Verhandelingen (Leiden)* 211: 1-98

BREURE, A.S.H. & J.R. SCHOUTEN. 1985. Notes on and descriptions of Bulimulidae (Mollusca, Gastropoda), 3. *Zoologische Verhandelingen (Leiden)* 211: 1-98.

BREWER, S. W. & M. A. H. WEBB. 2002. A seasonal evergreen forest in Belize: unusually high tree species richness for northern Central America. Botanical Journal of the Linnean Society. 138: 275-296

BURCH, J. B. 1955. Some ecological factors of the soil affecting the distribution and abundance of land snails in eastern Virginia. Nautilus 69(2):26-29.

BURCH, J. B. 1962. *How to know the eastern land snails.* Wm. C. Brown Publishing Company. Dubuque, Iowa. 214 pp.

BURCH, J. B. & T. A. PEARCE. 1990. *Terrestrial Gastropoda.* in D. L. Dindall (ed.), Soil Biology Guide. Wiley and Sons, Inc., New Jersey. Pp. 201-309.

CAMERON, R. A. D. 1970. Differences in the distributions of three species Helicid land snails in the limestone district of Derbyshire. Proceedings of the Royal Society of London. 176:130-159.

CHASE, R., & K. C. BLANCHARD. 2006. The snail's love-dart delivers mucus to increase paternity. Proceedings of the Royal Society Biology. 273:1471-1475.

COCHRAN, J. 2007. Diet, habitat, and ecomorphology of cichlids in the upper Bladen river, Belize. Master's thesis. Texas A. & M.

CONEY, C. C., W. A. TARPLEY, J. C. WARDEN, & J. W. NAGEL. 1982. Ecological studies of land snails in the Hiwassee River Basin of Tennessee, USA. Malacological Review. 15:69-106.

DALLINGER, R. 1993. Strategies of metal detoxification in Terrestrial invertebrates. *in*: Dallinger, R. and Rainbow, P. S. (eds.). Ecotoxicology of Metals in Invertebrates. Lewis Publishers. Boca Raton, Florida. Pp. 245-289.

DALLINGER, R. & W. WEISER. 1984a. Molecular fractionation of Zn, Cu, Cd, and Pb in the midgut gland of *Helix pomatia* L. Comparative Biochemistry and Physiology. 79:125-129.

DALLINGER, R., B. BERGER, C. GRUBER, P. HUNZIKER & S. STUZENBAUM. 2000. Metallothioneins in terrestrial invertebrates: structural aspects, biological significance, and implications for their use as biomarkers. Cellular Molecular Biology 46:331-346.

DAY, M. J. 2007. Chapter 5. *Karst Landscapes:* In Bundschuh, J. and Alvarado, G. eds. Central America: Geology, Resources & Hazards. Taylor & Francis, London.

DEVRIES G. W. 2003. Enhancing collaboration for conservation & development in Southern Belize. Master's Thesis. University of Michigan. pp 183-206.

DOUGLAS, D. 2011. Land snail species diversity and composition among different disturbance regimes in central and eastern Kentucky forests. Master's thesis. Eastern Kentucky University. Richmond, KY.

DOURSON, D. 2007. A selected land snail compilation of the Central Kentucky Knobstone escarpment on Furnace Mountain in Powell County, Kentucky, USA. Journal of the Kentucky Academy of Sciences. 68 (2):119-131.

DOURSON, D. 2008. The feeding behavior and diet of an endemic West Virginia land snail, *Triodopsis platysayoides.* American Malacological Bulletin. 26:153-159.

DOURSON, D. 2009. A natural history of the Bladen Nature Reserve and its gastropods. Goatslug Publications, Bakersville, NC, USA. 148 pp.

DOURSON, D. 2012. Four new land snail species from the southern Appalachian Mountains. Journal of the North Carolina Academy of Sciences. 128:1-10.

DOURSON, D. 2012. Biodiversity of the Maya Mountains: a focus on the Bladen Nature Reserve, Belize, Central America. Goatslug Publications, Bakersville, NC, USA. 343 pp.

DOURSON, D. & J. BEVERLY. 2009. Diversity, substrata divisions and biogeographical affinities of land snails at Bad Branch State Nature Preserve, Letcher County, KY, USA. Journal of the Kentucky Academy of Science. 68(2):119-131.

DOURSON, D., R. CALDWELL, K. LOUKES, & W. GARCIA. 2011. Land snails of the Bladen Nature Reserve with notes on their significance to other organisms in southern Belize (Toledo District), Central America. Mesoamericana. 15(3): 31-42.

FOOTE, B. A. 1959. Biology and life history of the snail-killing flies belonging to the genus Sciomyza. Annals of the Entomological Society of America. 52:31-32.

FOURNIE, J. & M. CHETAIL. 1984. Calcium dynamics in land gastropods. American Zoologist. 24:857-870.

GEIGER, R. 1965. The climate near the ground. Harvard University Press. Cambridge, MA, USA.

GERVAIS, J., A. TRAVESET, AND M. F. WILSON. 1998. The potential for seed dispersal by the banana slug (*Ariolimax columbanus*). American Midland Naturalist. 140:103-110.

GETZ, L. L. 1974. Species diversity of terrestrial snails in the Great Smoky Mountains, USA. Nautilus. 88:6-9.

GITTENBERGER, B. 2006. Nodes of large degree in random trees and forests. Random Structures and Algorithms. 28: 374–385.

GOODRICH, C., AND H. VAN DER SCHALIE. 1937. Mollusca of Peten and North Alta Vera Paz, Guatemala. University of Michigan Museum of

326

Zoology. Miscellaneous Publications. 34:1-47.

GOSZ, J. R., G. E. LIKENS, & F. H. BORMANN. 1973. Nutrient release from decomposing leaf and branch litter in the Hubbard Brook Forest, New Hampshire. Ecological Monographs. 43:173-191.

GRAVELAND, J. R., J. H. VAN DER WAL, J. H VAN BALEN, & A. J. VAN NOORDWIJK. 1994. Poor reproduction in forest passerines from decline of snail abundance on acidified soils. Nature. 368:446-448

GRAVELAND, J. R. 1996. Avian eggshell formation in calcium-rich and calcium-poor habitats: importance of snail shells and anthropogenic calcium sources. Canadian Journal of Zoology. 74:1035-1044.

GUHA, M. M & R. L. MITCHELL. 1966. The trace and major element composition of some deciduous trees: Chapter 2. Seasonal Changes. Plant and Soil. 24:90-112.

HAAS. F., & A. SOLEM. 1960a. Non-marine mollusks from British Honduras. Nautilus. 73(4):129-131.

HALL, I. H. & J. H. BATESON. 1972. Late Paleozoic lavas in Maya Mountains, British Honduras, and their possible regional significance. The American Association of Petroleum Geologists Bulletin. 56(5): 950-963.

HAMES, R., K. V. ROSENBURG, J. D. LOWE, S. E. BARKER, & A. A. DHONDT. 2002. Adverse effects of acid rain on the distribution of the woodthrush *Hylocichla mustelina* in North America. Proceedings of the National Academy of Sciences. 99 No. 16.

HAMMOND, N. 1982. The prehistory of Belize. Journal of Field Archaeology. 9:349-362.

HENDERSON, J. B. & P. BARTSCH. 1920. A classification of the American operculate land mollusks of the family Annulariidae. Proceedings of the United States National Museum. 58: 49-82.

HINKLEY, A. A. 1920. Guatemala Mollusca. Nautilus, 34: 37-55.

HOGUE, C. L. 1985. Cultural Entomology. Annual Review of Entomology. Vol. 32: 181-199.

HOGUE, C. L. 1993. *Latin American insects and entomology*. University of California Press.

HOLDRIDGE, L. R., W. C. GRENKE, W. H. HATHEWAY, T. LIANG, & J. A. TOSI, JR. 1971. *Forest environments in tropical life zones: a pilot study*. Pergamon Press, Oxford, UK.

HUBRICHT, L. 1985. *The distributions of the Native Land Mollusks of the Eastern United States*. Fieldiana. FMN. Chicago, Illinois. 191 pp.

IREMONGER, S. AND N. BROKAW. 1995. Vegetation classification and mapping methodology as a basis for gap analysis of protected area coverage in Belize *in*: *Towards a National Protected Area Systems Plan for Belize*. Programme for Belize & Inter-American Development Bank Synthesis Report.

JACOT, A. P. 1935. Mollscan populations of old growth forests and re-wooded fields in the Asheville Basin in North Carolina. Ecology. 16:603-605.

JENKINS, M. A., S. JOSE & P. S. WHITE. 2007. Impacts of an exotic disease and vegetation change on foliar calcium cycling in Appalachian forests. Ecological Applications. 17(3):869-881.

JOCHUM, A., A. M. WEIGAND, E. BOCHUD, T. INABNIT, D. D. DORGE, B. RUTHENSTEINER, A. FARRE, G. MARTELS & M. KAMP-SCHULTE. Three new species of Carychium O. F. Müller, 1773 from the southeastern USA, Belize, and Panama are described using computer tomography CT (Eupulmonata, Ellobioidea, Carychiidae). Zookeys 675:97-127.

KALISZ, P. J. & J. E. POWELL. 2003. Effect of calcareous road dust on land snails and millipedes in acid forest soils of the Daniel Boone National Forest. Forest Ecology and Management. 186:177-183.

KARLIN, E. J. 1961. Ecological relationships between vegetation in the distribution of land snails in Montana, Colorado, and New Mexico. American Midland Naturalist. 65:60-66.

KAVOUNTZIS, E. 1989. Thesis (Ph. D.) An investigation of Maya ritual cave use with special reference to Naj Tunich, Peten, Guatemala. University of California, Los Angeles, USA.

KING, D. T. AND L. W. PETRUNY. 2003. *Stratigraphy and sedimentology of coarse impactoclastic breccia units within Cretaceous-Tertiary boundary section, Albion Island, Belize.* in: Koeberl, C. and F. Martinez-Ruiz eds. Impact markers in the stratigraphic record (impact studies). Springer-Verlag, Berlin. p 203-228.

KOENE, J. M., T. S. LIEW, K. MONTAGNE-WAJER, & M. SCHILTHUIZEN. 2013. A syringe-like love dart injects mal accessory gland products in a tropical hermaphrodite. PLoSOne. 8e69968.

KOTTEK, M., J. GRIESER, C. BECK, B. RUDOLF, & F. RUBEL. 2006. World Map of the Köppen-Geiger climate classification updated. Meteorologische Zeitschrift .15(3): 259-263.

KÖPPEN, W. 1900. Versuch einer Klassifikation der Klimate, vorzugsweise nach ihren Beziehungen zur Pflanzenwelt. Geographie Zeitschriften 6: 593 –611, 657–679.

LAWRENCE, J. F. & E. B. BRITTON. 1994. Australian beetles. Melbourne University Press.

LEE, J. C. 1994. *The Amphibians and Reptiles of the Yucatán Peninsula.* Cornell University Press, Ithaca. New York.

LODI, M. AND J. M. KOENE. 2016. The love-darts of land snails: integrating physiology, morphology and behaviour, *Journal of Molluscan Studies.* 82 (1):1–10.

MACHENSTED, U, AND K. MARKEL. 2001. Chapter 4: Radular structure and function. Pp. 213-236 in: G. M. Barker ed. *The biology of terrestrial mollusks.* CABI. New York, NY.

MARSHALL, J. S. 2007. Chapter 3. *Geomorphology and Physiographic Provinces of Central America:* In Bundschuh, J. and Alvarado, G. eds. Central America: Geology, Resources, and Hazards. Taylor and Francis, London. pp. 75-122.

MARTENS, E. v. 1890-1901. *Biologia Centrali-Americana. Mollusca.* 1-706. British Museum Natural History, London.

MARTIN, A. C., H. S. ZIM, & A. L. NELSON. 1951. American Wildlife and Plants: A Guide to Wildlife Food Habits. Dover, New York. 500 pp.

MCHARGUE, J. S. & W. R. ROY. 1932. Mineral and nitrogen content of the leaves of some forest trees at different times in the growing season. Botanical Gazette. 94:381-393.

MEERMAN, J. C. & W. SABIDO. 2001. Central American ecosystems map: Belize. Vol. I & II. Programme for Belize, Belize City, Belize.

MILLER, T. E. 1996. Geologic and hydrologic controls on karst and cave development in Belize. Journal of Cave and Karst Studies. 58(2): 100-120.

MIURA, O., M. E. TORCHIN, E. BERMINGHAM, D. K. JACOBS & R. F. HECHINGER. 2012. Flying shells: historical dispersal of marine snails across Central America. Proceedings of the Royal Society Biological Sciences. 279:1061-1067.

MONTGOMERY, C. E., J. M. RAY, A. H. SAVITZKY, E. J. RODRIGUEZ-GRIFFITH, H. L. ROSS & K. R. LIPS. 2007. *Sibon longifrenis* diet. Herpetological Review. 38(3):343.

MORITZ, C., K. S. RICHARDSON, S. FERRIER, J. STANISIC, S. E. WILLIAMS & T. WHIFFIN. 2001. Biogeographic concordance and efficiency of taxon indicators for establishing conservation priority in a tropical rainforest biota. Proceedings of the Royal Society London. 268:1875-1881.

NASKRECKI, P. & K. NISHIDA. 2007. Novel trophobiotic interactions in lantern bugs (Insecta: Auchenorrhyncha: Fulgoridae). Journal of Natural History. 41(37–40): 2397–2402.

NATION, R. 2005. The influence of soil calcium on land snail diversity in the Blue Ridge Escarpment of South Carolina. Dissertation presented to Clemson University. Clemson, South Carolina, USA.

NEKOLA, J. C. 1999. Terrestrial gastropod richness of carbonate cliff and associated habitats in the Great Lakes region of North America. Malacologia 41(1):231-252.

PEARCE, T. A. 2008. When a snail dies in the forest, how long will the shell persist? Effect of dissolution and micro-bioerosion. American Malacological Bulletin. 26:111-117.

PETERS, J. A. 1960. The snakes of the subfamily Dipsadine. Miscellaneous Publication of the Museum of Zoology, University of Michigan. 114:1-228.

PETRANKA, J. G. 1982. The distribution and diversity of land snails on Big Black Mountain, Kentucky. (A thesis). University of Kentucky. Lexington, KY.

PETRANKA, J. W. 1998. *Salamanders of the United States and Canada.* Smithsonian Institution Press, Washington, D. C.

PHILIPPI, R. A. 1842-1844. Abbildungen und beschreibungen neuer oder wennig gekannter Conchylien, (I): 1-204. Cassel.

PHILLIPS, R. A., J. MEERMAN, T. BOOMSMA, R. HOWE, R. MARTINEZ, R. BARBOUR, B. MARTINICO, A. BRADSHAW, L. POP, & I. MAI. 2014. Breeding Biology of the Hook-billed Kite in Belize: A Successful Triple Brooding. Poster at Raptor Research Conference. Sacromento, California, USA. PDF.

PILSBRY, H. A. 1891. Proceedings Academy of Natural Sciences Philadelphia. 315. Plate 15 fig. 14.

PILSBRY, H. A. 1918-1920. Manual of Conchology: Pupillidae: Gastrocoptinae, Vertigininae. Academy of Natural Sciences Philadelphia. Ser. 2. Vol. 25: 1-401.

PILSBRY, H. A. 1920. Review of the *Thysanophora plagioptycha* group. Nautilus. 33: 93-96.

PILSBRY, H. A. 1920b. Mollusca from Central America and Mexico. Proceedings of the Academy of Natural Sciences of Philadelphia, 71: 212-223.

PILSBRY, H. A. 1920e. Mollusca from Central America and Mexico. Proceedings of the Academy of Natural Sciences of Philadelphia. 72: 195-202.

PILSBRY, H. A. 1920-1921. Manual of Conchology: Pupillidae: Vertigininae, Pulpillinae. Academy of Natural Sciences Philadelphia. Ser. 2. Vol. 26: 1-254.

PILSBRY, H. A. 1926b. The land mollusks of the Republic of Panama and the Canal Zone. Proceedings of the Academy of Natural Sciences of Philadelphia. 78: 57-126.

PILSBRY, H. A. 1927-1935. Manual of Conchology: Geographical distribution of Pupillidae; Strobilopsidae, Valloniidae and Pleurodiscidae. Academy of Natural Sciences Philadelphia. Ser. 2, Vol. 28: 1-26.

PILSBRY, H. A. 1939. Land Mollusca of North America (north of Mexico). Vol. 1, Pt. 1: 1-573. Monographs of the Academy of Natural Science of Philadelphia.

PILSBRY, H. A. 1940a. Land Mollusca of North America (north of Mexico). Vol. 1, Pt. 2: 575-994. Monographs of the Academy of Natural Science of Philadelphia.

PILSBRY, H. A. 1946. Land Mollusca of North America (north of Mexico). Vol. II, Pt. 1: 1-520. Monographs of the Academy of Natural Science of Philadelphia.

PILSBRY, H. A. 1948. Land Mollusca of North America (north of Mexico). Vol. II, Pt. 2: 521-1113. Monographs of the Academy of Natural Science of Philadelphia.

PLATT, S. G., T. R. RAINWATER, A. G. FINGER, J. B. THORBJARNARSON, T. A. ANDERSON, & S. T. MCMURRAY. 2006. Food habits, ontogenetic dietary partitioning and observations of foraging behavior of Morelet's crocodile (*Crocodylus moreletii*) in northern Belize. Herpetological Journal. 16: 281-290.

POLLARD, E. 1975. Aspects of the ecology of *Helix pomatia* L. Journal of Animal Ecology. 44:305-329.

POPE, K. O. 2005. Chicxulub impact ejecta deposits in southern Quintana Roo, Mexico and Central Belize. Geological Society of America. No. 384.

POTTER, C. S., H. L. RAGSDALE & C. W. BERISH. 1987. Reabsorption of foliar nutrients in a regenerating southern Appalachian forest. Oecologia. 73:268-271.

RAWLS, H. & R. YATES. 1971. Fluorescence in Endontid snails. Nautilus. 85:17-20.

REEDER, P., R. BRINKMANN & E. ALT. 1996. Karstification on the northern Vaca Plateau, Belize. Journal of Cave and Karst Studies. 58(2): 121-130.

REID, F. A. 2006. *Mammals of North America, 4th edition*. Peterson Field Guide. Houghton-Mifflin, New York, NY.

RICHARDS, P. W. 1996. *The tropical rain forest, 2nd edition*. Cambridge University Press, Cambridge.

RICKLEFS, R. E. & K. K. MATTHEWS. 1982. Chemical characteristics of the foliage of some deciduous trees in southeastern Ontario. Canadian Journal of Botany. 60:2037-2045.

RICHTER, K. O. 1980. Evolutionary aspects of mycophagy in *Ariolimax columbianus* and other slugs. *In*: D. L. Dindal, ed., *Soil Ecology as Related to Land Use Practices*. Proceedings of the VII International Colloquium of Soil Biology, Washington D. C.

RICHLING, I. 2004. Classification of the Helicinidae: Review of Morphological Characteristics Based on a Revision of the Costa Rican Species and Application to the Arrangement of the Central American Mainland Taxa (Mollusca: Gastropoda: Neritopsina). Malacologia. 45 (2):195-440.

RIMMER, C. C., K. P. MCFARLAND, D. C. EVERS, E. K. MILLER, Y. AUBRY, D. BUSBY & R. J. TAYLOR. 2005. Mercury concentrations in Bicknell's thrush and other insectivorous passerines in Montane forests of northeastern North America. Ecotoxicology. 14:223-240.

SAY, T. 1829a. Descriptions of new species of shells. The Disseminator of Useful Knowledge [New Harmony]: 25-26.

SCHEIFLER, R., C, SCHWARTZ, G. ECHEVARRIA, A. DE VAUFLEURY, P. M. BADOT, & J.L. MORE. 2003. "Nonavailable" soil cadmium is bioavailable to snails: evidence from isotopic dilution experiments. Environmental Science and Technology. 37(1):81-86.

SHEARER A.& J. W. ATKINSON. 2001. Comparative analysis of food-finding behavior of an herbivorous and a carnivorous land snail. Invertebrate Biology. 120:199-205.

SYMONDSON, W. O. C. 2004. Chapter 2: Coleoptera (Carabidae, Staphylinidae, Lampyridae, Drilidae, and Silphidae) as predators of terrestrial gastropods. pp. 37-84 in *Natural Enemies of Terrestrial Mollusks*. CABI Publishing. Oxford, UK.

THABAH, A., G. LI, Y. WANG, B. LIANG, K. HU, S. ZHANG, & G. JONES. 2007. Diet, echolocation calls, and phylogenic affinities of the great evening bat (*Ia Io*; Vespertilionidae): another carnivorous bat. Journal of Mammalogy. 88(3):728–735.

THOMPSON, F. G. 1957. A collection of land and fresh water mollusks from Tabasco, Mexico. Nautilus. 70(3): 97-102.

THOMPSON, F. G. 1958. The land snail genus *Microconus*. Nautilus. 72: 5-10.

THOMPSON, F. G. 1959. Two new pleurocerid snails from eastern Mexico. Occasional papers of the Museum of Zoology. University of Michigan. 600: 1-8

THOMPSON, F. G. 1963a. Two Mexican species of *Guillarmodia.* Nautilus. 76:95-99.

THOMPSON, F. G. 1963b. Systematic notes on the land snails of the genus *Tomocyclus* (Cyclophoridae). Breviora of the Museum of Comparative Zoology. 181:1-9.

THOMPSON, F. G. 1963c. New land snails from El Salvador. Proceedings of the Biological Society of Washington. 79:19-32.

THOMPSON, F. G. 1964. Systematic studies on Mexican land snails of the genus *Holospira*, subgenus *Bostrichocentrum* (STYLOMMATOPHORA, UROCOPTIDAE). Malacologia. 2: 131-143.

THOMPSON, F. G. 1966. A new pomatiasid from Chiapas, Mexico. Nautilus. 80: 24-28.

THOMPSON, F. G. 1967c. The land and freshwater snails of Campeche. Bulletin of the Florida State Museum. 11:221-250.

THOMPSON, F. G. 1968a. Some Mexican land snails of the family Urocoptidae. Bulletin of the Florida State Museum. 12:125-183; fig. 1-29.

THOMPSON, F. G. 1969b. *Bulimulus unicolor* and *Bulimulus ocraspira* Nautilus. 82:106-107.

THOMPSON, F. G. 1976. The genus *Epirobia* in Chiapas, Mexico. Nautilus. 90: 41-46.

THOMPSON, F. G. 1987a. Giant carnivorous land snails from Mexico and Central America. Bulletin of the Florida State Museum. 30:29-52.

THOMPSON, F. G. 1995. New and little-known land snails of the family Spiraxidae from Central America and Mexico (Gastropoda, Pulmonata). Bulletin of the Florida Museum of Natural History. 39:45-85.

THOMPSON, F. G. 1998. *Holospira* Martens, 1850 (Mollusca, Gastropoda, Urocoptidae): proposed designation of *Cylindrella goldfussi* Menke, 1947 as the type species. Bulletin Zoological Nomenclature. 55: 87-87.

THOMPSON, F. G. 2011. *An annotated checklist and bibliography of the land and freshwater snails of Mexico and Central America.* Florida Museum of Natural History. On-line publication in pdf format. fgt@flmnh.ufl.edu

THOMPSON, F. G. & A. CORREA-SANDOVAL. 1994. Land snails of the genus *Coelocentrum* from northeastern México. Bulletin of the Florida Museum of Natural History. 36:41-173.

THOMPSON, F. G. & D. C. DOURSON. 2013. A new land snail of the genus Eucalodium from Belize (Gastropoda:Pulmonata:Urocoptoidea: Eucalodiidae). Nautilus. 127:153-155.

TORRES, C. DE LA, P. BARTSCH, & J. P. E. MORRISON. 1942. The cyclophorid operculate land mollusks of America. United States National Museum, Smithsonian Institution Bulletin. 181:204-205.

TREWARTH, G. T. 1981. *The Earth's Problem Climates,* 2nd edition. University of Wisconsin Press.

TRYON, G. W., JR., 1865-1872. American Journal of Conchology Volumes 1-7. *Ed.*

TURNER, B. L. II & P. D. HARRISON. 1983. Pulltrouser Swamp: Ancient Maya habitat, agriculture, and settlement in northern Belize. 310 pp. University of Texas Press. Austin, Texas.

UIT DE WEERD, D.R. 2008. Delimitation and phylogenetics of the highly diverse land snail family Urocoptidae (Gastropoda, Pulmonata) based on 28S rRNA sequence data: A reunion with *Cerion*. Journal of Molluscan Studies. 74: 317-329.

VESTERDAL, I., & K. RAULUND-RASMUSSEN. 1998. Forest floor under seven tree species along a soil fertility gradient. Canadian Journal of Forest Reserves. 28:1636-1647.

WADA, S., K. KAWAKAMI & S. CHIBA 2012. Snails can survive passage through a bird's digestive system. Journal of Biogeography. 39: 69–73.

WATTERS, G. T. 2006. The Caribbean land snail family Annulariidae: a revision of higher taxa and catalog of the species. 1-577. Leiden. Backhuys Publishing.

WATTERS, G. T. 2014. A revision of the Annulariidae of Central America (Gastropoda: Littorinoidea). Zootaxa. 3878(4):301-350. DOI: 10.11646/zootaxa.3878.4.1.

WHITSON, M. 2005. Cepaea nemoralis (Gastropoda, Helicidae): the invited invader. Journal of the Kentucky Academy of Science. 66:82-87.

WORLD RESOURCES INSTITUTE. 1992. *Global Biodiversity Strategy: A Policy Maker's Guide.* World Resource Institute, International Union for the Conservation of Nature and United Environmental Protection, Washington, D. C.

WRIGHT, A. C. S., D. H. ROMNEY, R. H. ARBUCKLE, & V. E. VIAL. 1959. Land in British Honduras. Colonial Research Publication N. 24. Her Majesty's Stationery Office, London.

# Index of Scientific Names

# Y

*youngii, Miradiscops* 273
*yucatanea, Polygyra* 95

# Z

*Zonitoides* 267

---

Codes to Index

(A) = Aquatic
(E) = Exotic
(U) = Undetermined

* Not yet documented in Belize but pictured in book for comparison

ABOUT THE AUTHORS

**Daniel C. Dourson** is a biologist, naturalist, wildlife illustrator and photographer who has been studying the land snails of Belize since 2006 when he and his wife, Judy, moved to Belize to assist Belize Foundation for Research and Environmental Education (BFREE), a non-profit organization dedicated to environmental education and the conservation of the Bladen Nature Reserve. From 2007-2016, Dan participated in numerous expeditions into the Bladen Nature Reserve and the surrounding Maya Mountains. In 2016, he was the co-principal investigator along with James Rotenberg (UNCW) funded by National Geographic/Waitt Grants Program titled "Harpy Eagles, Snails and a Sinkhole: Discovering How Ecosystems Work by Looking Through Three Unknown Portals of the Bladen Nature Reserve, Belize. The late Dr. Fred Thompson, Curator Emeritus, of the Florida Museum of Natural History and the leading expert on Central and South American land snails served as a mentor, assisting in initial identification confirmation. Dan spent many hours with Fred as he learning the fauna of Central America. In 2013, Dr. Thompson and Dan described a new species of land snail from Belize, *Eucalodium belizensis.*

Dan has shared his passion for the natural world and in particular, land snails with biologists and the general public throughout Belize. He self-funded collecting excursions and trained local Belizeans to become field technicians in order to obtain samples from as many habitats and locations in Belize as possible. He continues to research and write books about the natural world, both in Belize and the USA. Dan is currently writing a series of books about the biodiversity of the Red River Gorge in Kentucky, USA.

His books include: *Wild Yet Tasty (editions I and II); Rare Land Snails of the Cherokee National Forest, TN, USA (co-authored with Ron Caldwell); Land Snails of Kentucky; Land Snails of the Great Smoky Mountains and the southern Appalachians; Land Snails of West Virginia; Land Snails of the Bladen Nature Reserve; Biodiversity of Belize with a focus on the Bladen Nature Reserve, Reptiles and Amphibians of the Red River Gorge and Wildflowers and Ferns of Red River Gorge.*

Email: theroguebiologist@gmail.com

**Dr. Ronald S. Caldwell** is a retired biology professor, having taught for 38 years at Lincoln Memorial University in Tennessee, USA . There, he founded and became the first director of Cumberland Mountain Research Center (CMRC). He also established the Powell River Aquatic Research Station. Ron was instrumental in the establishment of the Rainforest Science Cooperative Lab at BFREE that provided collecting materials as well as lab equipment for the study of land snails. The lab has proven to be a useful space for visiting scientists from all over the world. He has worked on land snails of Belize for several years with his very good friend and colleague Dan Dourson. During that time, Ron promoted and supervised undergraduate research projects on land snails that contributed valuable information for this book. Ron received grants to survey land snails from a variety of habitats throughout Belize which contributed significantly to the book. He also organized the first land snail training workshop in the BFREE lab with Dan. Ron's research both in Belize and the U.S. also explores the use of micro-snails as indicators of forest health.

**Judy Dourson** is a retired teacher, having taught in public schools for 28 years. From 2006-2013, she served as the Director of Educational Programs for Belize Foundation for Research and Environmental Education (BFREE) in Belize where she facilitated field courses for university and high school students and assisted researchers from all over the world. For this project, Judy served as field assistant (spending countless hours alongside Dan surveying the land snails of Belize throughout the country), research assistant and editor. With only 24 species of land snails documented from Belize prior to 2006, Judy was responsible for the retrieval of out-of-print literature, museum specimen procurement for identification purposes and general research assistant while working in the Florida Museum of Natural History molluscan collection. She continues to work as a field assistant and researcher with Dan. They are currently working on a series of books about the biodiversity of the Red River Gorge in Kentucky, USA. Judy and Dan plan to continue their research of land snails in Central America with the possibility of expanding to other countries in the region.

Email: doursonbio@gmail.com

www.ingramcontent.com/pod-product-compliance
Lightning Source LLC
Chambersburg PA
CBHW041043331016

41458CB00103B/6465